Energy and Water Cycles in the Third Pole

Energy and Water Cycles in the Third Pole

Editors

Yaoming Ma
Zhongbo Su
Lei Zhong

MDPI • Basel • Beijing • Wuhan • Barcelona • Belgrade • Manchester • Tokyo • Cluj • Tianjin

Editors

Yaoming Ma
Institute of Tibetan Plateau
Research, Chinese Academy of
Sciences
Beijing
China

Zhongbo Su
Faculty of Geo-Information
Science and Earth Observation
(ITC)
University of Twente
Enschede
The Netherlands

Lei Zhong
School of Earth and Space
Sciences
University of Science and
Technology of China
Hefei
China

Editorial Office
MDPI
St. Alban-Anlage 66
4052 Basel, Switzerland

This is a reprint of articles from the Special Issue published online in the open access journal *Water* (ISSN 2073-4441) (available at: www.mdpi.com/journal/water/special_issues/energy_cycle_water_remote).

For citation purposes, cite each article independently as indicated on the article page online and as indicated below:

LastName, A.A.; LastName, B.B.; LastName, C.C. Article Title. *Journal Name* **Year**, *Volume Number*, Page Range.

ISBN 978-3-0365-3928-7 (Hbk)
ISBN 978-3-0365-3927-0 (PDF)

Cover image courtesy of Lei Zhong

Contents

Preface to "Energy and Water Cycles in the Third Pole"

As the most prominent and complicated terrain on the globe, the Tibetan Plateau (TP) is often called the "Roof of the World", "Third Pole"or "Asian Water Tower". The energy and water cycles in the Third Pole have great impacts on the atmospheric circulation, Asian monsoon system and global climate change. On the other hand, the TP and the surrounding higher elevation area are also experiencing evident and rapid environmental changes under the background of global warming. As the headwater area of major rivers in Asia, the TP's environmental changes—such as glacial retreat, snow melting, lake expanding and permafrost degradation—pose potential long-term threats to water resources of the local and surrounding regions. To promote quantitative understanding of energy and water cycles of the TP, several field campaigns, including GAME/Tibet, CAMP/Tibet and TORP, have been carried out. A large amount of data have been collected to gain a better understanding of the atmospheric boundary layer structure, turbulent heat fluxes and their coupling with atmospheric circulation and hydrological processes. The focus of this reprint is to present recent advances in quantifying land–atmosphere interactions, the water cycle and its components, energy balance components, climate change and hydrological feedbacks by in situ measurements, remote sensing or numerical modelling approaches in the "Third Pole"region.

This reprint was funded by the Second Tibetan Plateau Scientific Expedition and Research Program (Grant No. 2019QZKK0103)

Yaoming Ma, Zhongbo Su, and Lei Zhong
Editors

Editorial

Energy and Water Cycles in the Third Pole

Yaoming Ma [1,2,3,4,5,6], **Lei Zhong** [7,8,9,10,*] and **Zhongbo Su** [11,12]

1   State Key Laboratory of Tibetan Plateau Earth System, Resources and Environment (TPESRE), Institute of Tibetan Plateau Research, Chinese Academy of Sciences, Beijing 100101, China; ymma@itpcas.ac.cn
2   College of Earth and Planetary Sciences, University of Chinese Academy of Sciences, Beijing 100049, China
3   College of Atmospheric Science, Lanzhou University, Lanzhou 730000, China
4   National Observation and Research Station for Qomolongma Special Atmospheric Processes and Environmental Changes, Dingri 858200, China
5   Kathmandu Center of Research and Education, Chinese Academy of Sciences, Beijing 100101, China
6   China-Pakistan Joint Research Center on Earth Sciences, Chinese Academy of Sciences, Islamabad 45320, Pakistan
7   School of Earth and Space Sciences, University of Science and Technology of China, Hefei 230026, China
8   CAS Center for Excellence in Comparative Planetology, USTC, Hefei 230026, China
9   Frontiers Science Center for Planetary Exploration and Emerging Technologies, University of Science and Technology of China, Hefei 230026, China
10   Jiangsu Collaborative Innovation Center for Climate Change, Nanjing 210023, China
11   Faculty of Geo-Information Science and Earth Observation (ITC), University of Twente, 7514 AE Enschede, The Netherlands; z.su@utwente.nl
12   School of Environmental Science and Engineering, Chang'an University, Xi'an 710064, China
*   Correspondence: zhonglei@ustc.edu.cn

Citation: Ma, Y.; Zhong, L.; Su, Z. Energy and Water Cycles in the Third Pole. *Water* **2022**, *14*, 1175. https://doi.org/10.3390/w14071175

Received: 1 April 2022
Accepted: 2 April 2022
Published: 6 April 2022

Publisher's Note: MDPI stays neutral with regard to jurisdictional claims in published maps and institutional affiliations.

The energy and water cycles in the Third Pole have great impacts on the atmospheric circulation, Asian monsoon system and global climate change. On the other hand, as the largest high-elevation part of the cryosphere outside the polar regions, with vast areas of mountain glaciers, permafrost and seasonally frozen ground, the Third Pole is characterized as an area sensitive to global climate change [1–3]. The Tibetan Plateau (TP) and the surrounding higher elevation area are experiencing evident and rapid environmental changes, such as glacial retreat, snow melting, lake expanding and permafrost degradation. All these changes pose potential long-term impacts to water resources of local and surrounding regions. A better understanding of the water and energy cycles is essential for assessing and understanding the causes of changes in the cryosphere and hydrosphere in relation to changes of plateau atmosphere in the Asian monsoon system and for predicting the possible changes in water resources in South and East Asia [3].

To this end, the aim of this Special Issue was to present recent advances in quantifying (1) land–atmosphere interactions, (2) the water cycle and its components, (3) energy balance components, (4) climate change, and (5) hydrological feedbacks by in-situ measurements, remote sensing or numerical modelling approaches in the TP.

Ten articles (nine research articles and one review) are published in this Special Issue, covering the quantitative assessments of land surface radiation fluxes, evapotranspiration, water vapor transport and runoff, as well as the distinct surface processes over lake and glacier driven by warming climate. Besides, the coupling mechanism between the vertical motion of air with the near-surface meteorological variables is analyzed. Additionally, the pollution characteristics and possible sources of PFASs (per- and poly-fluoroalkyl substances) in both surface water and precipitation are also discussed.

Analysis of long-term, ground-based radiation budgets on the TP help to enhance scientific understanding of land-atmosphere interactions and their influence on weather and climate change in this region. Wang et al. (2021) [4] systematically analyzed the in situ measurements from 2006 to 2019 at six research stations over the TP. Despite the differences in climate and land cover, both land surface albedo and upwelling shortwave radiation decreased at all sites, while the downwelling and upwelling longwave radiation, net radiation, surface temperature and air temperature presented increasing trends at most stations.

Evapotranspiration (ET) is a key parameter in the surface energy and water balance, whose accurate estimation is important but still challenging. The ET estimates from one satellite merged dataset (EB), one assimilation product (GLDAS), two reanalysis datasets (ERA5 and ERA-Interim) and two WRF simulations (DDM and CPM) were intercompared by Dan et al. (2021) [5] in detail. The temporal and spatial variations and driving factors of ET were discussed by Song et al. (2022) [6]. As the air temperature, precipitation and NDVI (normalized difference vegetation index) increased at a rate of 0.07 °C/decade, 24.73 mm/decade and 0.02/decade in Qinghai Province from 2000 to 2020, respectively, the actual ET presented a significant increasing rate of 37.26 mm/decade, which is nearly three times of the increasing rate in China. The results also revealed that the air temperature is the dominant driving factor of ET, followed by NDVI and precipitation. Additionally, ET increased by 2.84 mm/100 m with increasing altitude, which also exhibited distinct heterogeneity in its driving factors across the altitude gradient, where the influence of precipitation, NDVI, and air temperature dominated at 1700–2600 m, 2600–3800 m, and 3800–6000 m, respectively.

Changes in the surface energy and water fluxes can induce the transport of water vapor. The variation patterns of water vapor budget and its relationship with precipitation were reported by Qi et al. (2021) [7], and the relationships between the sensible/latent heat and the water vapor flux divergence were revealed by Li et al. (2021) [8]. The height of the water vapor transportation channel of the western air flow was found to be higher than 3000 m, while that for the southwestern and southeastern air flows was about 2000 m. A negative correlation between the surface fluxes and the water vapor flux divergence was depicted. The southwest boundary of southeast TP was found to be the key area affecting the water vapor flux divergence. Determination of the location and density of water vapor sources is of great importance to the improvement of extreme precipitation forecasts. The relation of the atmospheric vertical motion with the climate was discussed by Tian et al. (2021) [9] via climate diagnosis and statistical analysis. Moreover, Li et al. (2021) [10] evaluated the runoff simulation skills via WRF-Hydro, and achieved an improvement of 6.6% in root mean square error against in situ measurements. The enhanced WRF-Hydro simulated an increase in latent heat flux, but a decrease in sensible heat flux and soil surface temperature due to the moist soil.

The TP preserves a sufficient amount of lakes and glaciers, which are both experiencing hydrological change in the context of climate warming. The surface processes and ice phenology of lakes on the TP were investigated by Lang et al. (2021) [11], and an increasing trend in lake surface temperature and latent heat flux, as well as a decreasing trend in sensible heat flux and ice thickness, were illustrated. Based on surface mass balance parameterizations, the mean glacier volume loss of Da Anglong Glacier during 2016–2098 was simulated by Zhao et al. (2022) [12] to be 38% and 83% of the volume in 2016 under RCP2.6 and RCP8.5.

Finally, the characteristics and possible sources of PFASs in surface water and precipitation in China were reported by Wang et al. (2022) [13]. The concentration of PFASs in the surface water in different areas of China varied from 0.775 ng/L to 1.06×10^6 ng/L, while that in precipitation was lower, ranging from 4.2 ng/L to 191 ng/L. Although the concentrations of PFASs in surface water and precipitation in the TP were lower comparably (0.115–6.34 ng/L and 0.115–1.24 ng/L, respectively), influenced by the southeast monsoon in summer, PFASs can reach the TP through long-distance transportation and finally enter the surface water of this area through depositions.

In summary, this Special Issue mainly presents the up-to-date advances on the quantitative assessments of surface radiation fluxes, ET, water vapor transport, runoff, and the typical surface processes over lake and glacier on the TP. These selected papers are novel and timely for the understanding of land–atmosphere interactions driven by climate warming over the TP.

We trust that the collation of these papers will provide quantitative references for better assessment and prediction of the energy and water cycle processes in the "Third Pole".

Author Contributions: Conceptualization, Y.M. and Z.S.; writing—original draft preparation, L.Z.; writing—review and editing, Y.M. and L.Z. All authors have read and agreed to the published version of the manuscript.

Funding: This research was jointly funded by the Second Tibetan Plateau Scientific Expedition and Research Program (Grant No. 2019QZKK0103), the Strategic Priority Research Program of the Chinese Academy of Sciences (Grant No. XDA20060101), the National Natural Science Foundation of China (Grant Nos. 41875031 and 91837208), and CLIMATE-TPE (ID 58516) in the framework of the ESA-MOST Dragon 5 program.

Conflicts of Interest: The authors declare no conflict of interest.

References

1. Ma, Y.; Hu, Z.; Xie, Z.; Ma, W.; Wang, B.; Chen, X.; Li, M.; Zhong, L.; Sun, F.; Gu, L.; et al. A long-term (2005–2016) dataset of hourly integrated land-atmosphere interaction observations on the Tibetan Plateau. *Earth Syst. Sci. Data* **2020**, *12*, 2937–2957. [CrossRef]
2. Zhong, L.; Ma, Y.; Xue, Y.; Piao, S. Climate change trends and impacts on vegetation greening over the Tibetan Plateau. *J. Geophys. Res. Atmos.* **2019**, *124*, 7540–7552. [CrossRef]
3. Su, Z.; Ma, Y.; Chen, X.; Dong, X.; Du, J.; Han, C.; He, Y.; Hofste, J.; Li, M.; Li, M.; et al. Monitoring water and energy cycles at climate scale in the Third Pole Environment (CLIMATE-TPE). *Remote Sens.* **2021**, *13*, 3661. [CrossRef]
4. Wang, C.; Ma, Y.; Wang, B.; Ma, W.; Chen, X.; Han, C. Analysis of the radiation fluxes over complex surfaces on the Tibetan Plateau. *Water* **2021**, *13*, 3084. [CrossRef]
5. Dan, J.; Gao, Y.; Zhang, M. Detecting and attributing evapotranspiration deviations using dynamical downscaling and convection-permitting modeling over the Tibetan Plateau. *Water* **2021**, *13*, 2096. [CrossRef]
6. Song, Z.; Feng, Q.; Gao, Z.; Cao, S.; Cao, G.; Wang, Z. Temporal and spatial differences and driving factors of evapotranspiration from terrestrial ecosystems of the Qinghai Province in the past 20 years. *Water* **2022**, *14*, 536. [CrossRef]
7. Qi, D.; Li, Y.; Zhou, C. Variation characteristics of summer water vapor budget and its relationship with the precipitation over the Sichuan Basin. *Water* **2021**, *13*, 2533. [CrossRef]
8. Li, M.; Wang, L.; Chang, N.; Gong, M.; Ma, Y.; Yang, Y.; Chen, X.; Han, C.; Sun, F. Characteristics of the water vapor transport in the canyon area of the southeastern Tibetan Plateau. *Water* **2021**, *13*, 3620. [CrossRef]
9. Tian, R.; Ma, Y.; Ma, W. Vertical motion of air over the Indian Ocean and the climate in East Asia. *Water* **2021**, *13*, 2641. [CrossRef]
10. Li, G.; Meng, X.; Blyth, E.; Chen, H.; Shu, L.; Li, Z.; Zhao, L.; Ma, Y. Impact of fully coupled hydrology-atmosphere processes on atmosphere conditions: Investigating the performance of the WRF-Hydro model in the Three River source region on the Tibetan Plateau, China. *Water* **2021**, *13*, 3409. [CrossRef]
11. Lang, J.; Ma, Y.; Li, Z.; Su, D. The impact of climate warming on lake surface heat exchange and ice phenology of different types of lakes on the Tibetan Plateau. *Water* **2021**, *13*, 634. [CrossRef]
12. Zhao, W.; Zhao, L.; Tian, L.; Wolovick, M.; Moore, J. Simulating the evolution of Da Anglong Glacier, western Tibetan Plateau over the 21st century. *Water* **2022**, *14*, 271. [CrossRef]
13. Wang, F.; Zhuang, Y.; Dong, B.; Wu, J. Review on per- and poly-fluoroalkyl substances' (PFASs') pollution characteristics and possible sources in surface water and precipitation of China. *Water* **2022**, *14*, 812. [CrossRef]

Article

Analysis of the Radiation Fluxes over Complex Surfaces on the Tibetan Plateau

Chunxiao Wang [1,2,3], Yaoming Ma [1,2,3,*], Binbin Wang [1,2], Weiqiang Ma [1,2], Xuelong Chen [1,2] and Cunbo Han [1,2]

1 Land-Atmosphere Interaction and Its Climatic Effects Group, State Key Laboratory of Tibetan Plateau Earth System, Resources and Environment (TPESRE), Institute of Tibetan Plateau Research, Chinese Academy of Sciences, Beijing 100101, China; wangchx202109@163.com (C.W.); wangbinbin@itpcas.ac.cn (B.W.); wqma@itpcas.ac.cn (W.M.); x.chen@itpcas.ac.cn (X.C.); hcb@itpcas.ac.cn (C.H.)
2 College of Earth and Planetary Sciences, University of Chinese Academy of Sciences, Beijing 100049, China
3 Lanzhou University, Lanzhou 730000, China
* Correspondence: ymma@itpcas.ac.cn

Abstract: Analysis of long-term, ground-based observation data on the Tibetan Plateau help to enhance our understanding of land-atmosphere interactions and their influence on weather and climate in this region. In this paper, the daily, monthly, and annual averages of radiative fluxes, surface albedo, surface temperature, and air temperature were calculated for the period of 2006 to 2019 at six research stations on the Tibetan Plateau. The surface energy balance characteristics of these six stations, which include alpine meadow, alpine desert, and alpine steppe, were then compared. The downward shortwave radiation at stations BJ, QOMS, and NAMORS was found to decrease during the study period, due to increasing cloudiness. Meanwhile, the upward shortwave radiation and surface albedo at all stations were found to have decreased overall. Downward longwave radiation, upward longwave radiation, net radiation, surface temperature, and air temperature showed increasing trends on inter-annual time scales at most stations. Downward shortwave radiation was maximum in spring at BJ, QOMS, NADORS, and NAMORS, due to the influence of the summer monsoon. Upward shortwave radiation peaked in October and November due to the greater snow cover. BJ, QOMS, NADORS, and NAMORS showed strong sensible heat fluxes in the spring while MAWORS showed strong sensible heat fluxes in the summer. The monthly and diurnal variations of surface albedo at each station were "U" shaped. The diurnal variability of downward longwave radiation at each station was small, ranging from 220 to 295 W·m^{-2}. The diurnal variation in surface temperature at each station slightly lagged behind changes in downward shortwave radiation, and the air temperature, in turn, slightly lagged behind the surface temperature.

Keywords: Tibetan Plateau; surface characteristic parameter; radiation fluxes; observation data; land-atmosphere interaction

Citation: Wang, C.; Ma, Y.; Wang, B.; Ma, W.; Chen, X.; Han, C. Analysis of the Radiation Fluxes over Complex Surfaces on the Tibetan Plateau. *Water* **2021**, *13*, 3084. https://doi.org/10.3390/w13213084

Academic Editor: Teresa Afonso do Paço

Received: 7 August 2021
Accepted: 29 October 2021
Published: 3 November 2021

Publisher's Note: MDPI stays neutral with regard to jurisdictional claims in published maps and institutional affiliations.

1. Introduction

With a mean elevation over 4000 m, the Tibetan Plateau is considered as 'the roof of the world' or 'third pole' and has the world's most complex mountain topography [1]. The high and undulating endorheic hinterland of the Tibetan Plateau is surrounded by a chain of steeply descending marginal mountains, including the eight highest peaks of the world, including Mount Everest, in the south [2]. This extensive plateau lies between 26°00′ N and 39°47′ N, 73°19′ E, and 104°47′ E [3]. The complex and high-elevation topography, and the solar radiation absorbed by the ground in summer, lead to significant land-atmosphere interactions across the Tibetan Plateau. Consequently, the region's energy and water circulation processes have important effects on the Asian monsoon, the East Asian general circulation, and global climate change [4–6].

Solar radiation is the basic energy source driving a diversity of physical processes in the atmosphere, and it is also an important meteorological element characterizing the thermal condition of the Tibetan Plateau. At the same time, the free atmosphere is subject to various thermal as well as dynamic effects, which propagate from the plateau surface, through the near-earth layer and into the boundary layer [7–9]. Therefore, the study of the various radiation fluxes that affect the development of the boundary layer is particularly important. The ground gains heat due to the absorption of downward shortwave radiation and downward longwave radiation emitted by the atmosphere. The ground can also lose heat via the emission of upward longwave radiation and reflection of incoming shortwave radiation. In the absence of other modes of heat exchange, the net radiation determines the change in surface temperature. Surface temperature is an indicator that characterizes the variability of heat sources [10] and is an important parameter that describes the material exchange and energy balance between the surface and the atmosphere. Moreover, the air temperature is directly influenced by the surface temperature, as the surface emits upward longwave radiation to heat the near-surface air. Changes in air temperature can thus reflect the influence of the surface on the near-surface layer of the plateau.

The vast area, complex subsurface type, high altitude, and uneven distribution of a small number of observation stations on the Tibetan Plateau limit our understanding of land-atmosphere interactions in this region. Many studies in the past were based either on satellite data and reanalysis data or on short time series of observations. Moreover, many studies focused on the analysis of solar radiation or net radiation and did not analyze the radiation components. There is a lack of detailed analysis of long-term observation data over complex surfaces in highland areas.

Ma et al. [11,12] first analyzed the pre-monsoon, mid-monsoon, and post-monsoon radiation characteristics of the Nagqu region using the radiometric observations of GAME/Tibet during the 1998 Intensification Observation Period (IOP), and then compared the observations with the results obtained from remote sensing parameterization.

Philipona et al. [13] showed profiles of solar and terrestrial radiation measured with balloon-borne radiometers. They revealed the solar absorption in the free atmosphere and strong reflection in clouds and albedo effects on the ground and the atmosphere above. They also revealed that the longwave upward radiation is partly absorbed and reemitted by water vapor and other greenhouse gases. Obregón et al. [14] used the satellite data during the period 2000–2018. They found that water vapor and aerosols reduce solar radiation reaching the surface. This reduction ranges between 2% and 8% for aerosol optical thickness, 11.5% and 15% for precipitable water vapor, and 14% and 20% for the combined effect. Wang et al. [15] also pointed out that aerosols and total clouds attenuate surface solar radiation using the second Modern-Era Retrospective Analysis for Research and Applications (MERRA-2) reanalysis product. Jandaghian et al. [16] used the online Weather Research and Forecasting model coupled with Chemistry (WRF-Chem) to simulate the effects of albedo enhancement on aerosol, radiation, and cloud interactions in the Greater Montreal Area during the 2011 heatwave period. They found that albedo enhancement led to a net decrease in radiative balance at solar noon by 25 W/m^2.

You et al. [17] analyzed the annual and seasonal variations of all-sky and clear-sky surface solar radiation in the eastern and central Tibetan Plateau during the period 1960–2009, based on surface observational data, reanalysis, and ensemble simulations with the global climate model ECHAM5-HAM. They found a decreasing trend in the mean annual all-sky surface solar radiation, at a rate of -1.00 W m^{-2} decade^{-1}. A stronger decrease of -2.80 W m^{-2} decade^{-1} was found in the mean annual clear-sky surface solar radiation series. They also indicated that both NCEP/NCAR and ERA-40 reanalysis do not capture the decadal variations of the all-sky and clear-sky surface solar radiation.

Neither the satellite data nor the reanalysis data accurately reflect the true surface characteristics. Meanwhile, the short time series of observation data cannot accurately capture longer-term trends. Therefore, we need to statistically analyze data on observed surface characteristics over long time periods to more accurately quantify the changes and relationships amongst surface radiation fluxes, surface temperature, and air temperature on the Tibetan Plateau. This is important for understanding land-atmosphere interactions their influence on weather and climate in this region.

2. Materials and Methods

2.1. Study Area

The observation sites used in this paper comprised 6 field stations operated by the Chinese Academy of Science on the Tibetan Plateau, namely, the Qomolangma Atmospheric and Environmental Observation and Research Station (QOMS), the Southeast Tibetan Observation and Research Station for the Alpine Environment (SETORS), the BJ site of Nagqu Station of Plateau Climate and Environment (BJ), the Nam Co Monitoring and Research Station for Multisphere Interaction (NAMORS), the Ngari Desert Observation and Research Station (NADORS), and the Muztagh Ata Westerly Observation and Research Station (MAWORS). The stations were at altitudes in the range 3327 m to 4730 m, and included several surface types (e.g., alpine meadow, alpine desert, and alpine steppe). Figure 1 shows the instrument setup and subsurface conditions at the observation site.

Figure 1. Instrument setup and subsurface conditions at the observation site [16].

QOMS is located in Tingri County, Tibet, about 40 km from Everest Base Camp. The station is built in a river valley, with relatively flat topography and an open area around it, and the surface is mainly bare ground with sparse and small vegetation [18]. SETORS is located in Linzhi County, Tibet. The station is built in a valley with a relatively flat topography, surrounded by woodland, and the surface type is an alpine meadow with good growth conditions, and the grass height can reach 30–40 cm in summer [18]. BJ is located in Amdo County, Tibet. The station is flat and open all around, the surface is mainly sandy soil with sparse distribution of fine stones, and alpine meadows with a height of 10–20 cm grow unevenly in summer [10]. NAMORS is located on the southeastern shore of Namucuo Lake in Tibet, Dangxiong County, backed by the snowy peaks of the Nyingchi Tanggula mountain range, the lower cushion for alpine meadows [18]. The surface of NADORS and MAWORS are similar, both being desert, gravel and sparse short grass. NADORS is located in Ritu County, Ali Region, Tibet. MAWORS is located in Aktau County, Xinjiang, near Mushtag Mountain and Karakuri Lake, which is a typical westerly climate influence area. Figure 2 shows the Surface emissivity and station distribution in the observation area.

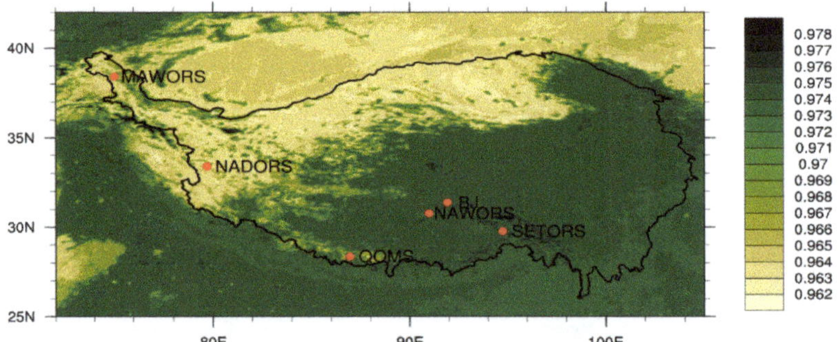

Figure 2. Surface emissivity and station distribution in the observation area.

Table 1 shows the geographical characteristics of the 6 field stations. Table 2 shows the detailed information on observation items.

Table 1. List of the geographic characteristics of the six sites [19].

Site	Latitude	Longitude	Elevation (m)	Land Cover
QOMS	28.36° N	86.95° E	4298	Alpine desert
SETORS	29.77° N	94.74° E	3327	Alpine meadow
BJ	31.37° N	91.90° E	4509	Alpine meadow
NAMORS	30.77° N	90.96° E	4730	Alpine steppe
NADORS	33.39° N	79.70° E	4270	Alpine desert
MAWORS	38.42° N	75.03° E	3668	Alpine desert

Table 2. List of the detailed information on observation items [19].

Site	Variables	Sensors Models	Manufacturers	Period
BJ	Air temperature	HMP45D	Vaisala	2006–2014
		HMP155	Vaisala	2016–2019
	Radiations	CM21 for shortwave radiation PIR for longwave radiation	Kipp and Zonen Eppley	2006–2019
QOMS	Air temperature	HMP45C-GM	Vaisala	2006–2019
	Radiations	CNR1	Kipp and Zonen	2006–2019
SETORS	Air temperature	HMP45C-GM	Vaisala	2006–2019
	Radiations	CNR1	Kipp and Zonen	2006–2019
NADORS	Air temperature	HMP45C	Vaisala	2006–2019
	Radiations	NR01	Kipp and Zonen	2006–2019
MAWORS	Air temperature	HMP155A	Vaisala	2006–2019
	Radiations	NR01	Kipp and Zonen	2006–2019
NAMORS	Air temperature	HMP45D	Vaisala	2006–2019
	Radiations	CMP6	Vaisala	2006–2019

2.2. Data and Methods

The observations used here were collected hourly from 2006 to 2019, and include downward shortwave radiation, upward shortwave radiation, upward longwave radiation, downward longwave radiation, and air temperature. The annual and monthly averages of downward shortwave radiation, upward shortwave radiation, and surface albedo were calculated using observation data from 8:00 to 20:00 Beijing time.

The upward longwave radiation, downward longwave radiation, and air temperature data at SETORS were problematic. The values were greater than the other stations, and the annual variation of upward longwave radiation and the diurnal variation in downward longwave radiation was not consistent with the variations of the rest of the stations, which was caused by the monitoring problem. The above data were rounded off and the calculation of the average value was not performed. Due to instrumental limitations, air temperatures were taken at 1.5 m at NAMORS, QOMS, and NADORS; 1.3 m at SETORS; and 1.9 m at MAWORS. Air temperatures were taken at 1.03 m at BJ from 2006 to 2014, and 1.5 m from 2015 to 2019. Although each radiometer observed the voltage value, the radiation data acquisition system has already calculated the radiation flux value according to the classical methodology and special controlling factor in each station of the Tibetan Plateau. We directly used the output radiation flux value of the observation system in our analysis. If the downward shortwave radiation or upward shortwave radiation value was less than 0, it would be revised to 0. If the surface albedo was greater than 0 and less than 1, it would be further averaged, otherwise, it would be excluded (Figure 3). When annual averages were calculated for each station, if the number of missing measurements in a given year was greater than 40% of the total number of data, the data for that year was rounded off, the annual average was no longer calculated and not represented in the graph. The monthly and daily averages were calculated by simply rounding off the missing measurements. Cloud data were selected from MOD08 product, NDVI data were selected from MOD13C2 product. Pearson's correlation coefficient r was calculated for the monthly average of cloud cover and the annual average of downward shortwave radiation, and also for the annual average of NDVI and the annual average of upward shortwave radiation. In addition, the observed surface temperatures were limited by instrumental functionality. To calculate accurate surface temperatures for each site, we selected the MOD11C3 product from 2006 to 2019. The MOD11C3 Version 6 product provides monthly Land Surface Temperature and Emissivity values in a 0.05 degree (5600 m at the equator) latitude/longitude Climate Modeling Grid. Each MOD11C3 product consists of LSTs, quality control assessments, observation times, view zenith angles, and the number of clear-sky observations, along with a percentage of land in the grid and emissivities from bands *20, 22, 23, 29, 31,* and *32*. Here we chose the emissivities from bands *31* and *32*.

The wide-band specific emissivity was obtained by linearly fitting the emissivity from bands *31* and *32* following the method proposed by Shunlin Liang [20], as follows:

$$\varepsilon = 0.261 + 0.314\varepsilon31 + 0.411\varepsilon32 \tag{1}$$

where ε is the wide-band emissivity, $\varepsilon31$ is the emissivity from band *31*, and $\varepsilon32$ is the emissivity from band *32*.

The total ground surface longwave irradiance includes the longwave radiation emitted from the ground and the downward longwave radiation reflected from the ground surface. The surface temperature can be calculated using the measured upward and downward longwave radiation and the surface emissivity:

$$Ts = \left(\frac{L_b^\uparrow - (1 - \varepsilon b)L_b^\downarrow}{\varepsilon b \sigma} \right)^{1/4} \tag{2}$$

where $L_b^\uparrow$ and $L_b^\downarrow$ are the upward longwave radiation and downward longwave radiation, respectively; εb is the surface emissivity; and σ is the Stephen Boltzmann constant (5.67×10^{-8} W·m^{-2} K^{-4}).

The fundamental equations governing the net energy budget of the Earth system are listed as follows:

$$Rn = Rsd + Rld - Rsu - Rlu \tag{3}$$

where Rsd is the downward shortwave radiation; Rld is the downward longwave radiation; Rsu is the upward shortwave radiation; and Rlu is the upward longwave radiation.

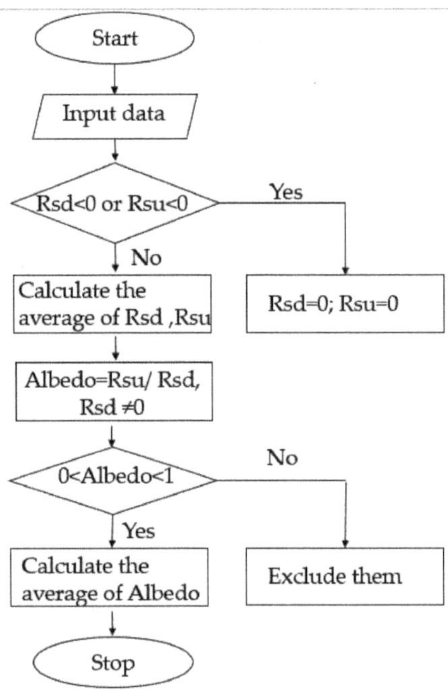

Figure 3. Flowchart representing the steps to process Rsd, Rsu, Albedo. Rsd is downward shortwave radiation, Rsu is upward shortwave radiation.

3. Data Analysis and Results

3.1. Downward Shortwave Radiation Flux

The radiation emitted by the sun may be absorbed and scattered by air molecules, water vapor, clouds, and dust in the atmosphere before it reaches the ground. The portion that eventually reaches the ground is called downward shortwave radiation [21], and is determined by both solar altitude angle and atmospheric transparency. The average altitude of the Tibetan Plateau is above 4000 m; here, the atmospheric cleanliness is high, thus the downward shortwave radiation in the Tibetan Plateau region is strong.

Atmospheric transparency is negatively correlated with cloud cover [22]. The higher the cloud cover is, the lower the atmospheric transparency is. In addition, atmospheric transparency can affect downward shortwave radiation [23]. The downward shortwave radiation reaching the ground decreased with decreasing atmospheric transparency. Therefore, an increase in cloudiness leads to a decrease in downward shortwave radiation to some extent, which can also be reflected in the relationship between the monthly average downward shortwave radiation and cloudiness at each station (Figure 4).

The interannual variability of downward shortwave radiation at each station (Figure 5) shows fluxes between 335 W·m^{-2} and 525 W·m^{-2} on the Tibetan Plateau. The downward shortwave radiation at BJ and QOMS had a tendency to decrease year by year, especially at BJ, which decreased by 1.55 W·m^{-2} per year, as previously attributed to an increase in convective clouds over the plateau [24,25]. However, there was an increasing trend in downward shortwave radiation at MAWORS, which may have been caused by decreasing cloud over MAWORS (Figure 6). The cloud cover at NADORS and SETORS showed an increasing trend, while the downward shortwave radiation at NADORS and SETORS showed no obvious change trend. In addition, the downward shortwave radiation at NAMORS decreased, and the cloud cover fluctuated. This may be due to changes in the atmosphere, such as water vapor and aerosols. However, the overall cloud cover at each station was negatively correlated with downward shortwave radiation can still be seen from Figure 4, and the Pearson correlation coefficient reached −0.79 and passed the 99% significance test.

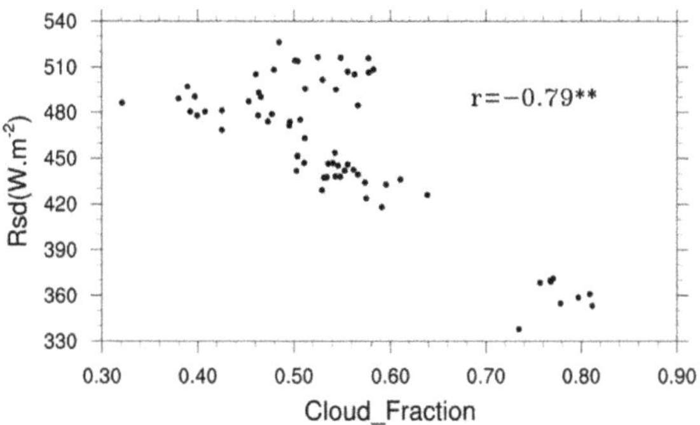

Figure 4. Relationship between the monthly average of downward shortwave radiation and the monthly average of cloud fraction. ** Significantly correlated at the 0.01 level (bilaterally).

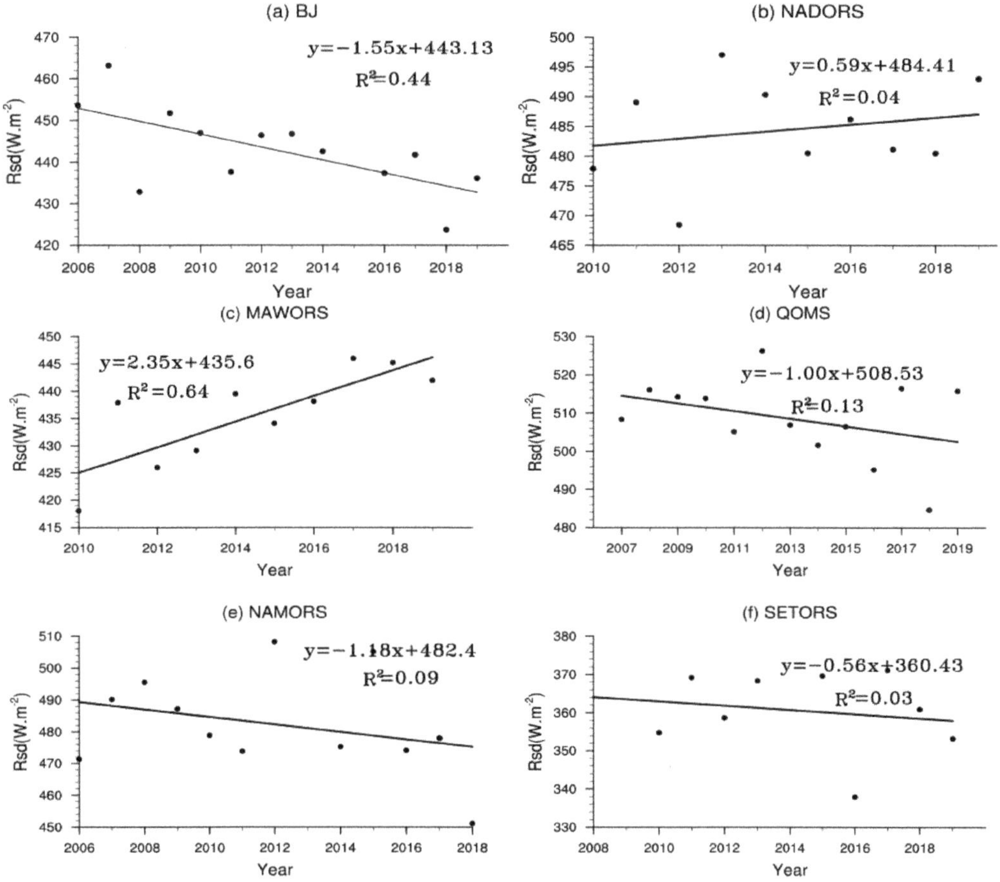

Figure 5. Interannual variation of downward shortwave radiation at each station.

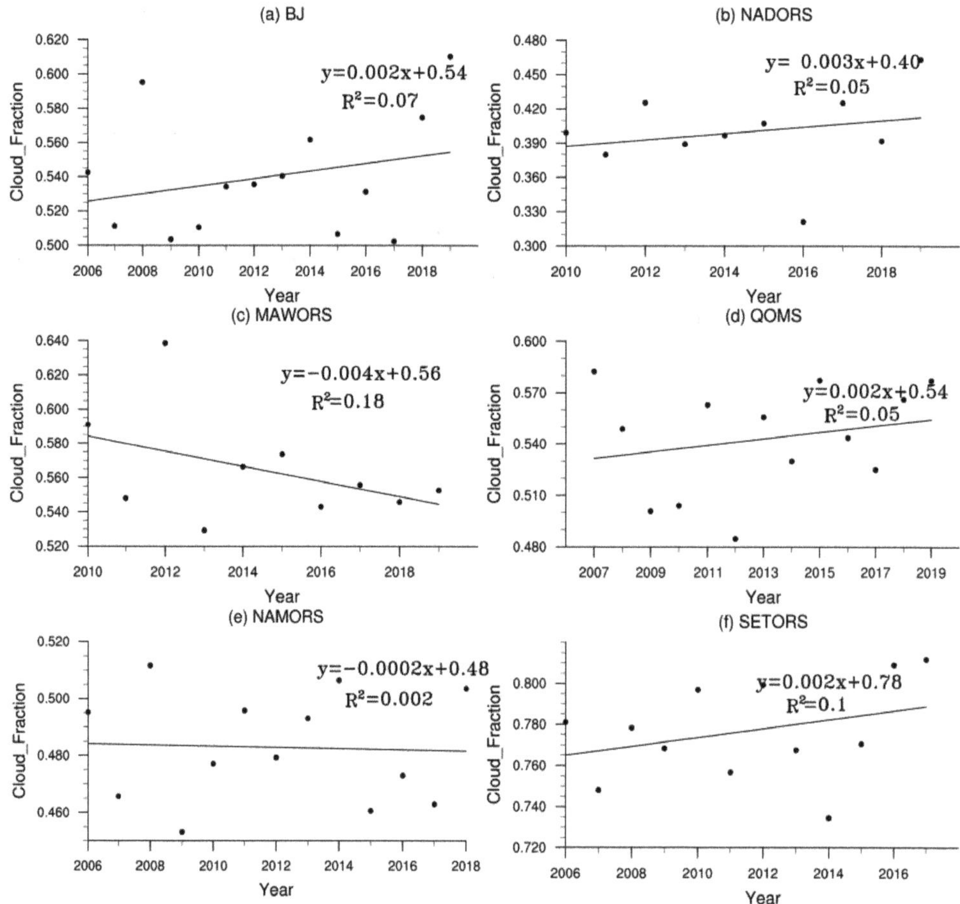

Figure 6. Interannual variation of cloud fraction at each station.

The intra-annual variations of downward shortwave radiation at each station (Figure 7a) showed a gradual increase with increasing solar altitude angle, starting from January. However, the downward shortwave radiation at BJ, QOMS, NAMORS, and NADORS did not reach their maximum in summer when the solar altitude angle was at its maximum, but in spring (when it reached a maximum of 620 $W \cdot m^{-2}$). Spring represents the early stage of the monsoon outbreak, characterized by low soil humidity, low cloudiness, and strong downward shortwave radiation. In summer, the monsoon is in its outbreak period: precipitation increases, the water vapor content increases, the air becomes more humid, and there is a corresponding decrease in atmospheric transparency, which results in less downward shortwave radiation reaching the ground. The downward shortwave radiation fluctuated between 450 $W \cdot m^{-2}$ and 550 $W \cdot m^{-2}$ at each station. In autumn and winter, the downward shortwave radiation decreased with the decreasing solar altitude angle. The annual variation of downward shortwave radiation at QOMS was obvious. The difference between the downward shortwave radiation in spring and summer reached 200 $W \cdot m^{-2}$. The annual variation of downward shortwave radiation at SETORS fluctuated relatively weakly, between 300 $W \cdot m^{-2}$ and 400 $W \cdot m^{-2}$. MAWORS is located in the northwestern part of the Tibetan Plateau, far inland, and is almost unaffected by the summer monsoon. At that site the downward shortwave radiation reached its maximum in July.

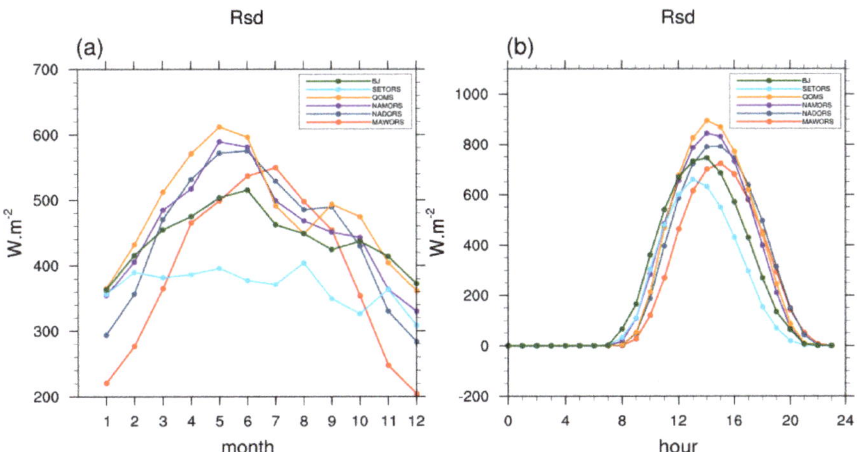

Figure 7. Annual variation of downward shortwave radiation (**a**) and diurnal variation of downward shortwave radiation (**b**) at each station.

The diurnal variation in downward shortwave radiation at each station (Figure 7b) shows similar patterns at all stations. From 8:00, the downward shortwave radiation increased above zero, reaching a maximum of 900 W·m^{-2} at around 14:00, and returned to negative values at around 21:00. QOMS had the largest diurnal range of downward shortwave radiation (0 to 950 W·m^{-2}). SETORS had the smallest diurnal range (0 to 700 W·m^{-2}).

3.2. Upward Shortwave Radiation Flux

Upward shortwave radiation comes from the reflection of downward shortwave radiation from the surface. Therefore, it is mainly controlled by the downward shortwave radiation and the surface conditions. Diurnal variations in upward shortwave radiation (Figure 8b) were similar to those of downward shortwave radiation: upward shortwave radiation increased with increasing solar altitude angle, to a maximum between 13:00 and 15:00, before decreasing again. The greatest diurnal range was at QOMS (0 to 230 W·m^{-2}), and the smallest was at SETORS (0 to 130 W·m^{-2}).

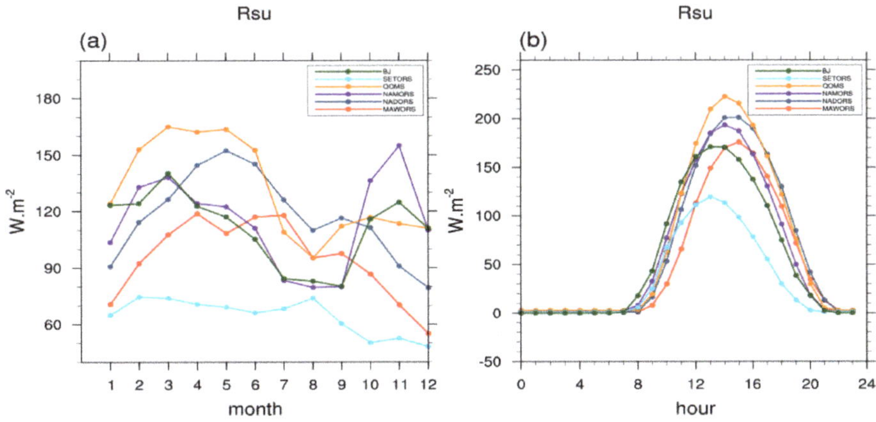

Figure 8. Annual variation of upward shortwave radiation (**a**) and diurnal variation of upward shortwave radiation (**b**) at each station.

The inter-annual variations in upward shortwave radiation at each station (Figure 9) ranged between 80 W·m^{-2} and 142 W·m^{-2}. The trend in upward shortwave radiation decreased year by year at all stations, most notably at NADORS, which decreased by 2.59 W·m^{-2} per year. On the one hand, the trend in upward shortwave radiation can be affected by downward shortwave radiation. On the other hand, it can be affected by vegetation density. An increase in vegetation density (NDVI) is generally expected to reduce upward shortwave radiation because of the strong absorptance in the Photosynthetically Active Radiation (PAR) region of the solar spectrum [26]. The downward shortwave radiation at BJ and NAMORS was decreasing year by year, and consequently, there was a decreasing trend in the upward shortwave radiation. The NDVI at these two stations showed an increasing trend(Figure 10), the vegetation cover increased, and the ground absorbed more downward shortwave radiation, which further led to a decrease in upward shortwave radiation. The downward shortwave radiation at NADORS and SETORS fluctuated, and the downward shortwave radiation at MAWORS increased, but the upward shortwave radiation at the above three stations increased, which was caused by the increase of ground vegetation coverage and the increase of NDVI. The NDVI at QOMS showed a decreasing trend, and the decreasing trend of upward shortwave radiation was mainly due to the decrease of downward shortwave radiation.

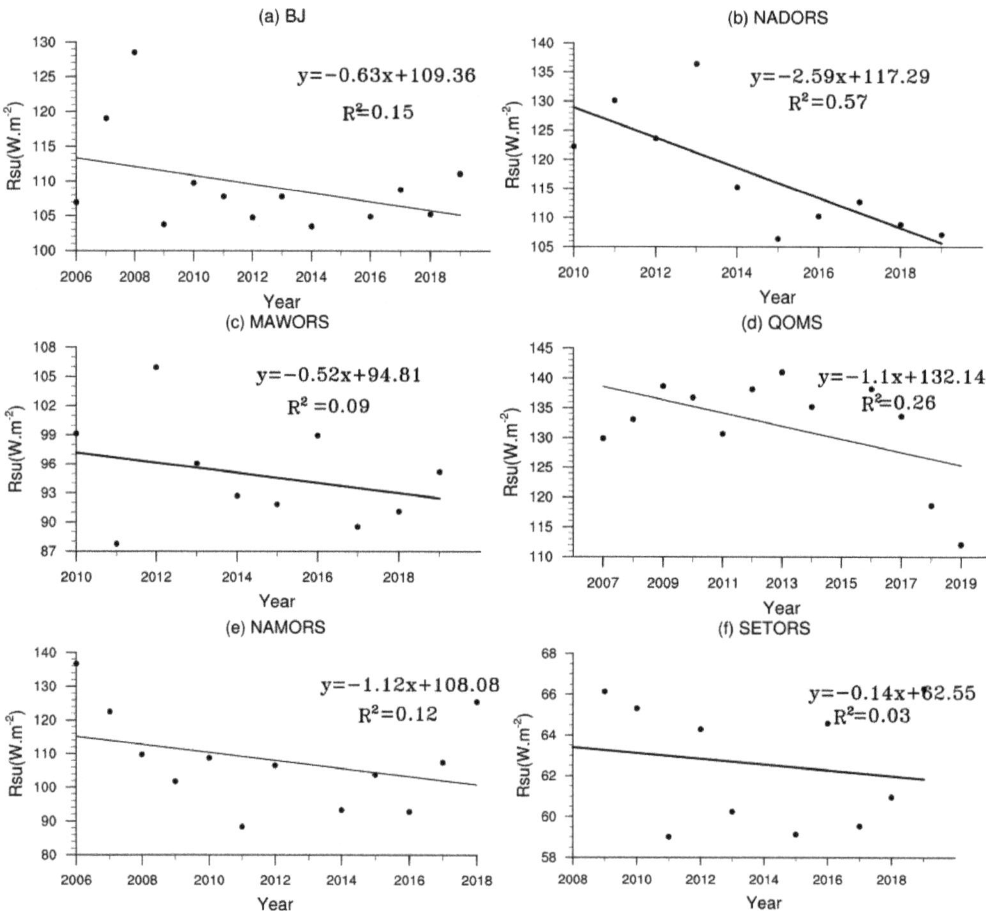

Figure 9. Interannual variation of upward shortwave radiation at each station.

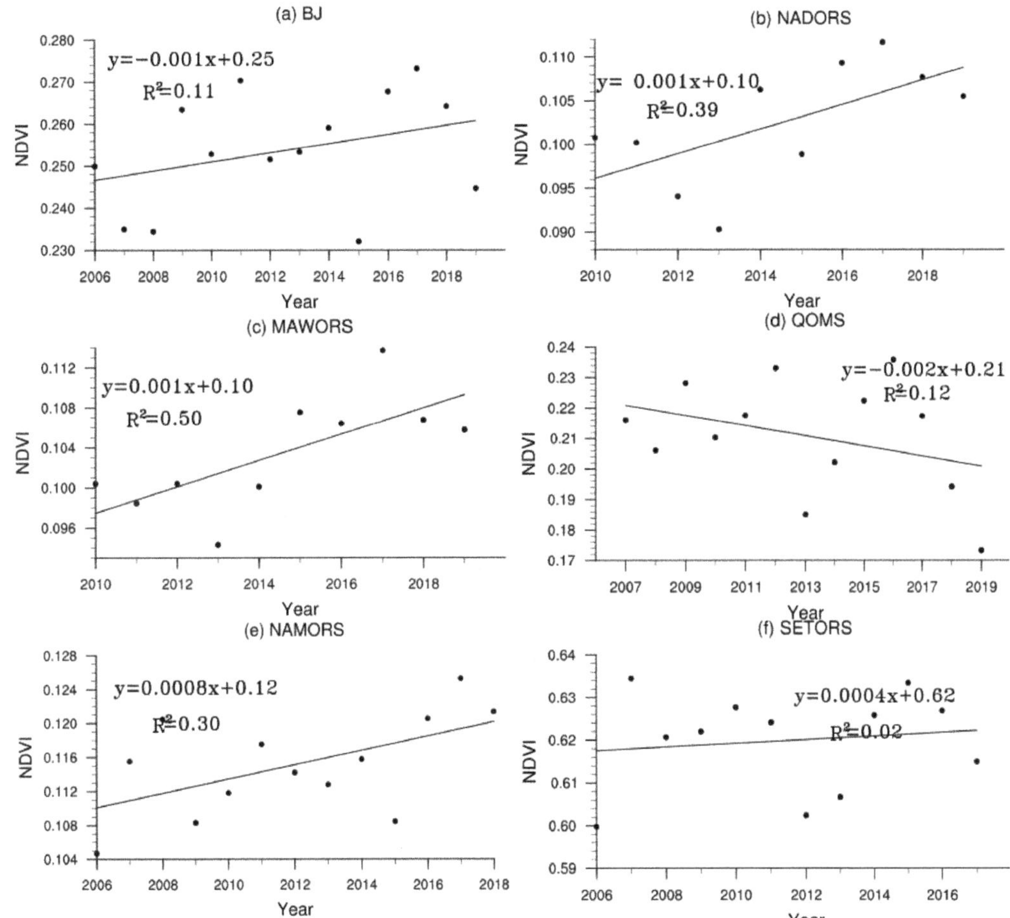

Figure 10. Interannual variation of NDVI at each station.

The annual variation of upward shortwave radiation (Figure 8a) showed trends at MAWORS and SETORS that were broadly consistent with those of downward shortwave radiation. There were notable annual variations in upward shortwave radiation at BJ, QOMS, NAMORS, and NADORS. In spring, the trend generally followed that of downward shortwave radiation. In summer, NDVI increased (Figure 11), vegetation became lush, and the ground absorbed more downward shortwave radiation, while the ground received less shortwave radiation. Therefore, the upward shortwave radiation decreased. In autumn, the air temperature dropped below 0 °C (Figure 12a), due to snow accumulation on the ground, an increasing trend of upward shortwave radiation was observed, especially at NAMORS, where the upward shortwave radiation in November reached 160 W·m^{-2}. In winter, the area of snow on the ground remained largely unchanged, thus the upward shortwave radiation flux decreased with the decreasing solar altitude angle.

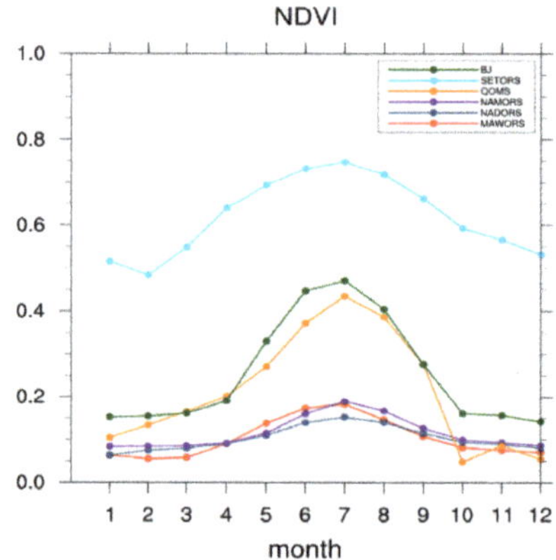

Figure 11. Annual variation of NDVI at each station.

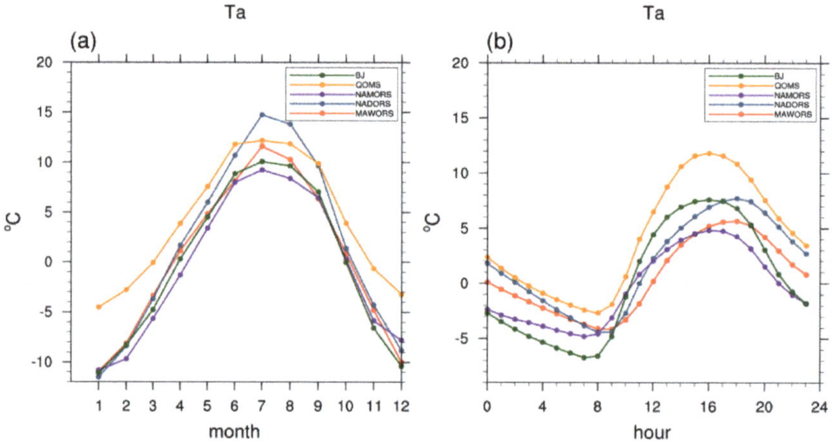

Figure 12. Annual variation of air temperature (**a**) and diurnal variation of air temperature (**b**) at each station.

3.3. Surface Albedo

Surface albedo is the ratio of upward shortwave radiation to downward shortwave radiation, and is an important factor affecting the energy balance of the surface. Surface albedo can be affected by solar altitude angle, atmospheric transparency, land cover, soil moisture, and weather conditions [27,28].

In general, snow melting or vegetation greening causes a typical decrease in surface albedo [29]. Therefore, surface albedo can broadly reflect changes in surface conditions, which is also reflected in the relationship between NDVI and surface albedo (Figure 13). The Pearson correlation coefficient between NDVI and surface albedo reached -0.73 and passed the significance test of 99%. We found the inter-annual variations in surface albedo (Figure 14) fluctuated between 0.18 and 0.33 at each station. From 2006 to 2016, the surface albedo at all stations decreased year by year. This trend was consistent with the findings of Li et al. [30], who used the MODIS shortwave white sky albedo product (MOD43A3) and found that the land surface albedo has been decreasing on the Tibetan Plateau from 2000 to 2013. This paper further supports their conclusion with real observation data. From 2006 to 2016, the surface albedo at all stations showed a decreasing trend, and the NDVI at all stations (except QOMS station) showed an increasing trend. From 2016 to 2019, the surface albedo at BJ, MAWORS, NAMORS, and SETORS showed an increasing trend, and the NDVI at these stations showed a decreasing trend. The interannual trends of surface albedo can broadly reflect the changes in surface conditions.

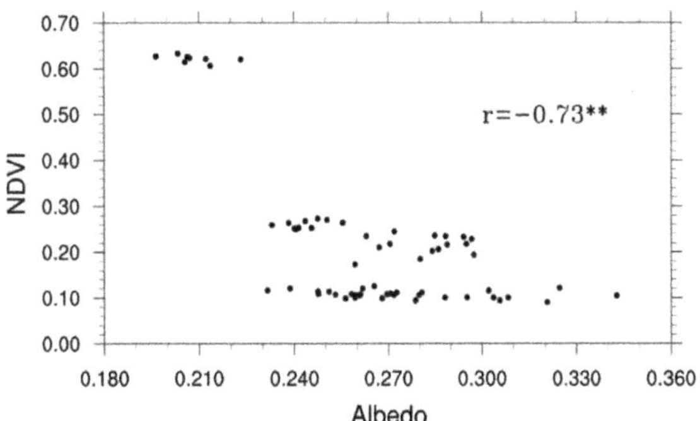

Figure 13. Relationship between the monthly average of NDVI and the monthly average of surface albedo. ** Significantly correlated at the 0.01 level (bilaterally).

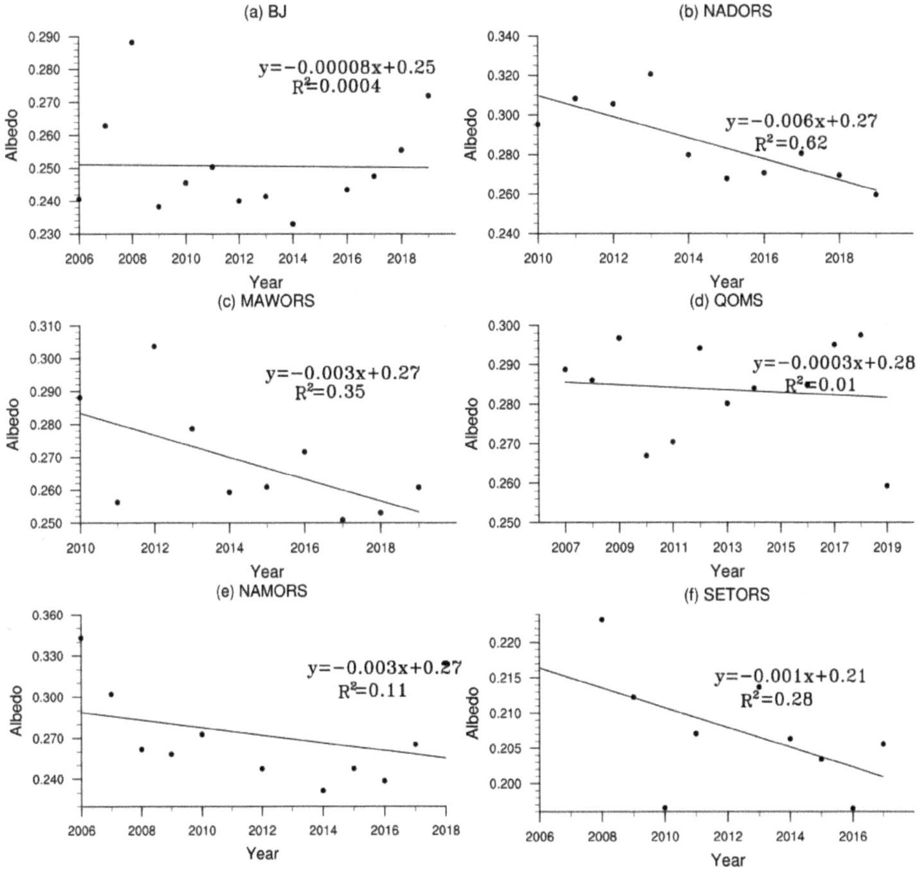

Figure 14. Interannual variation of surface albedo at each station.

The annual variation of surface albedo (Figure 15a) at each station generally follows a "U" shape. From January to March, the temperature at all stations was below 0 °C (Figure 12a), the ice was thicker, and the solar altitude angle was small, the surface albedo of each station was the largest, with a value around 0.35 (except SETORS). In April, the temperature at all stations was around 0 °C (Figure 12a), the snow and ice gradually melted, and the surface albedo gradually decreased. May to September is the plant growth period with large NDVI values at all stations (Figure 11). During this period, the vegetation cover increased, and the surface albedo continued to decrease, reaching a minimum in the range 0.2 to 0.25 during July or August. After October, the temperature at all stations was below 0 °C (Figure 12a), vegetation dieback, and increasing snow and ice cover caused the surface albedo to increase further. From October to December, the surface albedo at NAMORS was significantly higher than that of other stations due to the snow on the plateau, reaching a maximum of 0.48. The difference between surface albedo at SETORS and that at the other stations was not significant in summer, but in other seasons the surface albedo was significantly smaller than that at the rest of the stations, probably due to a lower snow and ice cover.

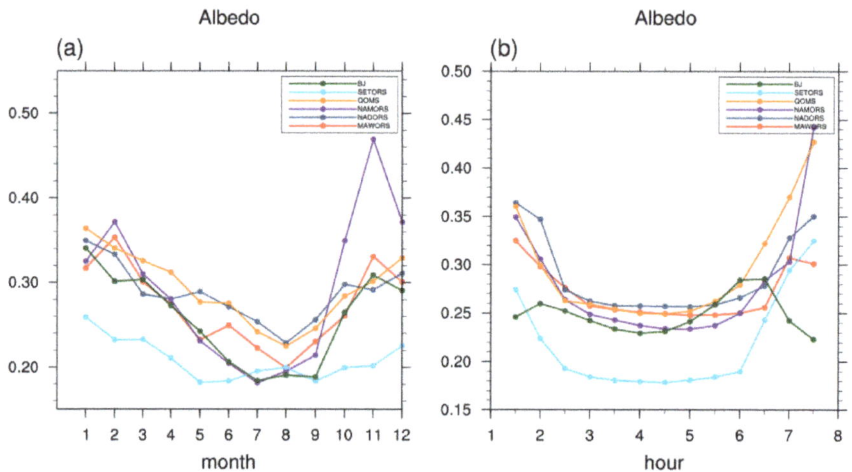

Figure 15. Annual variation of surface albedo (**a**) and diurnal variation of surface albedo (**b**) at each station.

The diurnal variations in surface albedo (Figure 15b) were similar at all stations, being highest in the morning and evening and lower at noon. This is mainly related to the change of solar altitude angle. When the solar altitude angle is low, a relatively greater proportion of the solar radiation reaching the ground is longwave, and the ground is very reflective to longwave radiation. When the solar altitude angle is high, the surface is less reflective to solar radiation. This effect is more evident when the solar altitude angle is low [31]. Therefore, the surface albedo at each station varied widely in the morning and evening but remained steady from 11:00 to 18:00. Surface albedo can broadly reflect changes in surface conditions. The ground surface at QOMS was covered with ice, sand, and gravel, and the vegetation was sparse. From 11:00 to 18:00, the surface albedo was around 0.27, which was the largest among the six stations. The ground surface at SETORS was forested grassland, with relatively high vegetation. The surface albedo from 11:00 to 18:00 was around 0.18, which was the smallest among the six stations.

3.4. Upward Longwave Radiation Flux

The ground surface temperature increases when the surface absorbs downward shortwave radiation. This leads to increased emission of longwave radiation back into the atmosphere. The inter-annual variations in upward longwave radiation (Figure 16) fluctuated between 321 W·m^{-2} and 368 W·m^{-2}. The upward longwave radiation showed overall increasing trends at BJ, MAWORS, QOMS, and NAMORS (most significant at MAWORS; weakest at NAMORS). Upward longwave radiation at NADORS showed a decreasing trend from 2011 to 2013, then an increasing trend from 2013 to 2019, with the initial decrease caused by the stronger upward longwave radiation in 2011. The overall increasing trend of upward longwave radiation at each station reflected changes in the surface climate of the plateau in the context of global warming.

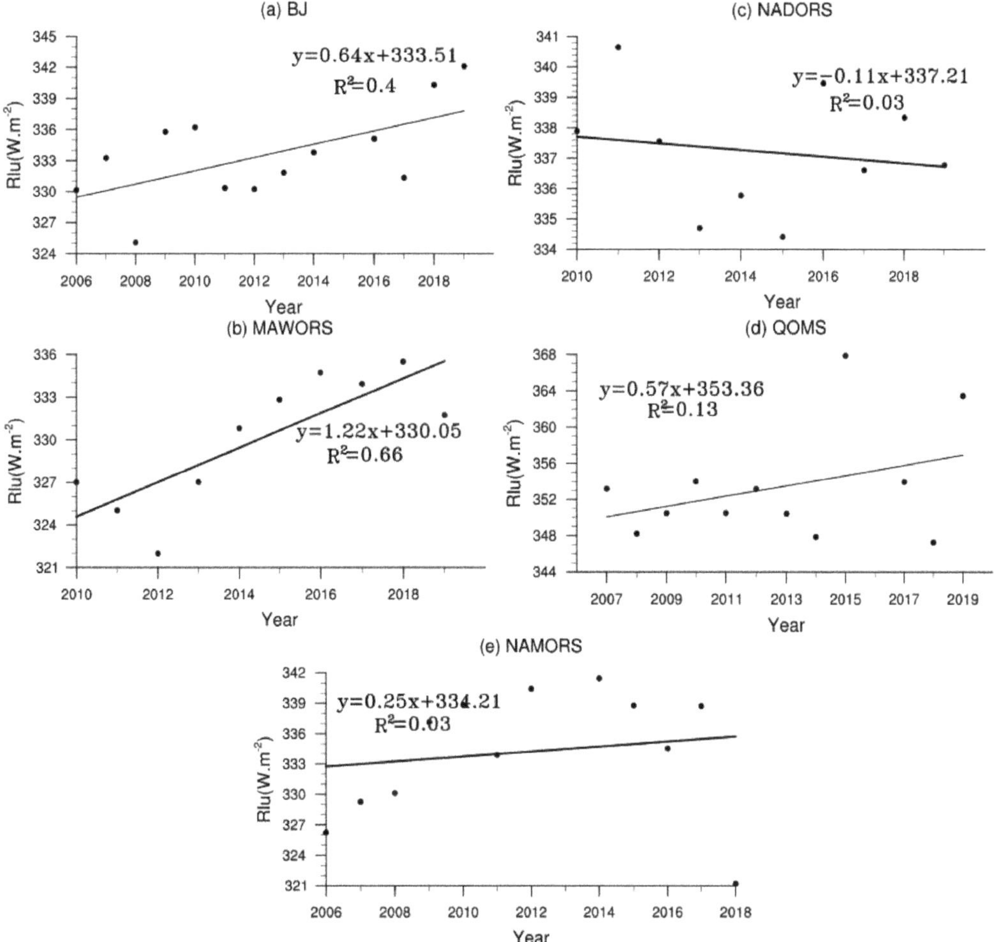

Figure 16. Interannual variation of upward longwave radiation at each station.

The annual variations in upward longwave radiation (Figure 17a) revealed obvious seasonal signals at each station, in which upward longwave radiation was significantly smaller in winter than in summer. Monthly maxima varied between the different stations: upward longwave radiation peaked at BJ in July and August, at QOMS and NAMORS in June, and at NADORS and MAWORS in July. The maximum varied between 380 W·m^{-2} and 410 W·m^{-2}. Due to the different surface characteristics and geographic locations, the upward longwave radiation changes showed varying patterns between different stations. However, in general, the upward longwave radiation at all stations reached the maximum in summer and the minimum in winter.

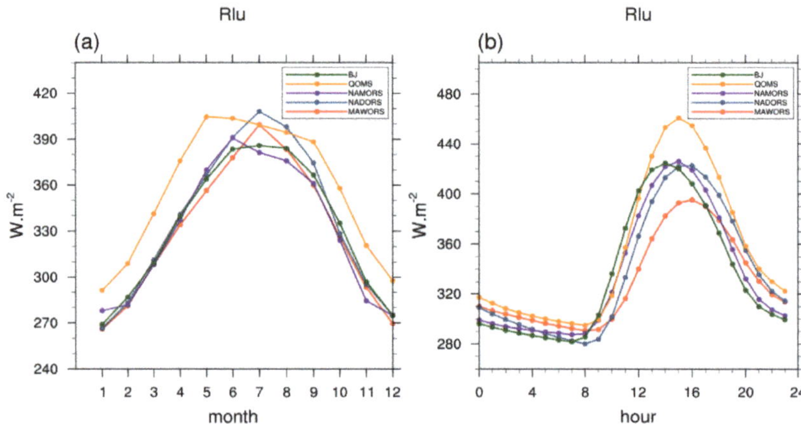

Figure 17. Annual variation of upward longwave radiation (**a**) and diurnal variation of upward longwave radiation (**b**) at each station.

Diurnal variations in upward longwave radiation (Figure 17b) show a peak between 15:00 and 16:00, slightly lagging the peak in downward shortwave radiation. The maximum upward longwave radiation at QOMS reached 470 W·m^{-2}, and the minimum was at MAWORS.

3.5. Downward Longwave Radiation Flux

Downward longwave radiation, also known as atmospheric inverse radiation, is longwave radiation emitted by the atmosphere to the surface. It is mainly influenced by temperature, atmospheric transparency, cloud amount, and cloud type. Geographical location also affects downward longwave radiation to some extent [32,33]. The Tibetan Plateau has a high altitude, and the overlying atmosphere holds less aerosol and water vapor than that of the lowlands. Therefore, the downward longwave radiation over the plateau is only 50%~70% of that in the lowlands [34].

The inter-annual variations of downward longwave radiation (Figure 18) ranged from 200 W·m^{-2} to 320 W·m^{-2} at each station and followed an increasing trend. MAWORS was mainly influenced by the increase of upward longwave radiation, NADORS was mainly influenced by the increase of convective clouds, while BJ, QOMS and NAMORS were influenced by a combination of both of those factors.

The annual variations of downward longwave radiation at each station (Figure 19a) showed an increase from January to a maximum in July or August and then a gradual decrease. In spring and winter, when temperatures were low, both upward and downward longwave radiation fluxes were small. In summer, at the time of monsoon break, water vapor was abundant, and upward longwave radiation was at its peak. Thus the downward longwave radiation was also greatest (at around 310 W·m^{-2}).

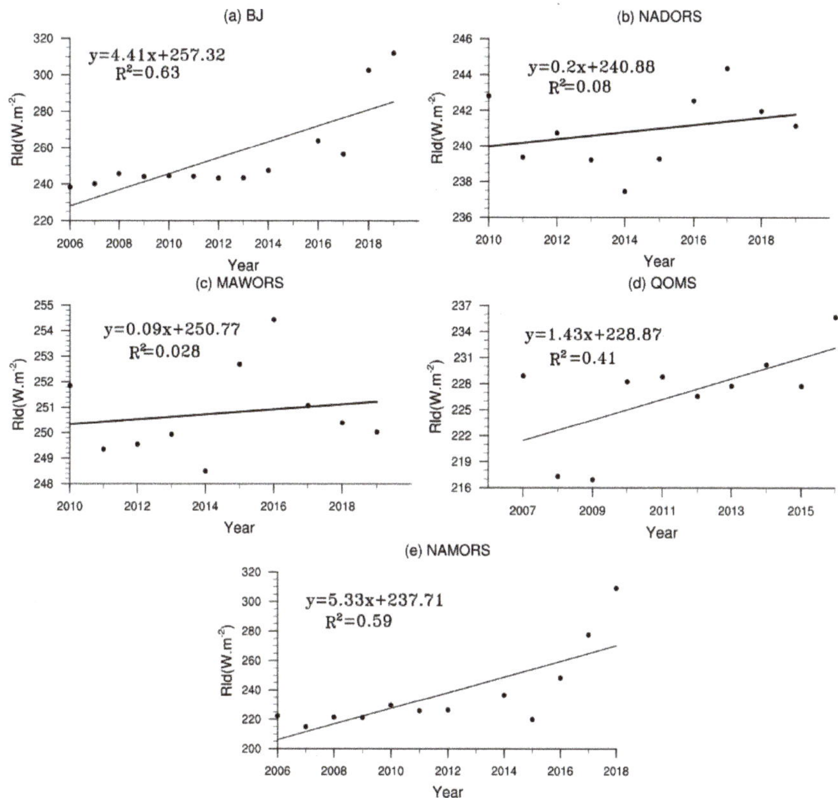

Figure 18. Interannual variation of downward longwave radiation at each station.

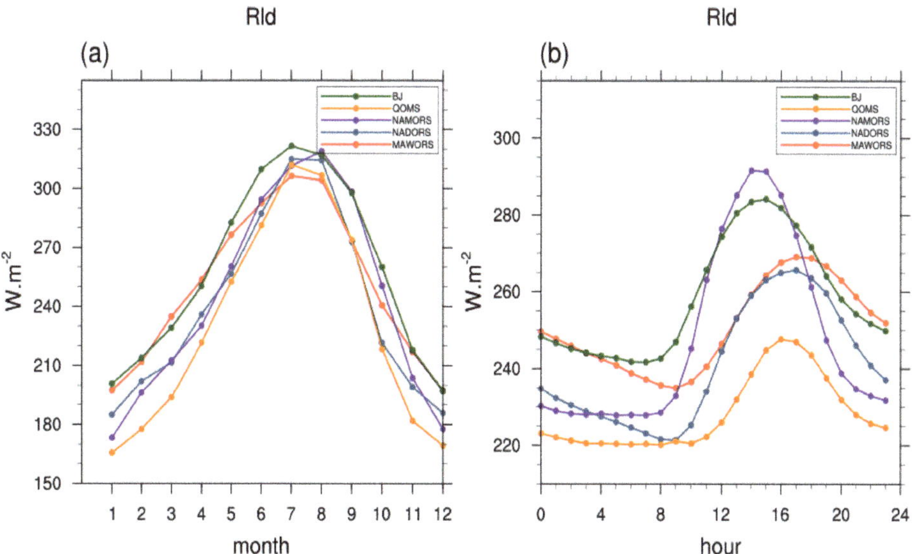

Figure 19. Annual variation of downward longwave radiation (**a**) and diurnal variation of downward longwave radiation (**b**) at each station.

Diurnal variations of downward longwave radiation (Figure 19b) were very small, ranging from 220 W·m^{-2} to 295 W·m^{-2}. The downward longwave radiation at NAMORS and BJ reached a minimum at around 8:00 and a maximum between 14:00 and 15:00. Respective minima and maxima at MAWORS, NADORS, and QOMS were at around 9:00 and 16:00 to 17:00. QOMS showed the least variation in downward longwave radiation.

3.6. Net Radiation Flux

The net radiation (Rn) is the difference between the downwards and upwards radiation fluxes at the ground surface. In the absence of other modes of heat exchange, the surface temperature increases when the net radiation is positive; remains constant when the net radiation is zero; and decreases when the net radiation is negative. Net radiation is jointly determined by each component of the radiation balance, and thus is affected by solar altitude angle, atmospheric transparency, altitude, underlying surface conditions, etc. [35].

The inter-annual variations in net radiation (Figure 20) ranged between 50 W·m^{-2} and 140 W·m^{-2}, with an overall increasing trend at each station. Although the downward shortwave radiation showed decreasing trends at BJ and NAMORS, the upward shortwave radiation reflected by the ground surface also decreased. This was due to increasing vegetation cover (Figure 10). Although the upward longwave radiation showed an increasing trend, the downward longwave radiation also increased. Overall, the net radiation showed an increasing trend. At QOMS, the upward shortwave radiation decreased, the downward longwave radiation increased. Thus, the net radiation showed an increasing trend. At NADORS, the downward shortwave radiation showed little change, the upward shortwave radiation decreased, and the downward longwave radiation increased. Thus, the net radiation showed an increasing trend. At MAWORS, the downward shortwave radiation increased and upward shortwave radiation decreased, and although the upward longwave radiation increased, the downward longwave radiation also increased, and overall the net radiation showed an increasing trend.

Annual variability in net radiation (Figure 21a) increased with the increasing total solar radiation at each station in spring. In summer, although the downward shortwave radiation at BJ, NADORS, QOMS, and NAMORS was slightly lower than that in spring (under the influence of monsoon), the upward shortwave radiation was lower than that in spring. Moreover, the downward longwave radiation was higher than that in spring. Therefore, the net radiation in summer at the above four stations was higher than that in spring. The net radiation at MAWORS continued to increase with the increasing total solar radiation. The net radiation flux at each station in summer ranged from 110 W·m^{-2} to 160 W·m^{-2}, and then decreased with the declining total solar radiation in autumn, reaching an annual minimum in winter.

Clear diurnal patterns of net radiation (Figure 21b) showed positive values from 8:00 to 20:00 (range from 0 to 500 W·m^{-2}) and negative at other times (range from 0 to -120 W·m^{-2}). The diurnal pattern generally followed downward shortwave radiation. NAMORS had the largest daily variation in net radiation, up to 600 W·m^{-2}.

From 8:00, the downward shortwave radiation increased above zero, reaching a maximum of 900 W·m^{-2} at around 14:00, and returned to negative values at around 22:00. QOMS had the largest diurnal range of downward shortwave radiation (0 to 950 W·m^{-2}). SETORS had the smallest diurnal range (0 to 700 W·m^{-2}).

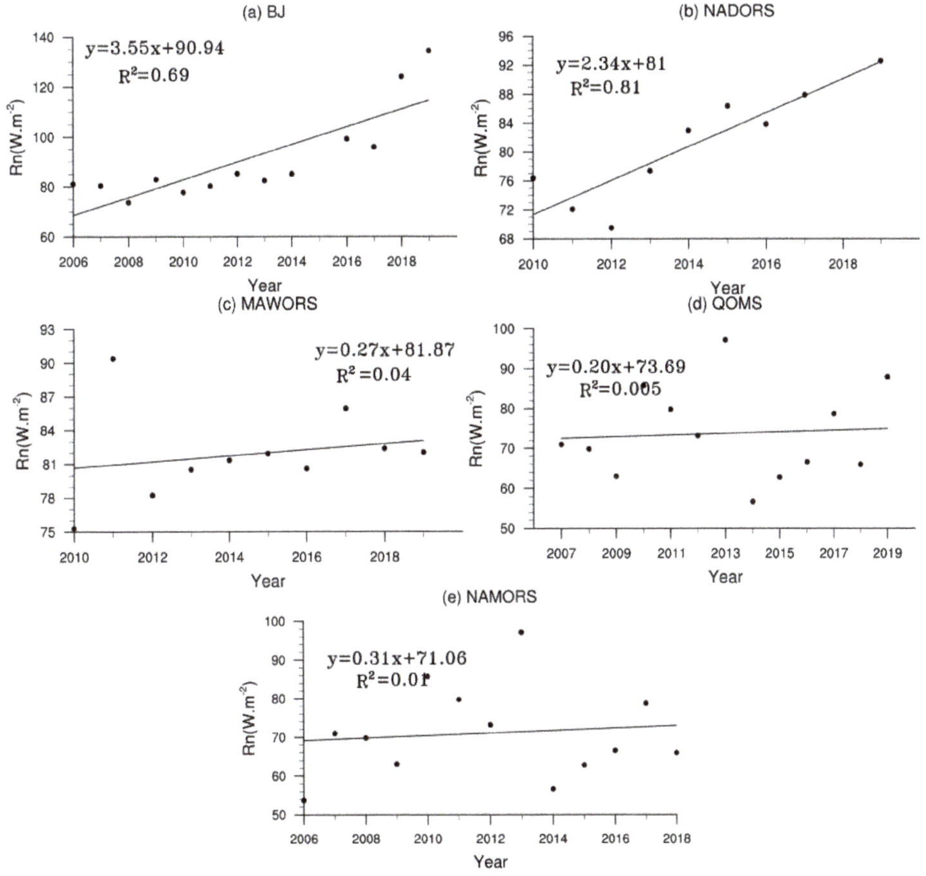

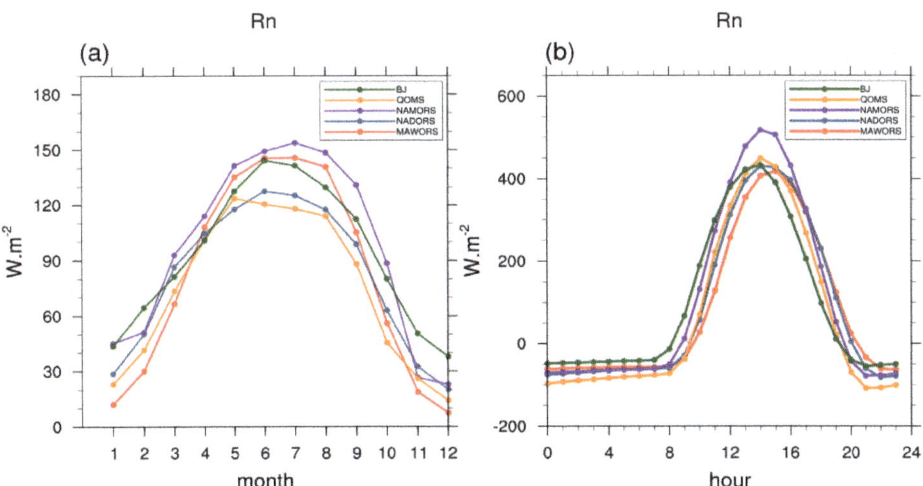

Figure 20. Interannual variation of net radiation of each station.

Figure 21. Annual variation of net radiation (**a**) and diurnal variation of net radiation (**b**) at each station.

3.7. Surface Temperature

The surface temperature represents the strength of the ground heat source [10]. The difference between the surface temperature and the air temperature above the surface directly affects the sensible and latent surface heat fluxes, which in turn affect the surface energy and water balances [36].

The inter-annual variations in surface temperature at each station (Figure 22) followed similar patterns to the upward longwave radiation and fluctuated between 1.2 °C and 5.5 °C. The surface temperature at BJ, MAWORS, QOMS, and NAMORS showed an increasing trend. The surface temperature at NADORS showed a decreasing trend from 2011 to 2013 and an increasing trend from 2013 to 2019.

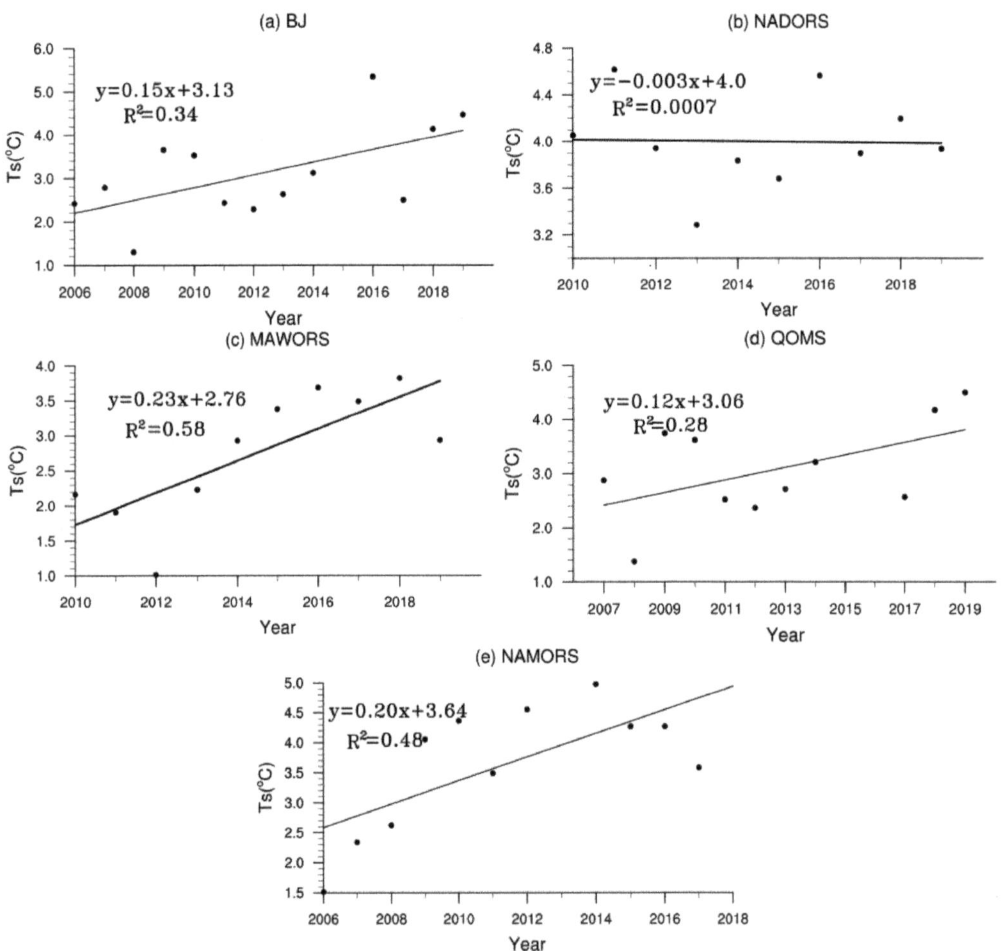

Figure 22. Interannual variation of surface temperature at each station.

The annual variation of surface temperature at each station (Figure 23a) was similar to those of the net radiation. Surface temperatures peaked in June and July, between 17 °C and 20 °C. The minimum reached in January was between −7 °C and −12 °C. From November to March, the surface temperature was less than 0 °C.

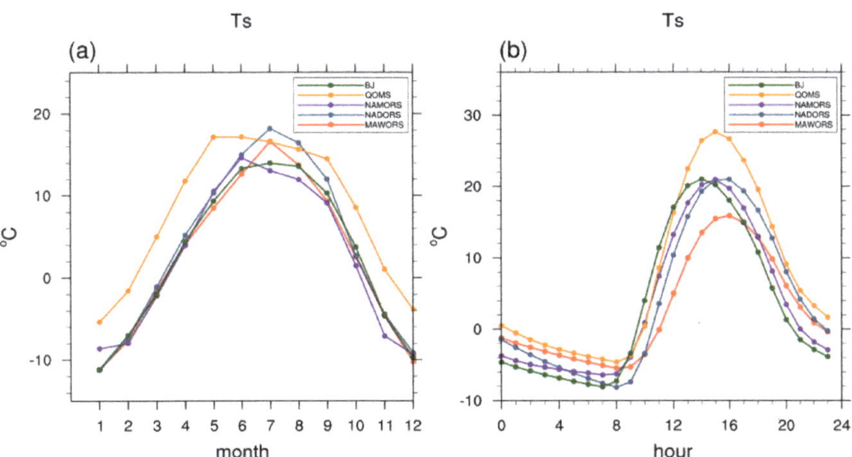

Figure 23. Annual variation of surface temperature (**a**) and diurnal variation of surface temperature (**b**) at each station.

Diurnal variations in surface temperature (Figure 23b) were similar at all stations and slightly lagged the downward shortwave radiation. The surface temperature peaked between 14:00 and 16:00 and reached its minimum at around 8:00. The surface temperature was below 0 °C from 0:00 to 10:00. Diurnal temperature ranges were relatively large, reaching 30 °C. The surface temperature was highest at QOMS, the southernmost station, and lowest at MAWORS, the northernmost station.

3.8. Air Temperature

Inter-annual variations in air temperature (Figure 24) fluctuated from −1 °C to 5.5 °C. The air temperature at BJ, MAWORS, and NAMORS showed an increasing trend, consistent with the trend of surface temperature. Air temperature at NADORS showed an increasing trend from 2011 to 2013 and a decreasing trend from 2013 to 2019. The increasing air temperature trends at the above stations were consistent with the global warming trend. There was no significant air temperature trend at QOMS.

Annual variations in air temperature were similar at each station (Figure 12a), reaching a maximum around July and a minimum around January. From October to March, the air temperature was below 0 °C. The MAWORS was less affected by the summer monsoon, and the difference between surface temperature and surface air temperature reached its maximum in July(Figure 25). The difference between surface temperature and surface air temperature at BJ, QOMS, NADORS, and NAMORS reached a maximum in spring(Figure 25), before the outbreak of the summer monsoon: at this time, surface vegetation growth had not yet commenced and the soil moisture was low. In addition, because the sensible heat flux is directly proportional to the difference between surface temperature and surface air temperature, the surface sensible heat transfer to the atmosphere was strong in spring. QOMS was significantly affected by the summer monsoon, the surface vegetation was sparse, soil moisture was low, and this station showed the strongest difference between surface temperature and surface air temperature.

Diurnal air temperature variations at each station (Figure 12b) reached a minimum between 8:00 and 9:00 and a maximum between 16:00 and 18:00. Following sunrise between 8:00 and 9:00, downward shortwave radiation became positive. At this time, the net radiation was still negative, and the air and surface temperatures at each station reached their daily minima. As the solar altitude angle increased, the downward shortwave radiation, surface temperature, and air temperature all increased. Between 13:00 and 15:00, the downward shortwave radiation and net radiation reached their diurnal maxima. Subsequently, the downward shortwave radiation started to weaken, but the energy gained at the surface was still more than that lost by the emission of upward longwave radiation.

Therefore, the surface temperature continued to rise while the net surface energy balance remained positive, reaching a peak between 16:00 and 18:00.

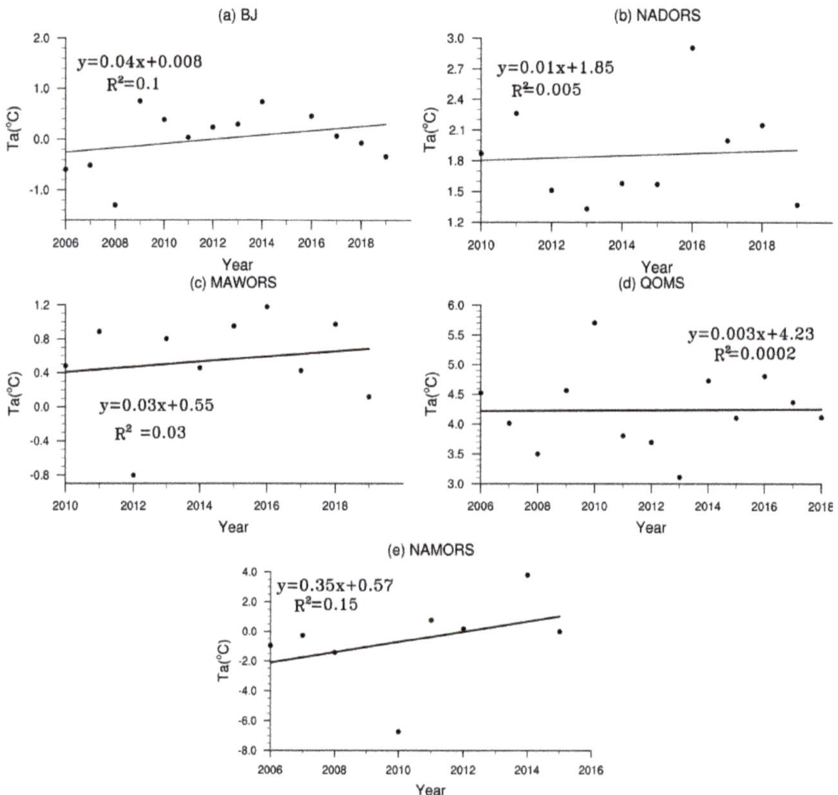

Figure 24. Interannual variation of air temperature at each station.

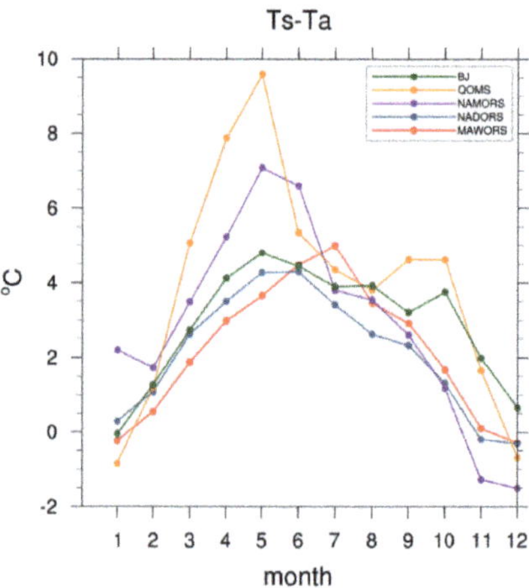

Figure 25. Annual variation of the difference between surface temperature and surface air temperature at each station.

4. Concluding Remarks

In this paper, the diurnal, annual, and inter-annual variations in downward shortwave radiation, upward shortwave radiation, downward longwave radiation, upward longwave radiation, surface albedo, net radiation, surface temperature, and air temperature of typical surfaces (alpine meadow, alpine desert, and alpine steppe) were analyzed using ground observations from 2006 to 2019 at QOMS, SETORS, BJ, NAMORS, MAWORS, and NADORS. The conclusions from our analysis are as followings.

1. The net radiation at each station (except SETORS) showed an increasing trend year by year. Although the upward longwave radiation at BJ, NADORS, MAWORS, QOMS, and NAMORS showed an increasing trend from 2013 to 2019, the downward longwave radiation also increased. Although the downward shortwave radiation at BJ and QOMS showed a decreasing inter-annual trend, probably due to an increase of cloudiness. The upward shortwave radiation increased due to decreasing downward shortwave radiation and increasing vegetation cover. Although the inter-annual downward shortwave radiation at NADORS and NAMORS fluctuated and decreased year by year, respectively. The upward shortwave radiation increased. At MAWORS, the downward shortwave radiation increased due to decreasing cloudiness and upward shortwave radiation decreased due to increasing vegetation cover.

2. The net radiation of each station is maximum in summer and minimum in winter. In summer, although the downward shortwave radiation at BJ, NADORS, QOMS, and NAMORS was slightly lower than that in spring (under the influence of monsoon), the upward shortwave radiation was lower than that in spring. The net radiation decreased with the declining total solar radiation in autumn, reaching an annual minimum in winter.

3. The diurnal pattern of net radiation generally followed downward shortwave radiation. It showed positive values from 8:00 to 20:00 (range from 0 to 500 $W \cdot m^{-2}$). Moreover, diurnal variations in upward shortwave radiation were similar to those of downward shortwave radiation. Diurnal variations in upward longwave radiation show a peak between 15:00 and 16:00, slightly lagging the peak in downward shortwave radiation. The daily variation of downward longwave radiation at each station was small, ranging from 220 to 295 $W \cdot m^{-2}$.

4. The surface albedo at each station was decreasing year by year due to stronger vegetation growth under global warming. The annual change in surface albedo at each station followed a "U" shape. In winter, the ground was frozen, and the surface albedo was large. During the plant growth period from May to September, the surface albedo was small and showed little change. The daily variation of surface albedo at each station was also "U" shaped. The grass height at SETORS can reach 30 to 40 cm in summer, thus, here, the surface albedo was the smallest of all the stations.

5. The surface temperature at BJ, MAWORS, QOMS, NAMORS, and NADORS showed an increasing trend from 2013 to 2019. The annual variation of surface temperature at each station generally followed the changes in net radiation. The diurnal variation at each station was generally consistent with changes in upward longwave radiation and slightly lagged the downward shortwave radiation. The southernmost station (QOMS) had the highest surface temperature, and the northernmost station (MAWORS) had the lowest surface temperature.

6. The air temperature at BJ, MAWORS, NAMORS, and NADORS showed an increasing trend from 2013 to 2019, consistent with the trend in surface temperature. The difference between surface temperature and surface air temperature at BJ, QOMS, NADORS, and NAMORS was maximum in spring before the monsoon outbreak. There was strong sensible heat transport from the surface to the atmosphere in spring. The difference between surface temperature and surface air temperature at MAWORS was maximum in summer. QOMS had the largest difference between surface temperature and surface air temperature. The diurnal variation in surface temperature at

each station slightly lagged behind changes in downward shortwave radiation, and the air temperature, in turn, slightly lagged behind the surface temperature.

In the absence of other modes of heat exchange, the net radiation determines the change in surface temperature. Surface temperature is an important parameter that describes the material exchange and energy balance between the surface and the atmosphere. Moreover, changes in air temperature can reflect the influence of the surface on the near-surface layer of the plateau. The variation characteristics of radiation fluxes, surface temperature, air temperature over the six stations in the Tibetan Plateau were derived in this study. The reasons for the variation of net radiation were analyzed in detail after analyzing each component of the radiation. This paper helps to enhance our understanding of land-atmosphere interactions and their influence on weather and climate in the Tibetan Plateau. However, due to the complex landscape of the Tibetan Plateau, we have to extend the results presented here to a broader perspective. In the other words, the results obtained in this study have to be compared to the analysis from more stations observation. All this research will be conducted in the coming days.

Author Contributions: Writing—original draft preparation, C.W.; formal analysis, C.W.; data curation, B.W., W.M., X.C., C.H.; supervision, Y.M. All authors have read and agreed to the published version of the manuscript.

Funding: This research was funded by the National Natural Science Foundation of China (91837208). The Second Tibetan Plateau Scientific Expedition and Research (STEP) program (2019QZKK0103), the Strategic Priority Research Program of Chinese Academy of Sciences (XDA20060101).

Institutional Review Board Statement: Not applicable.

Informed Consent Statement: Not applicable.

Data Availability Statement: The data presented in this study are openly available in National Tibetan Plateau Data Center. Ma, Y. (2020). A long-term dataset of integrated land-atmosphere interaction observations on the Tibetan Plateau (2005–2016). DOI: 10.11888/Meteoro.tpdc.270910. CSTR: 18406.11.Meteoro.tpdc.270910.

Acknowledgments: We would like to thank Zhipeng Xie, Ling Yuan of Institute of Tibetan Plateau Research, Chinese Academy of Sciences, and Jiankai Zhang of Lanzhou University for their suggestions and help in this study.

Conflicts of Interest: The authors declare no conflict of interest.

Abbreviations

The following abbreviations are used in this manuscript:

GAME/Tibet	GEWEX Asian Monsoon Experiment in the Tibetan Plateau
IOP	Intensification Observation Preriod
QOMS	Qomolangma Atmospheric and Environmental Observation and Research Station
SETORS	Southeast Tibetan Observation and Research Station for the Alpine Environment
BJ	BJ site of Nagqu Station of Plateau Climate and Environment
NAMORS	Nam Co Monitoring and Research Station for Multisphere Interaction
NADORS	Ngari Desert Observation and Research Station
MAWORS	Muztagh Ata Westerly Observation and Research Station

References

1. Qiu, J. The third pole. *Nature* **2008**, *454*, 393–396. [CrossRef]
2. Liu, J.; Ding, L.; Zeng, L.; Paul, T.; Yves, G. Large-scale terrain analysis of selected regions of the Tibetan Plateau: Discussion on the origin of plateau planation surface. *Earth Sci. Front.* **2006**, *13*, 285–299.
3. Zhang, Y.; Li, B.; Zheng, D. A discussion on the boundary and area of the Tibetan Plateau in China. *Geogr. Res.* **2002**, *21*, 128.
4. Ma, Y.; Yao, T.; Hu, Z.; Wang, J. The Cooperative Study on Energy and Water Cycle over the Tibetan Plateau. *Adv. Earth* **2009**, *24*, 1280–1284.
5. Yeh, T.; Lo, S.; Chu, P. The Wind Structure and Heat Balance in the Lower Troposphere over Tibetan Plateau. *Inst. Meteorol. Geophys.* **1957**, *28*, 108–121.

6. Yanai, M.; Li, C. Mechanism of heating and the boundary layer over the Tibetan Plateau. *Mon. Weather Rev.* **1994**, *122*, 305–323. [CrossRef]
7. Ma, Y.; Ma, W.; Hu, Z.; Li, M.; Wang, J.; Ishikawa, H.; Tsukamoto, O. Similarity Analysis of Atmospheric Turbulent Intensity over Grassland Surface of Qinghai-Xizang Plateau. *Plateau Meteorol.* **2002**, *21*, 514–517.
8. Ma, Y.; Osamu, T.; Wu, X.; Tamagawa, I.; Wang, J.; Ishikaw, H.; Hu, Z.; Gao, H. Characteristics of Energy Transfer and Micrometeorology in the Surface Layer of the Atmosphere above Grassy Marshland of the Tibetan Plateau Area. *Chin. J. Atmos. Sci.* **2000**, *24*, 715–722.
9. Ma, Y.; Osamu, T.; Wang, J.; Hirohiko, I.; Ichiro, T. Analysis of aerodynamic and thermodynamic parameters on the grassy marshland surface of Tibetan Plateau. *Prog. Nat. Sci.* **2002**, *12*, 38–42.
10. Huang, F.; Ma, W.; Li, M.; Ma, Y. Analysis on responses of land surface temperature on the Northern Tibetan Plateau to climate change. *Plateau Meteorol.* **2016**, *35*, 55–63. [CrossRef]
11. Ma, W.; Ma, Y.; Hu, Z.; Li, M.; Wang, J.; Qian, Z. Analyses on Surface Radiation Budget in Northern Tibetan Plateau. *Plateau Meteorol.* **2004**, *23*, 348–352.
12. Ma, W.; Ma, Y.; Hu, Z.; Li, M.; Sun, F.; Gu, L.; Wang, J.; Qian, Z. The Contrast Between the Radiation Budget Plus Seasonal Variation and Remote Sensing Over the Northern Tibetan Plateau. *J. Arid Land Resour. Environ.* **2005**, *19*, 109–115.
13. Philipona, R.; KRäUCHI, A.; Kivi, R.; Peter, T.; Wild, M.; Dirksen, R.; Fujiwara, M.; Sekiguchi, M.; Hurst, D.; Becker, R. Balloon-borne radiation measurements demonstrate radiative forcing by water vapor and clouds. *Meteorol. Z.* **2020**, *29*, 501–509. [CrossRef]
14. Obregón, M.; Costa, M.; Silva, A.; Serrano, A. Spatial and temporal variation of aerosol and water vapour effects on solar radiation in the mediterranean basin during the last two decades. *Remote Sens.* **2020**, *12*, 1316. [CrossRef]
15. Wang, Z.; Zhang, M.; Wang, L.; Qin, W.; Ma, Y.; Gong, W.; Yu, L. Investigating the all-sky surface solar radiation and its influencing factors in the Yangtze River Basin in recent four decades. *Atmos. Environ.* **2021**, *244*, 17888. [CrossRef]
16. Jandaghian, Z.; Akbari, H. Effects of increasing surface reflectivity on aerosol, radiation, and cloud interactions in the urban atmosphere. *Theor. Appl. Climatol.* **2020**, *139*, 873–892. [CrossRef]
17. You, Q.; Sanchez-Lorenzo, A.; Wild, M.; Folini, D.; Fraedrich, K.; Ren, G.; Kang, S. Decadal variation of surface solar radiation in the Tibetan Plateau from observations, reanalysis and model simulations. *Clim. Dyn.* **2013**, *40*, 2073–2086. [CrossRef]
18. Wu, R.; Ma, Y. Comparative analyses on radiation characteristics in different areas over the Tibetan Plateau. *Plateau Meteorol.* **2010**, *29*, 251–259.
19. Ma, Y.; Hu, Z.; Xie, Z.; Ma, W.; Wang, B.; Chen, X.; Li, M.; Zhong, L.; Sun, F.; Gu, L.; et al. A long-term (2005–2016) dataset of hourly integrated land-atmosphere interaction observations on the Tibetan Plateau. *Earth Syst. Sci. Data* **2020**, *12*, 2937–2957. [CrossRef]
20. Liang, S. *Quantitative Remote Sensing of Land Surfaces*, 1st ed.; John Wiley & Sons, Inc: Hoboken, NJ, USA, 2003; ISBN 9780471281665.
21. Gu, X.; Ma, Y.; Ma, W.; Sun, F. Climatic characteristics of surface radiation flux over the Qinghai-Tibetan Plateau. *Plateau Meteorol.* **2018**, *37*, 1458–1469. [CrossRef]
22. Srivastava, A.; Rodriguez, J.; Saco, P.; Kumari, N.; Yetemen, O. Global Analysis of At-mospheric Transmissivity Using Cloud Cover, Aridity and Flux Network Datasets. *Remote Sens.* **2021**, *13*, 1716. [CrossRef]
23. Baigorria, G.; Villegas, E.; Trebejo, I.; Carlos, J.; Quiroz, R. Atmospheric transmissivity: Distribution and empirical estimation around the central Andes. *Int. J. Climatol. A J. R. Meteorol. Soc.* **2004**, *24*, 1121–1136. [CrossRef]
24. Tang, W.J.; Yang, K.; Qin, J.; Cheng, C.C.K.; He, J. Solar radiation trend across China in recent decades: A revisit with quality-controlled data. *Atmos. Chem. Phys.* **2011**, *11*, 393–406. [CrossRef]
25. Yang, K.; Ding, B.; Qin, J.; Tang, W.; Lu, N.; Lin, C. Can aerosol loading explain the solar dimming over the Tibetan Plateau? *Geophys. Res. Lett.* **2012**, *39*, L20710. [CrossRef]
26. Bounoua, L.; Collatz, G.; Los, S.; Sellers, P.; Dazlich, D.; Tucker, C.; Randall, D. Sensitivity of climate to changes in NDVI. *J. Clim.* **2000**, *13*, 2277–2292. [CrossRef]
27. Chu, D.; Ma, W.; Zhaxi, D. Land Surface Albedo in the North Tibetan Plateau from Ground Observations and MODIS. *Remote Sens. Technol. Appl.* **2015**, *30*, 908–916. [CrossRef]
28. Li, Y.; Hu, Z. A Study on Parameterization of Surface Albedo over Grassland Surface in the Northern Tibetan Plateau. *Adv. Atmos. Sci.* **2009**, *26*, 161–168. [CrossRef]
29. Pang, G.; Chen, D.; Wang, X.; Lai, H. Spatiotemporal variations of land surface albedo and associated influencing factors on the Tibetan Plateau. *Sci. Total Environ.* **2021**, *804*, 150100. [CrossRef]
30. Tian, L.; Zhang, Y.; Zhu, J. Decreased surface albedo driven by denser vegetation on the Tibetan Plateau. *Environ. Res. Lett.* **2014**, *9*, 104001. [CrossRef]
31. Li, Y.; Zhao, L.; Zhou, H.; Xu, S.; Zhang, F. Changes in Reflected Radiation and Reflectivity for Growing Season of Alpine Swamp in the Northern Qinghai. *J. Glaciol. Geocryol.* **2007**, *29*, 137–143.
32. Held, I.; Soden, B. Water vapor feedback and global warming. *Annu. Rev. Energy Environ.* **2000**, *25*, 441–475. [CrossRef]
33. Stephens, G.L.; Wild, M.; Stackhouse, P.W., Jr.; Ecuyer, T.; Kato, S.; Henderson, D. The global character of the flux of downward longwave radiation. *J. Clim.* **2012**, *25*, 2329–2340. [CrossRef]
34. Wang, Y.; Sheng, P.; Liu, S. *Atmospheric Physics*, 1st ed.; Meteorological Press: Beijing, China, 1987.

35. Weng, D.; Chen, Y. Climatological Calculation of Downward Atmospheric Radiation for China and its Characteristic Distribution. *J. Nanjing Inst. Meteorol.* **1992**, *15*, 1–8. [CrossRef]
36. Ouyang, B.; Che, T.; Dai, L.; Wang, Z. Estimating Mean Daily Surface Temperature over the Tibetan Plateau Based on MODIS LST Products. *J. Glaciol. Geocryol.* **2012**, *34*, 296–303.

Article

Detecting and Attributing Evapotranspiration Deviations Using Dynamical Downscaling and Convection-Permitting Modeling over the Tibetan Plateau

Jingyu Dan, Yanhong Gao *  and Meng Zhang

Department of Atmospheric and Oceanic Sciences, Institute of Atmospheric Sciences, Fudan University, Shanghai 200438, China; 19213020003@fudan.edu.cn (J.D.); zhangm2012@outlook.com (M.Z.)
* Correspondence: gaoyh@fudan.edu.cn; Tel.: +86-21-31248844

Abstract: Terrestrial evapotranspiration (ET) over the Tibetan Plateau (TP) exerts considerable impacts on the local climate and the water cycle. However, the high-altitude, mountainous areas over the TP pose a challenge for field observations. To finely capture its ET characteristics, we employed dynamical downscaling modeling (DDM) with a 28 km resolution and convection-permitting modeling (CPM) with a 4 km resolution in a normal climatology year, 2014. The benchmark data were the surface energy balance–based global land ET dataset (EB). Other compared data included the Global Land-Surface Data Assimilation System (GLDAS) and two reanalysis datasets: ERA-Interim and ERA5. Results showed that EB exhibits a gradient from the southeastern to northwestern TP, which is in line with the precipitation pattern. GLDAS generally reproduces the annual mean magnitude and pattern but poorly represents the seasonal variations. DDM and CPM perform well in the monsoon season but underestimate ET in the non-monsoon season. The two reanalysis datasets greatly overestimate the ET in the monsoon season, but ERA-Interim performs well in the non-monsoon season. All five datasets underestimate the ET over tundra and snow/ice areas, both in the annual and seasonal means. ET deviations are dominated by precipitation deviations in the monsoon season and by surface net radiation deviations in the non-monsoon season.

Keywords: terrestrial evapotranspiration; Tibetan Plateau; convection-permitting modeling; monsoon season; non-monsoon season

Citation: Dan, J.; Gao, Y.; Zhang, M. Detecting and Attributing Evapotranspiration Deviations Using Dynamical Downscaling and Convection-Permitting Modeling over the Tibetan Plateau. *Water* **2021**, *13*, 2096. https://doi.org/10.3390/w13152096

Academic Editor: Gordon Huang

Received: 27 May 2021
Accepted: 27 July 2021
Published: 30 July 2021

Publisher's Note: MDPI stays neutral with regard to jurisdictional claims in published maps and institutional affiliations.

1. Introduction

The Tibetan Plateau (TP) is often referred to as "the roof of the world" and "the Asian water tower" because of the considerable impacts it exerts on the regional and global climate and water cycle. As the highest and widest plateau in the world, its complex and mountainous terrain is not only a barrier to the westerly belt at the same latitude but also strengthens the Indian monsoon through strong dynamic and thermodynamic effects, which accelerates the large-scale atmospheric circulation [1,2]. In addition, the TP possesses abundant water resources and is home to the sources of many of Asia's major rivers such as the Yangtze, Yellow, Tarim, Indus, Ganges, and Mekong [3,4]. The TP has attracted continuous attention amongst scientists owing to its sensitivity to climate change and its tendency to warm faster than the global average during the past few decades [5,6]. This enhanced warming has led to significant glacial and snow melt, permafrost degradation, and increasing precipitation [7,8], and these warming-induced changes also have marked effects on the processes of terrestrial evapotranspiration (ET) over the plateau, which accelerates the increase in ET [9].

As a critical component of water cycle, ET connects the land, ocean, and atmosphere; transports 60% of the world's precipitation over land [10]; affects the long-term evolution of vegetation [11,12]; and plays an important role in regulating regional droughts and floods [13]. As for the ET over the TP, research indicates that the mean ET ranges between

250 and 400 mm/year, is lowest in the northwestern TP and highest in the southeastern TP [9,14], and is limited by water in dry areas and by energy in wet areas at the annual scale [15]. Moreover, several studies have focused on the interannual variability of ET in the TP. For example, Zhang et al. estimated the ET in 16 catchments across the plateau and pointed out the rate of increase was 7 mm/(10 years) during 1966–2000 [16]. Li et al. investigated the ET of the Yellow, Yangtze, Qiangtang, and Qaidam basins from 1983 to 2006 and also found the ET showed an upward trend for all seasons [17]. Yin et al. pointed out that the increasing ET trend over the last 30 years is linked with increased precipitation [18].

The terrestrial ET is closely linked with land-surface characteristics (e.g., soil and land-use types). The altitude of the TP is mostly over 4000 m above the sea level, the general tendency is higher in the northwest TP and lower in the southeast TP. Restricted by the terrain height, the land-surface types over the TP are diverse, including grassland, wetland, tundra, snow, and so on [19,20]. Previous studies have found that the multi-year mean ET of low-covered grassland is much lower than that of medium- and high-covered grassland [21,22]; additionally, that of wetland over the TP has been increasing in recent years and may be responsible for wetland degradation [23]. The ET in the permafrost region of the TP has also been found to be affected by the freeze–thaw cycle and presents obvious seasonal changes [24]. Moreover, a recent study indicated that lake evaporation in the south is 1171.9 mm during the ice-free season, which is higher than that of northern TP (1059.7 mm), and the evaporated water amount is about 51.7 km^3 per year when all plateau lakes are included [25].

However, due to the complex topography and harsh natural environmental conditions, observation sites are sparsely distributed, and the representativeness of station observations is limited. Thus, considerable uncertainty exists with respect to the ET over the various land-surface types of the TP.

Given the sparseness of site observations, remote sensing and numerical climate simulations have provided alternative solutions for understanding ET at large scales and causal attributions for the deviations of ET over the TP using the above two ways has been addressed in some studies. Remote sensing usually combined with the traditional methods such as the Penman–Monteith (P-M) method [26], or with the surface energy balance methods, for example, the Surface Energy Balance System (SEBS) [27,28], these methods have a sound physical basis, but the uncertainty in the parameterization of environmental stress factors in the equation usually bring deviations of ET [29–32]. As for the numerical simulations, two kinds of simulation can be employed to model ET: "off-line" and "on-line". The former uses a land-surface model or hydrological model driven by near-surface meteorological variables, which is widely used in ET responses to climate changes. The latter uses a climate model coupled with a land-surface model or hydrological model. Compared to the former, the latter not only involves the ET responses to climate changes but also the ET feedbacks to regional climate. Therefore, it is a more physically reasonable way to simulate the land–atmosphere interaction, and both global and regional climate models can be used in on-line simulations. Although the land-surface model and global climate models are often used for climate studies, the ET over the TP remain poorly modeled, and several studies point out that the deviations of ET over TP were from the deviation of the storage of summer soil water storage, negative bias of precipitation, and the overestimation of downward shortwave radiation flux [15,33]. Some studies also indicated that these imperfections in global climate models (GCMs) may be induced by the coarse resolution and parameterizations for large-scale processes [34,35]. Regional climate models (RCMs) have higher spatial resolution and better depiction of physics parameterizations for meso- and micro-processes compared with GCMs [36–38]. Thus, RCMs are used for dynamic downscaling of the coupled coarse grid GCM output data or the input reanalysis data [39,40], so they provide data with high spatial and temporal resolution, which are to some extent more suitable for heterogeneous regions. Nowadays, this is called dynamical downscaling modeling (DDM) and has been widely applied to study regional climate and climate change in North America, Europe, Africa, and Asia [41–47]. As for the TP, Gao et al. showed that 30 km DDM can significantly reduce the biases in precipitation and give a more accurate net precipitation (precipitation minus ET) over the TP [48], this is in agreement with Lin et al., who indicated that the finer resolutions can greatly diminish the positive precipitation deviations over the TP, especially from 30 to 10 km [49]. However, because of the limitation of convection parameterization, the highest resolution of DDMs obtained over the TP can only reach 0.25° (≈28 km), which is still not high enough for the ET simulations over the TP [50,51].

Recently, the common calculation errors caused by the use of convection parameterization schemes were mentioned by some studies [52–56]. Meanwhile, Prein et al. and Ou et al. also pointed out that the convection parameterization schemes start to break down as deep convection starts to be resolved, explicitly when grid resolution becomes smaller than 10 km, and a so-called "gray zone" existed when grid spacing was between 10 and 4 km, at these scales, the individual convective cells cannot be resolved and may lead to insufficiently resolved deep convection [54,57]. Thus, convection-permitting modeling (CPM), which not only removes the convection parameterization scheme, but also greatly improves the data resolution and the representation of complex orography, is a more advanced method than DDM and has become a better choice for researchers studying the TP. Previous studies found that CPM usually gives more accurate results over the TP, especially for precipitation [49,58]. However, detailed comparison and evaluation of CPM-simulated ET over TP is rare. Therefore, in this study, we investigated (1) the performance of "on-line" simulations in ET over TP compared to remote sensing and "off-line" simulations; (2) whether dynamical downscaling (DDM and CPM) performs better in terms of the heterogeneity of ET—for instance, over different land-surface types—over the TP than global models; and (3) what causes the deviation of ET simulated by CPM and DDM over the TP, and whether it varies with the seasons.

The paper is structured as follows: Section 2 briefly introduces the Weather Research and Forecasting (WRF) model and model configurations, datasets, and methods. Section 3 evaluates the datasets mentioned above in terms of annual and seasonal means, as well as dominant land-use types, and then explores the factors contributing to the deviation in ET. Finally, the main conclusions and some further discussions are presented in Sections 4 and 5, respectively.

2. Model, Datasets, and Methods

2.1. Model and Configurations

The Weather Research and Forecasting (WRF) model is a next-generation mesoscale numerical weather prediction system (http://www.wrf-model.org/index.php, accessed on 20 June 2021) [59], which was initially developed in the late 1990s and is still being improved today by scientists all over the world. WRF includes several dynamic cores, such as a fully mass- and scalar-conserving flux from the mass coordinate version, as well as many different physical parameterizations (e.g., microphysics, cumulus parameterization, planetary boundary layer, shortwave and longwave radiation, and land-surface models) [51,60,61].

The DDM run (outer domain) covered the entire Eurasian continent with a 28 km horizontal grid spacing, while the CPM run (inner domain) covered the entire TP with a 4 km horizontal grid spacing. The topography of the TP is shown in Figure 1a,b. Clearly, the 4 km resolution topography shows far more detail in terms of mountain tops and valleys, especially in the southern TP.

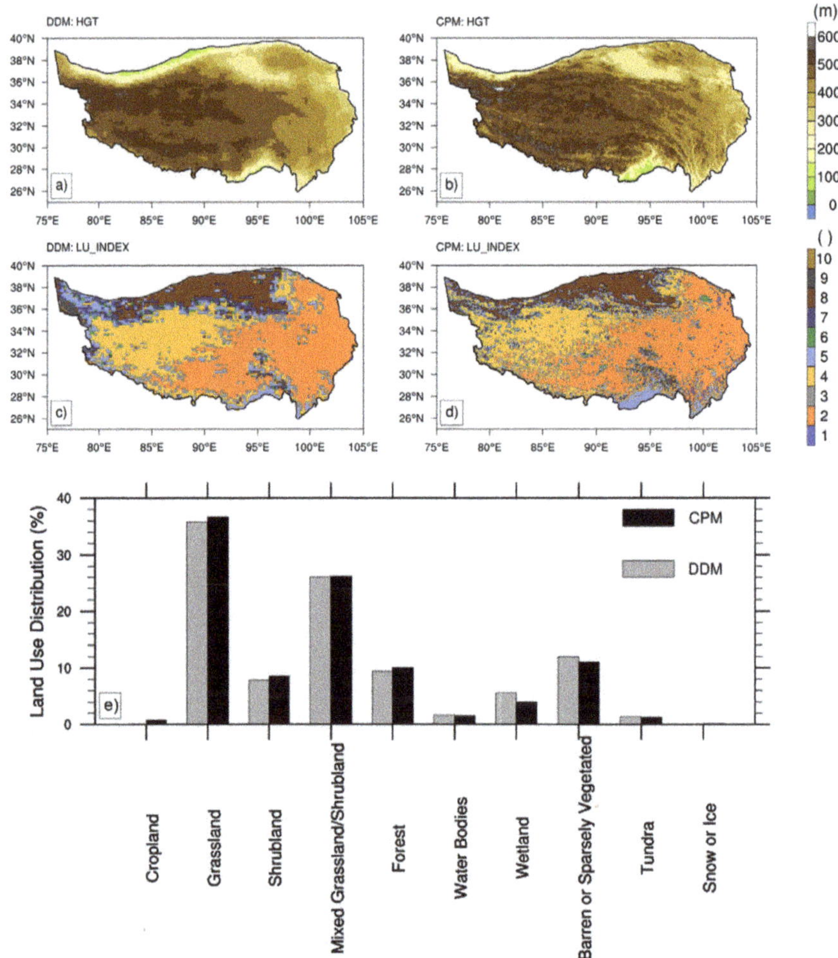

Figure 1. (**a,b**) The distribution of topography (unit: m), and (**c,d**) the dominant land-use categories over the Tibetan Plateau (TP). Panels (**a,c**) are from DDM (dynamical downscaling modeling), and panels (**b,d**) are from CPM (convection-permitting modeling). (**e**) The land-use distribution (unit: %) of ten dominant categories. In (**c–e**), the numbers 1–10 stand for cropland, grassland, shrubland, mixed grassland/shrubland, forest, water bodies, wetland, barren or sparsely vegetated, tundra, snow or ice, respectively.

Following our group's previous study on DDM and CPM [51], the NCAR (National Center for Atmospheric Research) CAM (Community Atmosphere Model) radiation scheme [62], the WSM6 (WRF Single-Moment 6-class microphysics scheme) [63], Yonsei University planetary boundary layer scheme [64], the Kain–Fritsch convection scheme [65] (used in the DDM run but not in the CPM run), and the Noah LSM (land-surface model) four-layer soil temperature and moisture model [66] were employed in this study too. The specific model configurations are shown in Table 1. We selected 2014 as the research period because the precipitation in this year was close to the climatological mean, which can help avoid the influence of certain extreme conditions and better test the ability of CPM and DDM to simulate ET under the climate mean state. The lateral boundary conditions and sea surface temperature (SST) were provided by the ERA-Interim reanalysis dataset at 6 h intervals. For both DDM and CPM, the simulations were separated into two stages: the first was initialized at 0000 UTC 1 October 2013, ended at 2300 UTC 31 May 2014, and archived in 3 h intervals; and the second was initialized at 0000 UTC 1 June 2014, ended at 2300 UTC 31 December 2014, and archived in 1 h intervals.

Table 1. The Weather Research and Forecasting (WRF) model configurations for dynamical downscaling modeling (DDM) and convection-permitting modeling (CPM).

WRF 3.8	DDM	CPM
Domain Region	Eurasian continent (centered at (92.8°E, 37.8°N))	Tibetan Plateau (75-105°E, 26-40°N)
Grid Cells	280 × 184	750 × 414
Horizontal Grid Spacing	28 km	4 km
Vertical Levels	27	
Radiation Scheme	CAM (Community Atmosphere Model) [62]	
Microphysics Scheme	WSM6 (WRF Single-Moment 6-class microphysics scheme) [63]	
Planetary Boundary Layer Scheme	Yonsei University [64]	
Convection Scheme	Kain-Fritsch [65]	Explicit
Land Surface Model	Noah [66]	
Boundary Conditions and SST	ERA-Interim reanalysis	

2.2. Datasets

Besides the DDM and CPM simulations, a gridded remote sensing dataset (EB), a dataset from off-line simulations (GLDAS), and two large-scale reanalysis datasets (ERA-Interim and ERA5) were adopted for comparison.

The EB dataset (http://data.tpdc.ac.cn, accessed on 20 June 2020) was derived from satellite data and a surface energy balance method for the global land area. It includes daily and monthly datasets from 2000 to 2017, with a 5 km horizontal resolution. This dataset has been proven to be robust across a variety of land-cover types and performs well in providing spatial and temporal information on the water cycle and land–atmosphere interactions for the Chinese landmass [67]. Therefore, EB was used as the reference ET in this study.

GLDAS [68] simulates the ET from four land-surface models driven by a series of conventional and satellite-derived observations. GLDAS products have a spatial resolution of 0.25° × 0.25° and a temporal resolution of 3 h. For a long time, GLDAS has been the only gridded dataset for ET available over the TP [60,69] and has been found to perform well in terms of surface air temperature and precipitation forcing [70].

ERA-Interim and ERA5 are commonly used reanalysis datasets released by the ECMWF (the European Center for Medium-Range Weather Forecasts). ERA-Interim [71] covers the period 1979 to 2019 and has a horizontal resolution of 0.7°. It was produced with a 12 h 4D-Var (four-dimensional variational) data assimilation scheme and an IFS (Integrated Forecast System) release Cy31r2 forecast model. Previous studies have shown that it is the outstanding performer among all reanalysis datasets in describing the water cycle over TP [48,69].

ERA5 [72] is the successor to ERA-Interim, with a temporal resolution of 1 h and horizontal resolution of 0.1°. With a more sophisticated hybrid incremental 4D-Var system [73] and advanced forecast model (IFS release Cy41r2), ERA5 covers a longer period from 1950 to the present day and outputs more meteorological elements. Moreover, its radiative transfer model, land-surface model, and snow data assimilation are all different from those of ERA-Interim. These improvements may have advantages in exploring regions with complex terrain, such as the TP.

2.3. Methods

To intercompare the six datasets with different temporal and spatial resolutions, all of the datasets were first monthly averaged and then interpolated into the same 0.25° grid through the local area averaging method. The comparison was conducted in terms of the annual and seasonal means as well as seasonal variabilities. According to the typical seasonal variation of precipitation over the TP [2], the monthly data were further averaged to the monsoon season (May–September) and the non-monsoon season (October–April).

Since the calculation of ET is directly land-use dependent, we compared the dominant land-use types in the DDM and CPM simulations. There are 24 land-use types for the USGS (United States Geological Survey) categories in the WRF model (http://www2.mmm.ucar.edu/wrf/users/docs/user_guide_V3.8/AWUsersGuideV3.8.pdf, accessed on 20 June 2021) [74]. However, not all of them exist over the TP. To present the categories clearly, we reclassified these land-use types over the TP into 10 categories (cropland, grassland, shrubland, mixed shrubland/grassland, forest, water bodies, wetland, barren or sparsely vegetated, tundra, and snow or ice) (provided in Table 2). The distributions of these 10 categories are shown in Figure 1c–e. Considering that the EB data do not include the underlying surface of water bodies, all comparisons in this study do not consider this surface type. The distributions according to DDM and CPM are quite similar, with very slight differences in percentages for each category (Figure 1c–e). Therefore, the distribution of the 10 dominant categories of CPM can be used in the comparison among the 0.25° grid cells via the nearest neighbor algorithm. To investigate the contributions from each land-use category to the TP average, the ET in the same category was averaged and intercompared among the six datasets.

Table 2. Reclassified 10-category land-use categories.

Land Use Category	Land Use Description	USGS (United States Geological Survey) [74]
1	Cropland	2–6
2	Grassland	7
3	Shrubland	8
4	Mixed Shrubland/Grassland	9-10
5	Forest	11–15
6	Water Bodies	16
7	Wetland	17-18
8	Barren or Sparsely Vegetated	19
9	Tundra	20–23
10	Snow or Ice	24

In addition, to better judge whether the ET deviation from the other five datasets over the dominate land-use categories of TP is significant compared to EB, we also conducted a 99% significance test with the Monte-Carlo method [75], which does not require the data to be normally distributed. The Monte-Carlo method regards the two datasets with sample sizes of n_a and n_b as a method of randomly extracting n_a samples from the total $(n_a + n_b)$ samples, it calculates the absolute value, D_i, of the sample mean difference from m resampled group as follows:

$$D_i = |\overline{x_{ai}} - \overline{x_{bi}}| \tag{1}$$

where $\overline{x_{ai}}$, $\overline{x_{bi}}$ represent the averaged value of the two new datasets in each group; $i = 1, 2, 3, \ldots, m$; m is re-sampling times, and m = 10,000 in this study. Then, the obtained D_i is sorted from smallest to largest and compared with the original sequence mean difference, D_{ori}. If the D_{ori} satisfies the following condition:

$$D_{ori} < D_{\frac{\alpha}{2}} \text{ or } D_{ori} > D_{1-\frac{\alpha}{2}} \tag{2}$$

where $\alpha = 0.01$ in this study, it is considered that the samples have passed the 99% Monte-Carlo significance test, and there is a significant difference between the two compared datasets.

Many factors play a key role in the simulation of ET, especially precipitation [76–78]. In addition, radiation provides the energy source for ET. However, the benchmark data EB is a satellite-derived dataset and has no other matching variables. Thus, it cannot be used as an attribution reference criterion. In order to explore the factors influencing the deviations of simulated ET, first, EB was used as the standard to find the nearest and farthest datasets to it, and then the nearest one was used as the reference in the comparison and the farthest to explore the source of deviations in attribution analysis (in Section 3.3).

Then, to explore the influence of precipitation on deviations in ET, the deviations of ET (dET) of the selected datasets were separated into those from precipitation (dP) and all other types of deviation (dET-dP), which were obtained from

$$\mathrm{d}A_i = A_{si} - A_{refi} \tag{3}$$

where dA_i is an abstract symbol, which can be replaced by deviations from ET (mm/d), precipitation (mm/d), net radiation flux (w m^{-2}), or soil temperature (K) in this equation; A_{si} is the corresponding variable (ET, precipitation, net radiation flux, or soil temperature) of the dataset furthest from the ET value of the EB data (mm/d); A_{refi} is the corresponding variable (ET, precipitation, net radiation flux, or soil temperature) of the dataset closest to the ET value of the EB data (mm/d); i is the land-use type range from 1 to 10 without 6 (the water bodies).

To further explore the contributions from radiation and the land-surface model in the non-monsoon season, the dET-dP was further analyzed by comparing deviations from surface net radiation (dR) and soil temperature (dST), which can be calculated using Equation (1). Considering the different units, a relative contribution was defined by normalized deviations, and defined as

$$\mathrm{d}B_{rj} = \frac{|dB_j|}{|dB_{max}|} \times 100\% \tag{4}$$

where dB_j is an abstract symbol, which can be replaced by the deviations from net radiation flux (w m^{-2}) or soil temperature (K) in this equation; dB_{rj} is the relative deviation (%) of the corresponding variable (net radiation flux or soil temperature); dB_{max} is the maximum deviation of the corresponding variable [net radiation flux (w m^{-2}) or soil temperature (K)] of all nine land-use types (without the water bodies); j is land-use type 9 (the tundra) or 10 (snow or ice).

The relative contributions from surface net radiation (dRr) and the land-surface model (dSTr - dRr) are represented by the magnitudes of the normalized deviations.

3. Results

3.1. Annual and Seasonal Mean ET

Figure 2 presents the annual mean distribution of ET in EB and its differences with those of the GLDAS, ERA-Interim, ERA5, DDM, and CPM datasets. EB exhibits a decrease in gradient from the southeastern TP to the northwestern TP. The daily ET reaches 2 mm d^{-1} in the southeastern TP and is almost down to zero in the northwestern TP (Figure 2a). The spatial distribution is generally in line with the pattern of precipitation over the TP [50,51]. Compared to EB, GLDAS shows a very similar annual mean spatial distribution (Figure 2b). ERA-Interim and ERA5 both significantly overestimate the ET almost for the entire TP (Figure 2c,d). As for the DDM and CPM, the simulations are very similar, both of them producing lower ET than EB in the central and southern regions of the plateau, with CPM being relatively lower than DDM (Figure 2e,f).

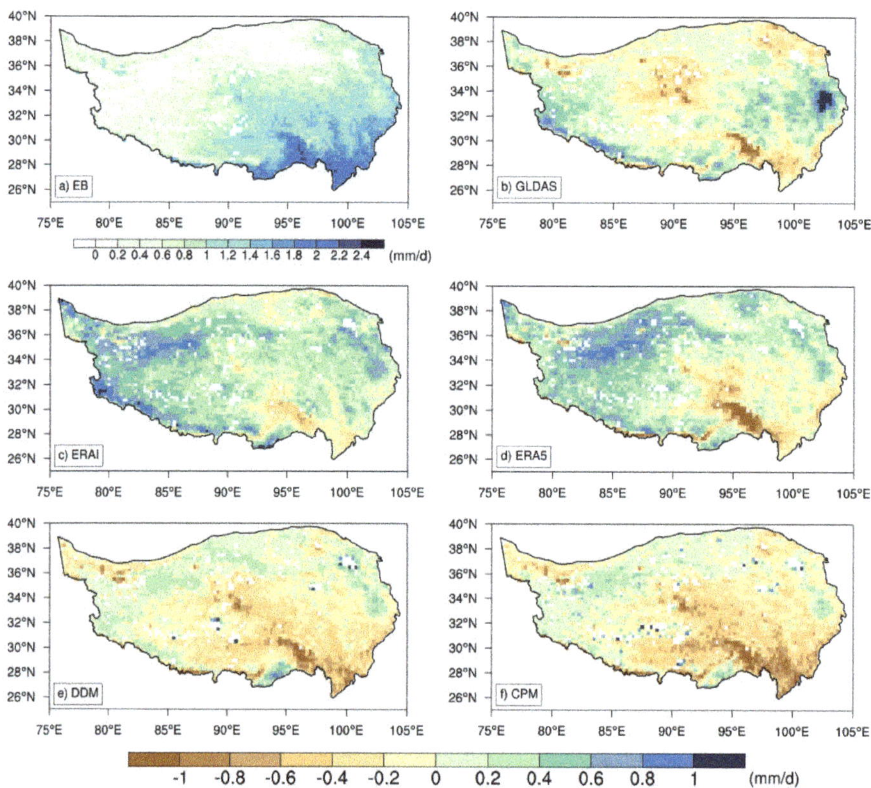

Figure 2. Distribution of the daily terrestrial evapotranspiration (ET) in 2014 in the (**a**) EB data (units: mm/d) and (**b**–**f**) the deviation from the EB distribution in the (**b**) GLDAS, (**c**) ERA-Interim, (**d**) ERA5, (**e**) DDM and (**f**) CPM data.

Figure 3 shows the annual and seasonal mean ET for the six datasets in 2014 averaged over the TP. Averaged over the TP, EB estimates around 0.79 mm of moisture was evaporated per day in 2014. Broken down, this figure is 1.12 mm d^{-1} for the monsoon season, which accounts for most of the annual ET, and 0.52 mm d^{-1} for the non-monsoon season. GLDAS shows obvious deviation in both monsoon season and non-monsoon season in comparison with EB. ERA-Interim and ERA5 have the largest annual deviation, which was mainly from the monsoon season. The DDM and CPM are the closest datasets to EB in the monsoon season (Figure 3 and Table 3), while CPM is the only one that underestimates the ET.

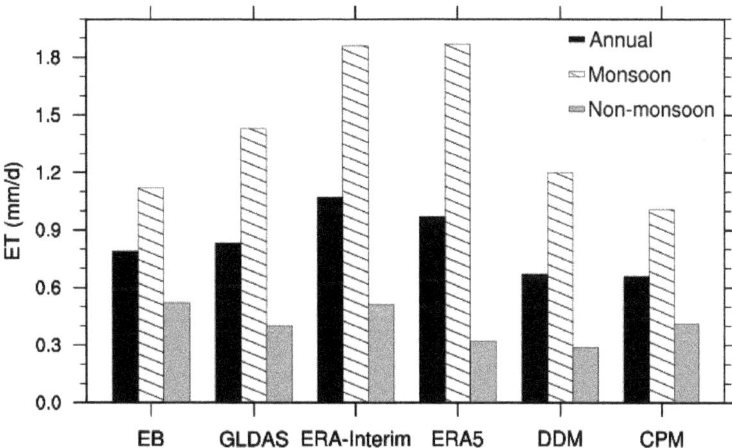

Figure 3. The annual and monsoon/non-monsoon seasonal mean ET in 2014 from the EB, GLDAS, ERA-Interim, ERA5, DDM, and CPM data averaged over the TP (units: mm/d).

Table 3. The mean ET from the EB, GLDAS, ERA-Interim, ERA5, DDM, and CPM data, and the standard deviation (SD), the root-mean-square error (RMSE), and spatial correlation coefficients (CORR) of GLDAS, ERA-Interim, ERA5, DDM, and CPM compared to EB.

		EB	GLDAS	ERA-Interim	ERA5	DDM	CPM
Value (unit: mm/d)	Ann	0.79	0.83	1.07	0.97	0.67	0.66
	Monsoon	1.12	1.43	1.86	1.87	1.20	1.01
	Non-Monsoon	0.52	0.40	0.51	0.32	0.29	0.41
SD	Ann	0.53	0.53	0.36	0.43	0.41	0.47
	Monsoon	0.73	0.94	0.61	0.63	0.76	0.63
	Non-Monsoon	0.41	0.30	0.25	0.38	0.20	0.45
RMSE	Ann	—	0.34	0.41	0.42	0.33	0.38
	Monsoon	—	0.65	0.89	0.95	0.46	0.47
	Non-Monsoon	—	0.30	0.21	0.37	0.38	0.38
CORR	Ann	—	0.94	0.95	0.91	0.95	0.93
	Monsoon	—	0.94	0.92	0.91	0.95	0.94
	Non-Monsoon	—	0.90	0.95	0.87	0.90	0.84

The seasonal mean ET of EB shows a great decrease in the non-monsoon season when compared to that in the monsoon season (Figure 3), which is mainly due to the large decrease in ET over the southern and central TP in the non-monsoon season in comparison with the monsoon season (Figure 4a,b). As for the GLDAS, it is relatively larger in the southeastern TP in the monsoon season and underestimated over the central and southern TP in the non-monsoon season (Figure 4c,d), which results in a balance in the annual mean. The overestimation of ERA-Interim and ERA5 exists almost across the whole TP but is largest in the northwestern TP in the monsoon season (Figure 4e,g). This is unsurprising given the known positive precipitation bias of ERA-Interim and ERA5 over the TP [79]. As for the non-monsoon season, ERA-Interim performs best compared to EB (Figure 4f and Table 3), while ERA5 shows an obvious underestimation especially for the southeastern TP (Figure 4h). DDM and CPM are the closest datasets to EB in the monsoon season (Figure 4i,k); however, in the non-monsoon season, DDM is significantly lower than EB for most of the TP but especially in the southeast (Figure 4j), while CPM ranks as the next best in comparison (Figure 4l and Table 3).

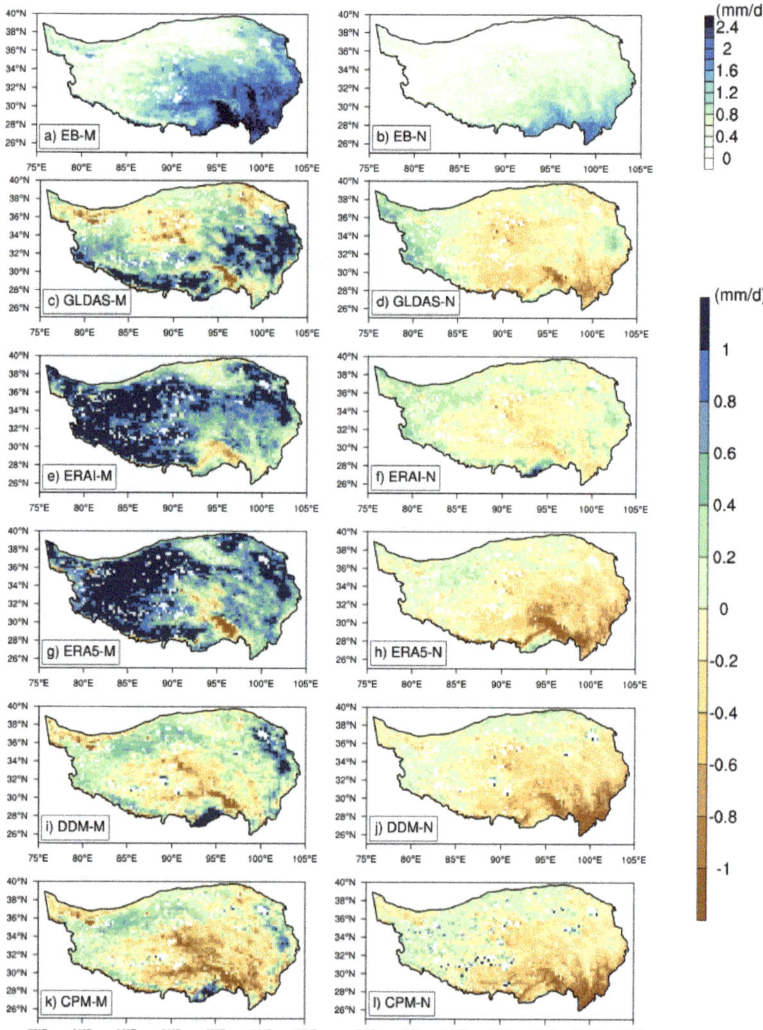

Figure 4. Distributions of the daily ET from the (**a**,**b**) EB data (units: mm/d), and (**c**–**l**) the differences of the (**c**,**d**) GLDAS, (**e**,**f**) ERA-Interim, (**g**,**h**) ERA5, (**i**,**j**) DDM, and (**k**,**l**) CPM data compared to EB, averaged over the monsoon (left-hand panels) and non-monsoon (right-hand panels) seasons.

3.2. ET over Dominant Land-Use Categories

Figure 5 presents box-and-whisker plots of the ET of nine land-use types based on the six datasets. The ET values averaged over the monsoon and non-monsoon seasons are given in Figures 6 and 7, respectively. In addition, we also used the Monte-Carlo test to detect whether the simulated ET is significantly different from EB, and the results are shown in Table 4.

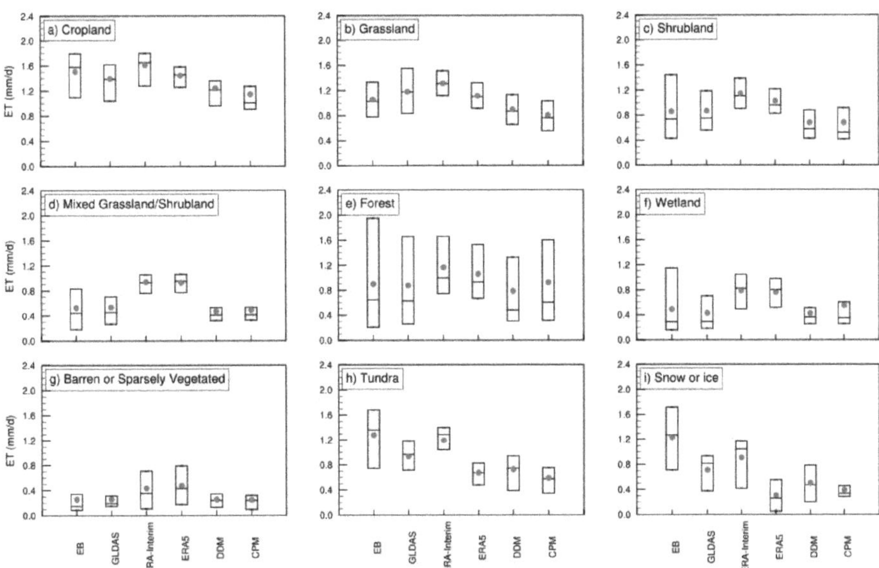

Figure 5. Box-and-whisker plots of ET of nine land-use categories based on six datasets averaged in 2014: (**a**) cropland; (**b**) grassland; (**c**) shrubland; (**d**) mixed grassland/shrubland; (**e**) forest; (**f**) wetland; (**g**) barren or sparsely vegetated; (**h**) tundra; (**i**) snow or ice (units: mm/d). The top and bottom of each box are the 20th and 80th percentiles in grid cells with the same land-use type. The lines in each box represent the median; red dots represent the mean.

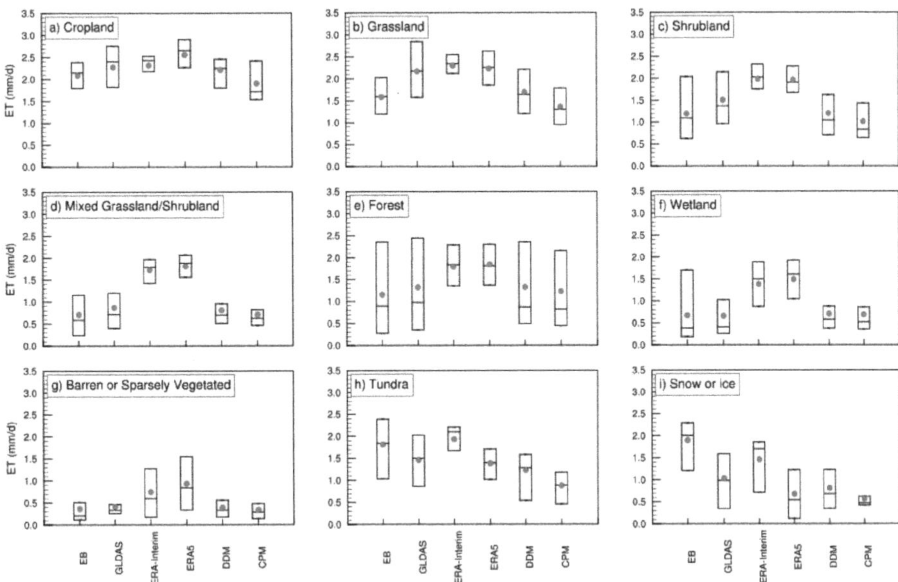

Figure 6. Box-and-whisker plots of ET of nine land-use categories based on six datasets for the monsoon season in 2014: (**a**) cropland; (**b**) grassland; (**c**) shrubland; (**d**) mixed grassland/shrubland; (**e**) forest; (**f**) wetland; (**g**) barren or sparsely vegetated; (**h**) tundra; (**i**) snow or ice (units: mm/d). The top and bottom of each box are the 20th and 80th percentiles in grid cells with the same land-use type. The lines in each box represent the median; red dots represent the mean.

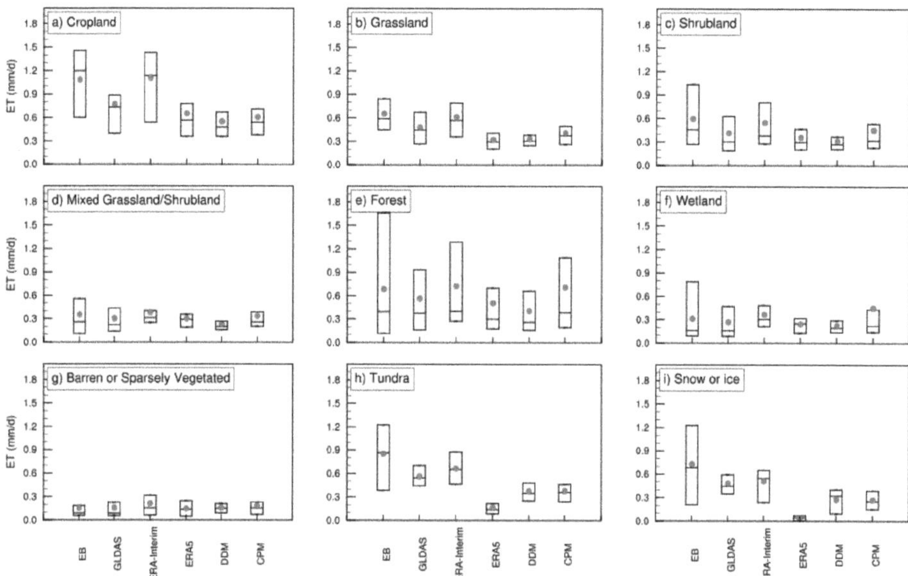

Figure 7. Box-and-whisker plots of ET of nine land-use categories based on six datasets for the non-monsoon season in 2014: (**a**) cropland; (**b**) grassland; (**c**) shrubland; (**d**) mixed grassland/shrubland; (**e**) forest; (**f**) wetland; (**g**) barren or sparsely vegetated; (**h**) tundra; (**i**) snow or ice (units: mm/d). The top and bottom of each box are the 20th and 80th percentiles in grid cells with the same land-use type. The lines in each box represent the median; red dots represent the mean.

Results from Figure 5 indicate that the ET of EB is generally larger over the vegetated ground of the TP. Cropland evaporates the most, with an annual mean value of 1.6 mm d^{-1}, while barren or sparsely vegetated ground only has 0.3 mm d^{-1}. The tundra and snow or ice surfaces present relatively high values for the annual mean. GLDAS shows relatively high consistency with EB in its annual mean over most land-use types, except grassland and tundra. Moreover, although the ET of GLDAS is also seen to be lower than that of EB over the snow/ice land surface, it is not significant (Table 4). The annual mean ET of ERA-Interim and ERA5 are relatively higher than those of EB, especially for the shrubland, mixed grassland/shrubland, forest, and wetland, while DDM and CPM produce lower annual mean ET over vegetated land.

Figure 6 presents the ET value of nine land-use types in the monsoon season. The ET of the EB dataset in the monsoon season is in line with its annual mean situation, the vegetated land evaporates generally more than the barren land. However, the performance of GLDAS in terms of the seasonal mean is unsatisfactory. In the monsoon season, clear overestimation exists for shrubland and grassland, which cover the largest area of the TP (Figure 1). The values provided by ERA-Interim and ERA5 are generally higher than those of other datasets, and this phenomenon exists and is significant for nearly all surfaces in the monsoon season (Table 4). The most obviously overestimated land-surface type of the reanalysis during the monsoon season is mixed grassland/shrubland, with the deviations almost reaching 1.1 mm d^{-1}. Moreover, for DDM and CPM, they both perform much better than other datasets and are closer to the EB data over most surfaces in the monsoon season.

Water 2021, 13, 2096

Table 4. The Monte-Carlo significance test of ET at confidence level of 99% over nine land-use categories of GLDAS, ERA-Interim, ERA5, DDM, and CPM compared to EB. * Indicates that the value has passed the 99% significance test, which means the ET is significantly different from that of EB over the corresponding land-surface type.

		99% Monte-Carlo Test				
		GLDAS	ERA-Interim	ERA5	DDM	CPM
Cropland	Ann					*
	Monsoon			*		
	Non-Monsoon	*		*	*	*
Grassland	Ann	*	*	*	*	*
	Monsoon	*	*	*	*	*
	Non-Monsoon	*	*	*	*	*
Shrubland	Ann		*	*	*	*
	Monsoon	*	*	*		*
	Non-Monsoon	*		*	*	*
Mixed Shrubland/Grassland	Ann		*	*	*	
	Monsoon	*	*	*	*	
	Non-Monsoon	*		*	*	
Forest	Ann		*	*		
	Monsoon		*	*		
	Non-Monsoon	*		*	*	
Wetland	Ann		*	*		
	Monsoon		*	*		
	Non-Monsoon					
Barren or Sparsely Vegetated	Ann		*	*		
	Monsoon		*	*		
	Non-Monsoon		*			
Tundra	Ann	*		*	*	*
	Monsoon	*		*	*	*
	Non-Monsoon	*	*	*	*	*
Snow or Ice	Ann			*	*	*
	Monsoon	*		*	*	*
	Non-Monsoon			*		

As for the non-monsoon season shown in Figure 7, the tundra and snow or ice surfaces of EB still present relatively high values. Apart from the monsoon season, GLDAS shows large underestimations over cropland in the non-monsoon season (Figure 7 and Table 4) and needs further study as well. ERA-Interim agrees well with EB for nearly all dominant land surfaces in comparison to the non-monsoon season, while the ET in ERA5 is generally lower than that of ERA-Interim, except for wetland and barren or sparsely vegetated land. The lower annual mean ET over vegetated land produced by DDM and CPM (shown in Figure 5) is mainly due to underestimation in the non-monsoon season, especially for DDM. As shown, DDM simulates considerably lower ET over cropland, grassland, shrubland, mixed shrubland/grassland, and forest, while CPM underestimates the ET over cropland, grassland as well as shrubland, which ultimately leads to the underestimation over the whole of the TP. Moreover, the underestimations for tundra and snow/ice in our simulations are significant, as Table 4 shows, and still need to be improved in future simulations.

3.3. Attribution of ET Deviations

The DDM- and CPM-simulated ET deviations compared to ERA-Interim and their counterparts from precipitation and other factors are presented in Figure 8. The solid lines represent a constant dET. Most points are located along the solid lines, which demonstrates that the ET deviation is equivalent for most land-cover categories, either in the monsoon or non-monsoon season. The dashed lines combined with the solid lines are used to separate the ET deviation. Points above or below the angle between the dashed and solid lines indicate that the ET deviation is more a result of the precipitation deviation. Points to the left or right indicate that the ET deviation is more a result of the other effects. Meanwhile, the points outside the red dashed box indicate that the deviations from precipitation and other effects are larger than one SD, which means the deviations are significant and cannot be ignored.

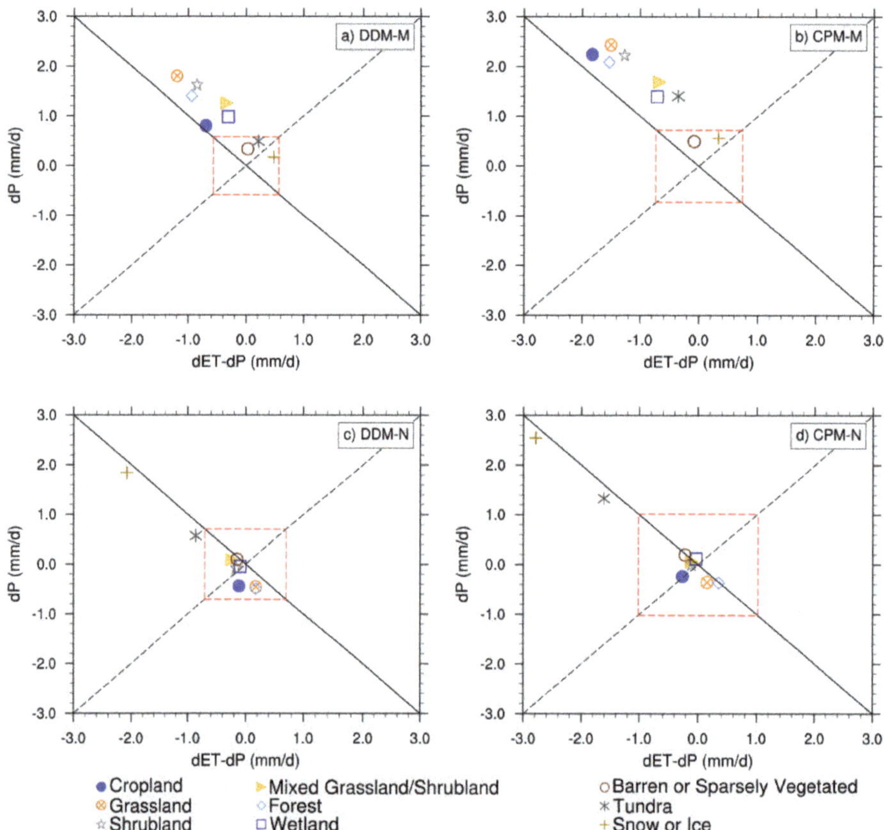

Figure 8. Deviations (monsoon: ERA-Interim minus DDM or CPM; non-monsoon: DDM or CPM minus ERA-Interim) in ET (units: mm/d) from precipitation and other factors for each land-use category in the (**a,b**) monsoon and (**c,d**) non-monsoon seasons. The ordinate is the ET deviation from precipitation, the abscissa is the ET deviation from the other effects. The black solid line indicates 1:1 correspondence of the first and third quadrants, and the gray dashed line indicates 1:1 correspondence of the second and fourth quadrants. The red dashed box indicates the SD from precipitation and other effects.

In the monsoon season, DDM and CPM agree well with EB. However, ERA-Interim and ERA5 show a large difference. Considering that the boundary conditions and SST in our simulations came from ERA-Interim, exploring its deviation from the two analog values can also lead to a better understanding of where the WRF mode has improved. Thus, DDM and CPM were selected as the reference data to explore the contributing factors to the ET deviations of ERA-Interim in the monsoon season. As shown in Figure 8a,b, all land-use categories in ERA-Interim, except snow or ice, show clear positive deviations in precipitation in comparison with DDM, which indicates that the overestimation of ET in ERA-Interim mainly comes from its large precipitation bias, especially for grassland. Importantly, the two WRF simulations greatly improve the result.

In the non-monsoon season, however, according to the previous analysis, the performances of DDM and CPM are not as good as those of the monsoon season. ERA-Interim shows the best performance and was, therefore, selected as the reference dataset in the non-monsoon season. The deviation from other factors dominates over most land surfaces of the TP, except cropland, grassland, and forest. As shown in Figure 8c,d, although most points are inside the red dashed box, the points for tundra and snow/ice show significant deviation. The other-factor deviations of the snow or ice surface type almost reach -3.0 mm d^{-1} in CPM. This is in line with the former analysis that precipitation in the monsoon season dominates the ET differences, while other effects such as surface radiation or deviations in the land-surface model play a relatively important role in the non-monsoon season.

As known, in the non-monsoon season, radiation is the dominant forcing that controls the energy balance of melting. To further explore the ET deviations over the tundra and snow/ice surfaces in the non-monsoon season, the relative contributions from surface net radiation deviation and land-surface process deviations were calculated and intercompared (Figure 9). All points are located above the 1:1 line, regardless of DDM or CPM, indicating that the lower ET of these two surfaces is mainly because of the underestimation of radiation, which may cause stronger freezing in the model. Specifically, radiation deviation of the tundra and snow/ice surfaces accounts for 90% and 62%, respectively, while the deviation from the land-surface processes is only 10% and 28%, in DDM. In CPM, the radiation deviation accounts for an even larger proportion, exceeding 90%.

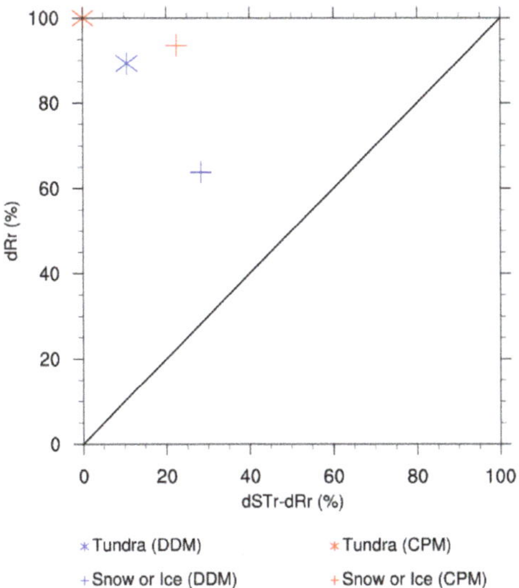

Figure 9. Relative contributions (normalized deviations between DDM (dynamical downscaling modeling) or CPM (convection-permitting modeling) and ERA-Interim) from surface net radiance deviation and land-surface processes deviation (unit: %) for tundra and snow or ice land-use categories in the non-monsoon season. The ordinate is the deviation from radiance; the abscissa is deviation from the land-surface processes. The black solid line indicates 1:1 correspondence of the first and third quadrants.

4. Discussion

The ET from remote sensing, off-line and on-line simulations was intercompared over the TP, and some significant deviations were found. There are, however, some drawbacks in this study.

4.1. Uncertainties of Usage Data

Site observations are generally considered to be the direct and most accurate measurement of ET, some atmospheric field experiments have gradually been carried out over the TP since 1970 [80–82], and several observation stations have been established [83]. However, due to the limited measurement space of one instrument, which can only measure a limited area within a few hundred to several kilometers [84], the existing observation sites on the TP are too few and insufficient to represent the ET over the whole TP.

Therefore, we selected the high-resolution remote sensing dataset EB, which performs well in land–atmosphere interactions over China, as the benchmark in this study. EB is based on the SEBS method, which requires parameterization of excessive resistance and is sensitive to the errors of air temperature (Ta) and land-surface temperature (Ts) [29]. Unlike the clear-sky conditions, the Ts used for ET calculation in EB can only be obtained by interpolation on cloudy days, and there is always error in interpolation, leading to uncertainty in the EB data. Moreover, many previous studies have noted that the uncertainty of lake parameterization and its impacts on regional climate are non-negligible [50,85,86] in current models; however, the EB data exclude the surfaces covered by lakes, meaning these areas were excluded in the present study, this is also a limitation of the EB dataset. Thus, multiple datasets from remote sensing can be considered for use in the following studies to reduce the uncertainties caused by a single dataset such as EB.

The other five datasets used in the study are all simulated from different models. GLDAS is the "off-line" simulation, which has wide applications in research. The data quality of GLDAS is mainly affected by the driving variables such as temperature, relative moisture, and wind speed [87]. Meanwhile, the lack of feedback mechanism also leads to the lack of the land–atmosphere interaction processes, which increases the uncertainties of GLDAS data as well.

The reanalysis datasets ERA-Interim and ERA5 are objective datasets obtained by conducting quality control on a series of observation data, including site observations, satellites, radar, soundings, aircraft, and ship, and then calculated through the numerical weather forecast data assimilation system. Therefore, these many sources of data bring to the reanalysis not only the advantages of a long-time scale and high spatial resolution but also the problem of poor data continuity [70,79]. The inherent uncertainties also come from the forecast model and data assimilation [88].

As for the "on-line" simulated DDM and CPM datasets, the atmospheric variables such as precipitation are also from simulation rather than being simply input as driving data. The selection of parameters in the model and the uncertainties of the parameterization scheme will influence the accuracy of ET, precipitation, and other variables simulated by DDM and CPM.

In addition, limited by the available computational resources, only a one-year simulation was conducted in this study. Uncovering whether the above results are of climatological significance would require further analysis with multi-year data.

4.2. Parameterizations of the Land-Surface Process

The analysis of the attribution of the ET deviation was only separated into precipitation and radiation deviation at present, with consideration of the water balance and energy balance. As is known, ET consists of soil evaporation, canopy transpiration, and canopy-intercepted water evaporation. Further in-depth study and analysis regarding the impacts of land-surface processes on ET deviations need to be carried out.

Land-surface processes can be solved by parameterization in models; however, previous studies indicated that despite the improvements offered by current simulation capabilities, there are still numerous uncertainties in the parameterization of land-surface processes. The DDM and CPM simulations underestimated the ET over vegetated ground, and the inaccurate representation of canopy light use, interception loss, and root water uptake processes resulting in underestimation of plant transpiration might be the reason for this [89,90]. Furthermore, the difference in the months when the coldest soil temperature appears between the reanalysis data and the WRF simulations (not shown) also indicates an insufficient consideration of snow parameterization, such as the higher snow albedo feedback mechanism, which results in lower incoming radiation and ultimately a cold deviation of the soil temperature [19,51].

Therefore, further investigations are necessary to improve the simulation effect of models and to obtain a more comprehensive and accurate understanding of the ET and water cycle process over the TP.

5. Conclusions

In this study, the ET from satellite merged data (EB), an off-line run (GLDAS), two global climate datasets (ERA5 and ERA-Interim) and two WRF dynamical downscaling simulations (DDM and CPM) were intercompared for a whole year, 2014, which is a normal year for precipitation, over the TP in terms of the annual mean, seasonal variation, and land-use types. Factors contributing to the deviations compared to EB were investigated in order to provide some insight into subsequent model improvements and land–atmosphere interaction simulations. The major conclusions are as follows:

Compared with the EB data, GLDAS generally reproduces the annual mean magnitude and spatial distribution, but seasonal variations are poorly presented. ERA-Interim and ERA5 generally overestimate ET, mainly in the monsoon season. ERA-Interim better reproduces the ET in the non-monsoon season. DDM and CPM perform well in the monsoon season but worse in the non-monsoon season.

ET is underestimated over tundra and snow or ice by all five datasets. GLDAS, ERA5, DDM, and CPM underestimate the ET over cropland, grassland, and shrubland in the non-monsoon season. GLDAS, ERA-Interim, and ERA5 overestimate the ET over grassland and shrubland in the monsoon season. ERA-Interim and ERA5 also overestimate the ET over the other four land-use types in the monsoon season.

Atmospheric forcing plays a dominant role in the simulation of ET. The considerable overestimation of precipitation dominates the ET deviation for the whole year, which is responsible for the particularly high results of ERA-Interim and ERA5 in the monsoon season. In the non-monsoon season, surface net radiation plays a secondary role after precipitation in the simulation of ET. The land-surface model exerts less impact than surface net radiation.

Author Contributions: Conceptualization, methodology: J.D. and Y.G.; validation: J.D., Y.G. and M.Z.; formal analysis, investigation: J.D.; data curation: Y.G.; writing—original draft preparation: J.D.; writing—review and editing: J.D., Y.G. and M.Z. All authors have read and agreed to the published version of the manuscript.

Funding: This research was jointly supported by the Second Tibetan Plateau Scientific Expedition and Research Program (grant 2019QZKK010314) and the Strategic Priority Research Program of the Chinese Academy of Sciences (grant XDA2006010202).

Institutional Review Board Statement: Not applicable.

Informed Consent Statement: Not applicable.

Data Availability Statement: Not applicable.

Acknowledgments: The authors thank Xuelong Chen for his valuable suggestions on this research. The authors thank the National Supercomputer Center in Tianjin for computational resources. The authors appreciate the National Tibetan Plateau Data Center for free access to the EB dataset (http://data.tpdc.ac.cn, 20 June 2021), the National Aeronautics and Space Administration (NASA) for free access to the GLDAS dataset, and the European Center for Medium-Range Weather Forecasts (ECMWF) for free access to the ERA5 and ERA-Interim datasets.

Conflicts of Interest: The authors declare no conflict of interest.

References

1. Bothe, O.; Fraedrich, K.; Zhu, X.H. Large-scale circulations and Tibetan Plateau summer drought and wetness in a high-resolution climate model. *Int. J. Climatol.* **2011**, *31*, 832–846. [CrossRef]
2. Yao, T.D.; Masson-Delmotte, V.; Gao, J.; Yu, W.S.; Yang, X.X.; Risi, C.; Sturm, C.; Werner, M.; Zhao, H.B.; He, Y.; et al. A review of climatic controls on δ18O in precipitation over the Tibetan Plateau: Observations and simulations. *Rev. Geophys.* **2013**, *51*, 525–548. [CrossRef]
3. Lu, C.X.; Yu, G.; Xie, G.D. Tibetan Plateau serves as a water tower. *IEEE Trans. Geosci. Remote Sens.* **2005**, *5*, 3120–3123.
4. Xu, X.D.; Lu, C.G.; Shi, X.H.; Gao, S.T. World water tower: An atmospheric perspective. *Geophys. Res. Lett.* **2008**, *35*, 1–5. [CrossRef]
5. Guo, D.L.; Wang, H.J. The significant climate warming in the northern Tibetan Plateau and its possible causes. *Int. J. Climatol.* **2011**, *32*, 1775–1781. [CrossRef]
6. Krause, P.; Biskop, S.; Helmschrot, J.; Flügel, W.-A.; Kang, S.; Gao, T. Hydrological system analysis and modelling of the Nam Co basin in Tibet. *Adv. Geosci.* **2010**, *27*, 29–36. [CrossRef]
7. Huntington, T.G. Evidence for intensification of the global water cycle: Review and synthesis. *J. Hydrol.* **2006**, *319*, 83–95. [CrossRef]
8. Yin, Y.H.; Wu, S.H.; Zhao, D.S. Past and future spatiotemporal changes in evapotranspiration and effective moisture on the Tibetan Plateau. *J. Geophys. Res. Atmos.* **2013**, *118*, 10850–10860. [CrossRef]
9. Kuang, X.X.; Jiao, J.J. Review on climate change on the Tibetan Plateau during the last half century. *J. Geophys. Res.* **2016**, *121*, 3979–4007. [CrossRef]
10. Oki, T.; Kanae, S. Global hydrological cycles and world water resources. *Science* **2006**, *313*, 1068–1072. [CrossRef]
11. Jung, M.; Reichstein, M.; Ciais, P.; Seneviratne, S.I.; Sheffield, J.; Goulden, M.L.; Bonan, G.; Cescatti, A.; Chen, J.Q.; de Jeu, R.; et al. Recent decline in the global land evapotranspiration trend due to limited moisture supply. *Nature* **2010**, *467*, 951–954. [CrossRef]
12. Wang, K.C.; Dickinson, R.E.; Wild, M.; Liang, S.L. Evidence for decadal variation in global terrestrial evapotranspiration between 1982 and 2002:1. Model development. *J. Geophys. Res.* **2010**, *115*, D20112. [CrossRef]
13. Douville, H.; Ribes, A.; Decharme, B.; Alkama, R.; Sheffield, J. Anthropogenic influence on multidecadal changes in reconstructed global evapotranspiration. *Nat. Clim. Change* **2013**, *3*, 59–62. [CrossRef]
14. Gao, G.; Chen, D.L.; Xu, C.Y.; Simelton, E. Trend of estimated actual evapotranspiration over China during 1960–2002. *J. Geophys. Res.* **2007**, *112*, D11120. [CrossRef]
15. Yang, K.; Ye, B.S.; Zhou, D.G.; Wu, B.Y.; Foken, T.; Qin, J.; Zhou, Z.Y. Response of hydrological cycle to recent climate changes in the Tibetan Plateau. *Clim. Chang.* **2011**, *109*, 517–534. [CrossRef]
16. Zhang, Y.Q.; Liu, C.M.; Tang, Y.H.; Yang, Y.H. Trends in pan evaporation and reference and actual evapotranspiration across the Tibetan Plateau. *J. Geophys. Res.* **2007**, *112*, D12110. [CrossRef]
17. Li, X.P.; Wang, L.; Chen, D.L.; Yang, K.; Wang, A.H. Seasonal evapotranspiration changes (1983–2006) of four large basins on the Tibetan Plateau. *J. Geophys. Res. Atmos.* **2014**, *119*, 13079–13095. [CrossRef]
18. Yin, Y.H.; Wu, S.H.; Zhao, D.S.; Zheng, D.; Pan, T. Modeled effects of climate change on actual evapotranspiration in different eco-geographical regions in the Tibetan Plateau. *J. Geogr. Sci.* **2013**, *23*, 195–207. [CrossRef]
19. Gao, Y.H.; Xiao, L.H.; Chen, D.L.; Chen, F.; Xu, J.W.; Xu, Y. Quantification of the relative role of land-surface processes and large-scale forcing in dynamic downscaling over the Tibetan Plateau. *Clim. Dyn.* **2017**, *48*, 1705–1721. [CrossRef]
20. Wang, X.D.; Zhong, X.H.; Fan, J.R. Spatial distribution of soil erosion sensitivity in the Tibet Plateau. *Pedosphere* **2005**, *15*, 465–472.
21. Cui, M.Y.; Wang, J.B.; Wang, S.Q.; Yan, H.; Li, Y.N. Temporal and spatial distribution of evapotranspiration and its influencing factors on Qinghai Tibet Plateau from 1982 to 2014. *J. Res. Ecol.* **2019**, *10*, 213–224.
22. Li, Z.Y.; Liu, X.H.; Ma, T.X.; Kejia, D.; Zhou, Q.P.; Yao, B.Q.; Niu, T.L. Retrieval of the surface evapotranspiration patterns in the alpine grassland-wetland ecosystem applying SEBAL model in the source region of the Yellow River, China. *Ecol. Modell.* **2013**, *270*, 64–75. [CrossRef]
23. Xue, Z.S.; Lyu, X.G.; Chen, Z.K.; Zhang, Z.S.; Jiang, M.; Zhang, K.; Lyu, Y.L. Spatial and temporal changes of Wetlands on the Qinghai-Tibetan Plateau from the 1970s to 2010s. *Chin. Geogr. Sci.* **2018**, *28*, 935–945. [CrossRef]
24. Zhang, Y.S.; Ohata, T.; Kadata, T. Land-surface hydrological processes in the permafrost region of the eastern Tibetan Plateau. *J. Hydrol.* **2003**, *283*, 41–56. [CrossRef]
25. Wang, B.B.; Ma, Y.M.; Su, Z.B.; Wang, Y.; Ma, W.Q. Quantifying the evaporation amounts of 75 high-elevation large dimictic lakes on the Tibetan Plateau. *Sci. Adv.* **2020**, *6*, eaay8558. [CrossRef] [PubMed]
26. Leuning, R.; Zhang, Y.Q.; Rajaud, A.; Cleugh, H.; Tu, K. A simple surface conductance model to estimate regional evaporation using MODIS leaf area index and the Penman-Monteith equation. *Water Resour. Res.* **2008**, *44*, 10. [CrossRef]
27. Su, Z. The surface energy balance system (SEBS) for estimation of turbulent heat fluxes. *Hydrol. Earth Syst. Sc.* **2002**, *6*, 85–99. [CrossRef]
28. Chen, X.; Su, Z.; Ma, Y.; Liu, S.; Yu, Q.; Xu, Z. Development of a 10-year (2001–2010) 0.1 data set of land-surface energy balance for mainland China. *Atmos. Chem. Phys.* **2014**, *14*, 13097–13117. [CrossRef]
29. Wang, K.C.; Dickinson, R.E. A review of global terrestrial evapotranspiration: Observation, modeling, climatology, and climatic variability. *Rev. Geophys.* **2012**, *50*, RG2005. [CrossRef]

30. Yan, H.; Wang, S.Q.; Billesbach, D.; Oechel, W.; Zhang, J.H.; Meyers, T.; Martin, T.A.; Matamala, R.; Baldocchi, D.; Bohrer, G. Global estimation of evapotranspiration using a leaf area index-based surface energy and water balance model. *Remote. Sens. Environ.* **2012**, *124*, 581–595. [CrossRef]

31. Li, X.L.; Liang, S.L.; Yuan, W.P.; Yu, G.R.; Chegn, X.; Chen, Y.; Zhao, T.B.; Feng, J.M.; Ma, Z.G.; Ma, M.G. Estimation of evapotranspiration over the terrestrial ecosystems in China. *Echohyrology* **2014**, *7*, 139–149. [CrossRef]

32. Chen, J.L.; Wen, J.; Kang, S.C.; Meng, X.H.; Yang, X.Y. The evapotranspiration and environmental controls of typical underlying surfaces on the Qinghai-Tibetan Plateau. *Sci. Cold Arid Reg.* **2021**, *13*, 53–61.

33. Xue, B.L.; Wang, L.; Li, X.P.; Yang, K.; Chen, D.L.; Sun, L.T. Evaluation of evapotranspiration estimates for two river basins on the Tibetan Plateau by a water balance method. *J. Hydrol.* **2013**, *492*, 290–297. [CrossRef]

34. Giorgi, F.; Francisco, R. Evaluating uncertainties in the prediction of regional climate change. *Geophys. Res. Lett.* **2000**, *27*, 1295–1298. [CrossRef]

35. Mearns, L.; Team, N. The North American Regional Climate Change Assessment Program (NARCCAP): Overview of phase II results. In Proceedings of the IOP Conference Series: Earth and Environmental Science, Copenhagen, Denmark, 10–12 March 2009.

36. Gao, Y.H.; Leung, L.R.; Salathe, E.P.; Dominguez, F.; Nijssen, B.; Lettenmaier, D.P. Moisture flux convergence in regional and global climate models: Implications for droughts in the southwestern United States under climate change. *Geophys. Res. Lett.* **2012**, *39*, L09711. [CrossRef]

37. Leung, L.R.; Kuo, Y.H.; Tribbia, J. Research needs and directions of regional climate modeling using WRF and CCSM. *Bull. Am. Meteorol. Soc.* **2006**, *87*, 1747–1752. [CrossRef]

38. Jacob, D.; Barring, L.; Christensen, O.B.; Christensen, J.H.; de Castro, M.; Deque, M.; Giorgi, F.; Hagemann, S.; Hirschi, M.; Jones, R.; et al. An inter-comparison of regional climate models for Europe: Model performance in present-day climate. *Clim. Chang.* **2007**, *81*, 31–52. [CrossRef]

39. Glisan, J.M.; Gutowski, W.J.; Cassano, J.J.; Higgins, M.E. Effects of spectral nudging in WRF on arctic temperature and precipitation simulations. *J. Clim.* **2013**, *26*, 3985–3999. [CrossRef]

40. Huang, Z.Y.; Zhong, L.; Ma, Y.M.; Fu, Y.F. Development and evaluation of spectral nudging strategy for the simulation of summer precipitation over the Tibetan Plateau using WRF (v4.0). *Geosci. Model Dev.* **2021**, *14*, 2027–2841. [CrossRef]

41. Deque, M.; Jones, R.G.; Wild, M.; Giorgi, F.; Christensen, J.H.; Hassell, D.C.; Vidale, P.L.; Rockel, B.; Jacob, D.; Kjellstrom, E. Global high resolution versus limited area model climate change projections over Europe: Quantifying confidence level from PRUDENCE results. *Clim. Dyn.* **2005**, *25*, 653–670. [CrossRef]

42. Gao, X.J.; Xu, Y.; Zhao, Z.C.; Pal, J.S.; Giorgi, F. On the role of resolution and topography in the simulation of East Asia precipitation. *Theor. Appl. Climatol.* **2006**, *86*, 173–185. [CrossRef]

43. Giorgi, F.; Bates, G.T.; Nieman, S.J. Simulation of the arid climate of the southern Great Basin using a regional climate model. *Bull. Am. Meteorol. Soc.* **1992**, *73*, 1807–1822. [CrossRef]

44. Laprise, R.; Caya, D.; Giguere, M.; Bergeron, G.; Boer, G.J.; McFarlane, N.A. Climate and climate change in western Canada as simulated by the Canadian regional climate model. *Atmos. Ocean* **1998**, *36*, 119–167. [CrossRef]

45. Li, S.S.; Lyu, S.H.; Gao, Y.H.; Zhang, Y. Effect of different horizontal resolution on simulation of precipitation in Qilian Mountain. *Plateau Meteor.* **2005**, *24*, 496–502.

46. Mearns, L.O.; Gutowski, W.; Jones, R.; Leung, R.; Yun, Q. A regional climate change assessment program for North America. *Eos. Trans. Am. Geophys. Union* **2009**, *90*, 311. [CrossRef]

47. Zhang, Y.X.; Duliere, V.; Mote, P.W.; Salathe, E.P. Evaluation of WRF and HadRM mesoscale climate simulations over the U.S. Pacific Northwest. *J. Clim.* **2009**, *22*, 5511–5526. [CrossRef]

48. Gao, Y.H.; Xu, J.W.; Chen, D.L. Evaluation of WRF mesoscale climate simulations over the Tibetan Plateau during 1979–2011. *J. Clim.* **2015**, *28*, 2823–2841. [CrossRef]

49. Lin, C.G.; Chen, D.L.; Yang, K.; Ou, T.H. Impact of model resolution on simulating the water vapor transport through the central Himalayas: Implication for models' wet bias over the Tibetan Plateau. *Clim. Dyn.* **2018**, *51*, 3195–3207. [CrossRef]

50. Gao, Y.H.; Chen, F.; Miguez-Macho, G.; Li, X. Understanding precipitation recycling over the Tibetan Plateau using tracer analysis with WRF. *Clim. Dyn.* **2020**, *55*, 2921–2937. [CrossRef]

51. Gao, Y.H.; Chen, F.; Jiang, Y.S. Evaluation of a convection-permitting modeling of precipitation over the Tibetan Plateau and its influences on the simulation of snow-cover fraction. *J. Hydrometeorol.* **2020**, *21*, 1531–1548. [CrossRef]

52. Arakawa, A. The cumulus parameterization problem: Past, present, and future. *J. Clim.* **2004**, *17*, 2493–2525. [CrossRef]

53. Brockhaus, P.; Lüthi, D.; Schär, C. Aspects of the diurnal cycle in a regional climate model. *Meteorol. Z.* **2008**, *17*, 433–443. [CrossRef]

54. Prein, A.F.; Langhans, W.; Fosser, G.; Ferrone, A.; Ban, N.; Goergen, K.; Keller, M.; Tolle, M.; Gutjahr, O.; Feser, F. A review on regional convection-permitting climate modeling: Demonstrations, prospects, and challenges. *Rev. Geophys.* **2015**, *53*, 323–361. [CrossRef] [PubMed]

55. Romps, D.M. A direct measure of entrainment. *J. Atmos. Sci.* **2010**, *67*, 1908–1927. [CrossRef]

56. Song, X.; Zhang, G.J. Convection parameterization, tropical pacifific double ITCZ, and upper-ocean biases in the NCAR CCSM3: Part I. Climatology and atmospheric feedback. *J. Clim.* **2009**, *22*, 4299–4315. [CrossRef]

57. Ou, T.H.; Chen, D.L.; Chen, X.C.; Lin, C.G.; Yang, K.; Lai, H.W.; Zhang, F.Q. Simulation of summer precipitation diurnal cycles over the Tibetan Plateau at the gray-zone grid spacing for cumulus parameterization. *Clim. Dyn.* **2020**, *54*, 3525–3539. [CrossRef]
58. Prein, A.F.; Gobiet, A.; Suklitsch, M.; Truhetz, H.; Awan, N.K.; Keuler, K.; Georgievski, G. Added value of convection permitting seasonal simulations. *Clim. Dyn.* **2013**, *41*, 2655–2677. [CrossRef]
59. Skamarock, W.C. Evaluating mesoscale NWP models using kinetic energy spectra. *Mon. Weather Rev. Mon.* **2004**, *132*, 3019–3032. [CrossRef]
60. Gao, Y.H.; Leung, L.R.; Zhang, Y.X.; Cuo, L. Changes in moisture flux over the Tibetan Plateau during 1979–2011: Insights from a high-resolution simulation. *J. Clim.* **2015**, *28*, 4185–4197. [CrossRef]
61. Huang, M.; Huang, B.; Gu, L.J.; Huang, H.L.A.; Goldberg, M.D. Parallel GPU architecture framework for the WRF single moment 6-class microphysics scheme. *Comput. Geosci.* **2015**, *83*, 17–26. [CrossRef]
62. Collins, W.D.; Rasch, P.J.; Boville, B.A.; Hack, J.J.; McCaa, J.R. Description of the NCAR community atmosphere model (CAM 3.0). *NCAR Tech.* **2004**, *226*, 1326–1334.
63. Hong, S.Y.; Lim, J.O.J. The WRF single-moment 6-class microphysics scheme (WSM6). *J. Korean Meteorol. Soc.* **2006**, *42*, 129–151.
64. Hong, S.Y.; Pan, H.L. Nonlocal boundary layer vertical diffusion in a medium-range forecast model. *Mon. Weather Rev.* **1996**, *124*, 2322–2339. [CrossRef]
65. Kain, J.S. The Kain–Fritsch convective parameterization: An update. *J. Appl. Meteor.* **2004**, *43*, 170–181. [CrossRef]
66. Chen, F.; Dudhia, J. Coupling an advanced land surface-hydrology model with the Penn State-NCAR MM5 modeling system. Part I: Model implementation and sensitivity. *Mon. Weather Rev.* **2011**, *129*, 569–585. [CrossRef]
67. Surface Energy Balance Based Global Land Evapotranspiration (EB-ET 2000-2017). Available online: http://data.tpdc.ac.cn (accessed on 25 June 2021).
68. Rodell, M.; Houser, P.R.; Jambor, U.; Gottschalck, J.; Mitchell, K.; Meng, C.J.; Arsenault, K.; Cosgrove, B.; Radakovich, J.; Bosilovich, M.; et al. The global land data assimilation system. *Bull. Am. Meteor. Soc.* **2004**, *85*, 381–394. [CrossRef]
69. Gao, Y.H.; Cuo, L.; Zhang, Y.X. Changes in moisture flux over the Tibetan Plateau during 1979–2011 and possible mechanisms. *J. Clim.* **2014**, *27*, 1876–1893. [CrossRef]
70. Wang, A.H.; Zeng, X.B. Evaluation of multi-reanalysis products with in situ observations over the Tibetan Plateau. *J. Geophys. Res.* **2012**, *117*, D05102.
71. Dee, D.P.; Uppala, S.M.; Simmons, A.J.; Berrisford, P.; Poli, P.; Kobayashi, S.; Andrae, U.; Balmaseda, M.A.; Balsamo, G.; Bauer, P.; et al. The ERA-Interim reanalysis: Configuration and performance of the data assimilation system. *Q. J. R. Meteor. Soc.* **2011**, *137*, 553–597. [CrossRef]
72. Hersbach, H.; Bell, B.; Berrisford, P.; Hirahara, S.; Horanyi, A.; Munoz-Sabater, J.; Nicolas, J.; Peubey, C.; Radu, R.; Schepers, D.; et al. The ERA5 global reanalysis. *Q. J. R. Meteor. Soc.* **2020**, *146*, 1999–2049. [CrossRef]
73. Bonavita, M.; Hólm, E.; Isaksen, L.; Fisher, M. The evolution of the ECMWF hybrid data assimilation system. *Q. J. R. Meteor. Soc.* **2016**, *142*, 287–303. [CrossRef]
74. NCAR. ARW version 3 modeling system user's guide. *Natl. Cent. Atmos. Res. Doc.* **2017**, *434*, 85–87.
75. Shi, N.; Wei, F.Y.; Feng, G.L.; Shen, T.L. Monte Carlo test method and application in correlation analysis and comprehensive analysis of meteorological field (in Chinese). *J. Nanjing Ins. Meteorol.* **1997**, *20*, 355–359.
76. Alves, M.D.; de Carvalho, L.G.; Vianello, R.L.; Sediyama, G.C.; de Oliveira, M.S.; de Sa, A. Geostatistical improvements of evapotranspiration spatial information using satellite land surface and weather stations data. *Theor. Appl. Climatol.* **2013**, *113*, 155–174. [CrossRef]
77. Brutsaert, W. Energy fluxes at the Earth's surface. In *Evaporation into the Atmosphere. Environmental Fluid Mechanics*; Springer: Dordrecht, The Netherlands, 1982; pp. 128–153.
78. Montaldo, N.; Oren, R. Changing seasonal rainfall distribution with climate directs contrasting impacts at evapotranspiration and water yield in the western Mediterranean region. *Earth Future* **2018**, *6*, 841–885. [CrossRef]
79. Bao, X.H.; Zhang, F.Q. Evaluation of NCEP-CFSR, NCEP-NCAR, ERA-Interim, and ERA-40 reanalysis datasets against independent sounding observations over the Tibetan Plateau. *J. Clim.* **2013**, *26*, 206–214. [CrossRef]
80. Tao, S.; Luo, S.; Zhang, H. The Qinghai-Xizang Plateau meteorological experiment (Qxpmex) May–August 1979. In Proceedings of the International Symposium on the Qinghai-Xizang Plateau and Mountain Meteorology, Beijing, China, 20–24 March 1986.
81. Wang, J. Land surface process experiments and interaction study in China-From HEIFE to IMGRASS and GAME-TIBET/TIPEX. *Plateau Meteorol.* **1999**, *18*, 280–294.
82. Ma, Y.; Yao, T.; Wang, J. Experimental study of energy and water cycle in Tibetan plateau—The progress introduction on the study of GAME/Tibet and CAMP/Tibet. *Plateau Meteorol.* **2006**, *25*, 344–351.
83. Ma, Y.M.; Hu, Z.Y.; Xie, Z.P.; Ma, W.Q.; Wang, B.B.; Chen, X.L.; Li, M.S.; Zhong, L.; Sun, F.L.; Gu, L.L.; et al. A long-term dataset of hourly integrated land–atmosphere interaction observations on the Tibetan Plateau. *Earth Syst. Sci. Data* **2020**, *12*, 2937–2957. [CrossRef]
84. Yao, Y.J.; Liang, S.L.; Li, X.L.; Chen, J.Q.; Wang, K.C.; Jia, K.; Cheng, J.; Jiang, B.; Fisher, J.B.; Mu, Q.Z.; et al. A satellite-based hybrid algorithm to determine the Priestley–Taylor parameter for global terrestrial latent heat flux estimation across multiple biomes. *Remote Sens. Environ.* **2015**, *165*, 216–233. [CrossRef]
85. Qiao, B.J.; Zhu, L.P.; Yang, R.M. Temporal-spatial differences in lake water storage changes and their links to climate change throughout the Tibetan Plateau. *Remote Sens. Environ.* **2019**, *222*, 232–243. [CrossRef]

86. Wang, B.B.; Ma, Y.M.; Wang, Y.; Su, Z.B.; Ma, W.Q. Significant differences exist in lake-atmosphere interactions and the evaporation rates of high-elevation small and large lakes. *J. Hydrol.* **2019**, *573*, 220–234. [CrossRef]
87. Dirmeyer, P.A.; Gao, X.; Zhao, M.; Guo, Z.C.; Oki, T.K.; Hanasaki, N. GSWP-2: Multimodel analysis and implications for our perception of the land surface. *Bull. Am. Meteorol. Soc.* **2006**, *87*, 1381–1397. [CrossRef]
88. Hodges, K.I.; Lee, R.W.; Bengtsson, L. A comparison of extratropical cyclones in recent reanalyses ERA-Interim, NASA MERRA, NCEP CFSR, and JRA-25. *J. Clim.* **2011**, *24*, 4888–4906. [CrossRef]
89. Lian, X.; Piao, S.L.; Huntingford, C.; Li, Y.; Zeng, Z.Z.; Wang, X.H.; Ciais, P.; McVicar, T.R.; Peng, S.S.; Ottle, C.; et al. Partitioning global land evapotranspiration using CMIP5 models constrained by observations. *Nat. Clim. Chang.* **2018**, *8*, 640–646. [CrossRef]
90. Tanaka, N.; Kume, T.; Yoshifuji, N.; Tanaka, K.; Takizawa, H.; Shiraki, K.; Tantasirin, C.; Tangtham, N.; Suzuki, M. A review of evapotranspiration estimates from tropical forests in Thailand and adjacent regions. *Agric. For. Meteorol.* **2008**, *148*, 807–819. [CrossRef]

Article

Temporal and Spatial Differences and Driving Factors of Evapotranspiration from Terrestrial Ecosystems of the Qinghai Province in the Past 20 Years

Zhiyuan Song [1,2,3,4], Qi Feng [5], Ziyi Gao [6], Shengkui Cao [1,2,3,4,*], Guangchao Cao [2,3,4] and Zhigang Wang [1,2,3,4]

[1] School of Geographical Science, Qinghai Normal University, Xining 810008, China; 202047341005@stu.qhnu.edu.cn (Z.S.); 201947331022@stu.qhnu.edu.cn (Z.W.)
[2] Academy of Plateau Science and Sustainability, People's Government of Qinghai Province and Beijing Normal University, Xining 810016, China; caoguangchao@qhnu.edu.cn
[3] Key Laboratory of Tibetan Plateau Land Surface Processes and Ecological Conservation (Ministry of Education), Qinghai Normal University, Xining 810008, China
[4] Qinghai Province Key Laboratory of Physical Geography and Environmental Process, Xining 810008, China
[5] Key Laboratory of Ecohydrology of Inland River Basin, Northwest Institute of Eco-Environment and Resources, Chinese Academy of Sciences, Lanzhou 730013, China; qifeng@lzb.ac.cn
[6] College of Tourism, Qinghai Nationalities University, Xining 810007, China; 2021078@qhmu.edu.cn
* Correspondence: 2025054@qhnu.edu.cn

Citation: Song, Z.; Feng, Q.; Gao, Z.; Cao, S.; Cao, G.; Wang, Z. Temporal and Spatial Differences and Driving Factors of Evapotranspiration from Terrestrial Ecosystems of the Qinghai Province in the Past 20 Years. *Water* **2022**, *14*, 536. https://doi.org/10.3390/w14040536

Academic Editors: Yaoming Ma, Zhongbo Su and Lei Zhong

Received: 13 January 2022
Accepted: 9 February 2022
Published: 11 February 2022

Abstract: As the "Asian Water Tower", understanding the hydrological cycles in Qinghai Province and its interior is critical to the security of terrestrial ecosystems. Based on Moderate Resolution Imaging Spectroradiometer (MODIS)16 evapotranspiration (ET) remote sensing data, we used least squares regression, correlation analysis, and t-test to determine the temporal and spatial changes and trends of ET in Qinghai Province and its five ecological functional regions, located on the Qinghai–Tibet Plateau (Plateau) Western China from 2000 to 2020. In addition, we discussed the main factors affecting the changes of ET in different regions of Qinghai Province over the first two decades of the 21st century along spatial as well as altitudinal gradients. The results showed that: (1) the average annual ET in Qinghai Province was 496.56 mm/a, the highest ET value appeared in the southeast of the study area (684.08 mm/a), and the lowest ET value appeared in the Qaidam region in the northwest (110.49 mm/a); (2) the annual ET showed an increasing trend with a rate of 3.71 mm/a ($p < 0.01$), the place where ET decreased most was in the Three-River Source region (−8–0 mm/a) in the southwest of the study area, and the ET increased the most in the Hehuang region in the east of the study area (9–34 mm/a); (3) temperature (T) was the dominant ET change factor in Qinghai Province, accounting for about 65.27% of the region, followed by the normalized difference vegetation index (NDVI) and precipitation (P) for 62.52% and 55.41%, respectively; and (4) ET increased significantly by 2.84 mm/100 m with increasing altitude. The dominant factors changed from P to NDVI and T as the altitude increased. The research is of practical value for gaining insight into the regional water cycle process on the Plateau under climate change.

Keywords: ET; Qinghai Province; driving factors; elevation-dependency

1. Introduction

On a global scale, the asymmetry of climate change leads to heterogeneity among ecosystems [1], affecting atmospheric water vapor content, and causing significant changes in water cycle systems, such as precipitation (P) and evapotranspiration (ET) [2]. As the central link in water, energy, and carbon cycles in the climate system, ET refers to the process by which surface water vapor enters the atmosphere in gaseous form [3], including soil evaporation and plant transpiration, and is also an active factor in any regional ecosystem. ET is a complex physical and biological process whose rates vary over time across the landscape as a function of differences in temperature (T), precipitation (P), and vegetative land cover. Studies have shown that global land ET provides about 60% of P to the atmosphere, and it may be even higher in arid regions [4]. Therefore, clarifying the spatial and temporal dynamic response of ET to climate change is of great significance for predicting regional and global hydrological cycles [5].

Most traditional methods for measuring ET are based on an on-site scale, including flux tower measurement [6], indirect estimation [7], empirical model [8], semi-empirical model [9], and process model [10]. However, the spatial heterogeneity of ET and the complexity of hydrological processes limit the application of these methods in regional and global scale assessments [11]. With the development of remote sensing, ET data can be obtained on a larger geographic scale, and with better timeliness, accuracy, and economy [12]. In particular, the Moderate Resolution Imaging Spectroradiometer (MODIS)16 global surface land ET data set, which is calculated based on the Penman–Monteith formula, can reach 86% simulation accuracy with high resolution and continuous coverage in different temporal and spatial scales around the world [7], has been widely used and its reliability of results have been verified. Liaqat et al. used flux tower measurements of semi-humid farmland and temperate forest in the Korean Peninsula to evaluate the accuracy of MODIS16 ET data, and found that MODIS16 data can provide reasonable accuracy for daily ET changes [13]. Xu et al. compared 12 types of ET data in the United States and found that MODIS16 data could capture seasonal changes in ET [14]. Studies of the Three Gorges Reservoir [15], North China Plain [16], and Taohe Basin [17] in China also showed that MODIS16 ET data could provide reasonable accuracy.

The Qinghai–Tibet Plateau (Plateau) has profoundly affected East Asia and the global atmospheric water cycle through special dynamic and thermal effects, and has become a critical sensitive region for global climate change [18]. Since the 1950s, the T of the Plateau has increased significantly, and P has increased slightly [19], which inevitably leads to ET changes in response [20]. Studies have shown that there are significant spatial differences in the interannual ET trends on the Plateau [21,22], and P is the main factor affecting the ET on the Plateau [2]. However, in Tibet's Ali, Lhasa River Valley, Qinghai Haibei region [23], and northeastern Plateau [24], ET was closely related to radiation and T. Meanwhile, the driving factors for the ET of different underlying surfaces were also different. The ET of wetlands and Kobresia humilis meadows was mainly affected by T [25,26], while that of alpine shrubs, alpine meadows, and alpine wetlands was determined by P [27,28]. In addition, factors, such as atmospheric water vapor, normalized difference vegetation index (NDVI), and soil moisture, also significantly impact ET [29–31]. These studies showed that the driving factors of ET were highly uncertain [32], depending on the difference in ecosystems caused by factors, such as the latitude and altitude of the study region, affecting the formation and feedback of the regional hydrological cycle [33]. The adaptation process of terrestrial hydrological evolution brought about by ET has a significant spatial connectivity. In practical applications, water resource managers and decisionmakers are more concerned about the temporal and spatial distribution and driving factors of ET at the administrative scale, such as provincial and city scales, rather than just watersheds or other natural subdivisions, yet, many current studies ignore this fact, which is not conducive to the implementation of unified water resources management and allocation.

Qinghai Province is the birthplace of Asian rivers, including the Yangtze River, the Yellow River, and the Lancang River. It is also the most populous region on the Plateau, and the study of its ET is of great significance to the ecological and hydrological systems of the Plateau. At present, studies on ET in Qinghai province have only focused on the watershed scale, such as Three-River Source [34] and Shaliu River Basin [35], but there is no comprehensive exploration from a spatial and altitudinal perspective. The purpose of this research is to: (1) evaluate the applicability of MODIS16 ET data in the terrestrial ecosystem of Qinghai Province; (2) analyze the temporal and spatial changes of ET in Qinghai Province from 2000 to 2020; and (3) assess the impact of climate factors (P, T) and environmental factors (NDVI) on ET over spatial and altitude gradients. The research is expected to provide a theoretical basis for exploring regional environmental protection, climate change, and agriculture and animal husbandry production in Qinghai Province.

2. Materials and Methods

2.1. Study Region

Qinghai Province is located in the northeastern part of the Plateau, between 89°25′–103°04′ E and 31°39′–39°11′ N, with a region of 717,500 square kilometers (Figure 1). Based on the ecological environment zoning management and control system, which from the perspective of ecological and environmental protection, promote differentiated and refined management in different regions, it has been divided into five ecological function zones in the province's administrative regions by the Qinghai Provincial government: (1) the Three-River Source region; (2) the Qaidam Basin region; (3) the Qilian Mountain region; (4) the region around Qinghai Lake; and (5) the Hehuang region. Among them, the Three-River Source region, which is the birthplace of the Yellow River, the Yangtze River, and the Lancang River, focuses on water conservation and biodiversity function maintenance to ensure the water source of the "Asian Water Tower"; the Qilian Mountain region, an important water source in the Hexi Corridor of China, such as the Shiyang River, Heihe River, and Shule River, focuses on the ecological management and restoration to improve the water conservation function; the region around Qinghai Lake focuses on maintaining biodiversity function to promote the positive cycle of the watershed, woodland, grassland and wetland ecosystems, and biodiversity ecosystems; the Qaidam Basin region focuses on coordinating the relationship between protection and development, protecting the original ecological surface and landforms, and rationally developing mineral resources; and the Hehuang region focuses on the restoration and rehabilitation of the environment.

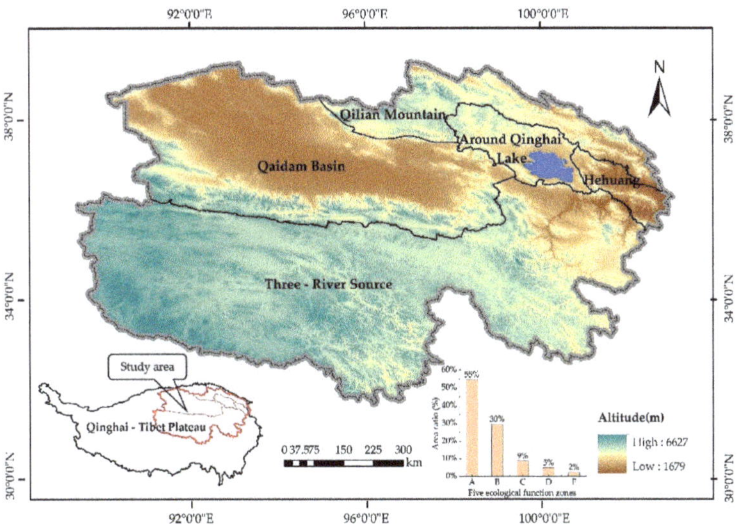

Figure 1. Location of Qinghai Province and its five ecological function zones.

The terrain of Qinghai Province is exceptionally complex, with an average altitude of over 3000 m. It has formed a unique "plateau climate" with cold winters and short summers. The annual average T ranges between −5 °C and 8 °C, with higher T distributed from Hehuang Region to Qaidam. P decreases from southeast to northwest, with annual P below 400 mm in most regions [36]. The main vegetation types in this region are alpine shrubs and alpine meadows [37].

2.2. Data Source and Processing

The ET discussed in this paper is the actual evapotranspiration, and the ET remote sensing product we used is the MODIS16 A3 global terrestrial ET product (Table 1), which is a remote sensing image data set based on the Penman–Monteith algorithm for calculating global surface ET. It has a spatial resolution of 500 m and an annual time scale. The estimated input data of MODIS16 includes remote sensing information such as leaf region index, albedo, vegetation coverage, and meteorological data, such as T, air pressure, relative humidity, and radiation. It includes ET, latent heat flux (LE), latent ET (PET), and latent heat flux (PLE) [7].

Table 1. Introduction table of research data sources.

Data Name	Data Sources	Time Resolution	Spatial Resolution
Evapotranspiration (ET)	https://search.earthdata.nasa.gov/ assessed on 2 June 2021	year	500 m
Normalized difference vegetation index (NDVI)	https://search.earthdata.nasa.gov/ assessed on 2 June 2021	month	1 km
Digital elevation model(DEM)	http://www.gscloud.cn/ assessed on 20 May 2021	—	90 m
Temperature (T), Precipitation (P)	http://www.geodata.cn assessed on 27 June 2021	month	0.0083333°

In order to study the response of ET to climate change on the spatial and altitudinal gradients, T and P were selected as climate variables in this study, and NDVI as the normalized vegetation index, which can reflect the coverage of vegetation and be also used as an essential factor affecting ET (Table 1). MODIS13A3 is a product of the global terrestrial vegetation index, the content is the grid normalized vegetation index (NDVI) with a spatial resolution of 1 km. The Digital Elevation Model(DEM) data are from the China Geospatial Data Cloud Platform, with a spatial resolution of 90 m. The T and P data set are China's monthly P data provided by the National Earth System Science Data Center (China's National Science & Technology Infrastructure). The data are based on the published the global 0.5° climate data from the Climatic Research Unit (http://www.cru.uea.ac.uk, accessed on 9 June 2021) and the global high-resolution climate data published by WorldClim (http://worldclim.org, accessed on 12 June 2021), which is formed in China through the Delta space downscaling program Downscale generation. The data of 496 independent meteorological observation points were used for verification, and the verification results are credible [38].

We used Python to convert T and P data into TIF format, cut and splice, NDVI, and meteorological data based on ArcGIS 10.6, then eliminated invalid values, projected transform NDVI and meteorological data, resampled according to 500 m resolution, and finally obtained the raster data of ET, NDVI, P, and T in Qinghai Province from 2000 to 2020.

2.3. Research Methods

The least-squares regression method estimates the linear trend of ET in Qinghai Province from 2000 to 2020. The calculation formula is as follows [39]:

$$SLOPE = \frac{n \times \sum_{i=1}^{n} i \times j_i - \sum_{i=1}^{n} i \sum_{i=1}^{n} i}{n \times \sum_{i=1}^{n} i^2 - \left(\sum_{i=1}^{n} i\right)^2} \tag{1}$$

In the formula, SLOPE is the slope of the regression equation, n is the length of the time series; if SLOPE > 0, it means ET increases; if SLOPE < 0, it means ET decreases.

Using correlation analysis to reflect the correlation between variables x and y, the calculation formula is [39]:

$$r_{xy} = \frac{\sum\limits_{i=1}^{n}(x_i - \overline{x})(y_i - \overline{y})}{\sqrt{\sum\limits_{i=1}^{n}(x_i - \overline{x})^2}\sqrt{\sum\limits_{i=1}^{n}(y_i - \overline{y})^2}} \tag{2}$$

In the formula, $\overline{x}$ is the multi-year average ET, $\overline{y}$ is the multi-year average NDVI, T, and P, i is the number of years ($i = 1, 2, 3, 4 \ldots 21$), n is the length of the research time series, is the ET value in the i-th year, is NDVI, T and P values in the year i. When $r > 0$, there is a positive correlation between the two, and when $r < 0$, there is a negative correlation between the two. The closer the absolute value is to 1, the stronger the correlation; the closer to 0, the weaker the correlation. The t-test completes the significance test of the correlation coefficient, and the calculation formula is [40]:

$$t = \frac{R_{xy}}{\sqrt{1 - R_{xy}^2}}\sqrt{n - m - 1} \tag{3}$$

Among them, n is the number of samples (the time series is 2001–2020, that is, $n = 21$), and m is the number of independent variables.

3. Results

3.1. Temporal and Spatial Characteristics of T, P, and NDVI

The spatial distribution characteristics of T, P and NDVI are shown in Figure 2a–c. In the study region, the ranges of T, P, and NDVI were 20.42–9.22 °C, 17.7–802.03 mm, and 0–0.86 (Figure 2a–c). The T in the Qaidam Basin and the Hehuang regions were relatively high, while the T in the Qilian Mountain and the Three-River Source regions were relatively low (Figure 2a). P and NDVI gradually decreased from the Three-River Source in the southeast to the Qaidam Basin(Figure 2b,c) in the northwest.

As shown in Figure 2d,e,f, T and P increased at a rate of 0.07 °C/decade and 24.73 mm/decade, respectively, from 2000 to 2020, but the increase was not significant ($p > 0.05$), while NDVI increased significantly at a rate of 0.02/decade ($p < 0.01$). By further analysis of the correlation between T and P with NDVI, we found that the P and NDVI were significantly correlated ($p < 0.05$), and precipitation determined the spatial distribution of NDVI (Figure 2b,c). Due to the significant terrain differences in Qinghai Province, although the overall increase in T and P was not significant, the trend of warming and humidification was still obvious in most regions (Figure 3b,c), which may be the climatic reason for the significant increase in NDVI. In addition, the ecological projects implemented by the Chinese government have a lot to do with the increase in NDVI. In 2003, the large-scale implementation of the "returning grazing land to grassland" ecological project began, and NDVI increased rapidly in 2004 (Figure 2c).

3.2. Characteristics of Interannual Variations in ET

Figure 4 shows that average ET variations in the study region during 2000–2020. ET has ranged from 405.88 to 549.3 mm in Qinghai Province over the 21 year period. The multi-year mean ET was 496.56 mm. The lowest and highest ET were 405.88 and 549.3 mm, occurring in 2000 and 2017, respectively. The average annual ET showed a significant increasing trend ($R^2 = 0.43$, $p < 0.01$) and increased at a rate of 37.26 mm/decade. In addition, the relative rate of average level distance showed that the annual average change of ET ranged from -18% to 11%. Before 2009, the rate of change was primarily negative, and after 2009, it was mostly positive.

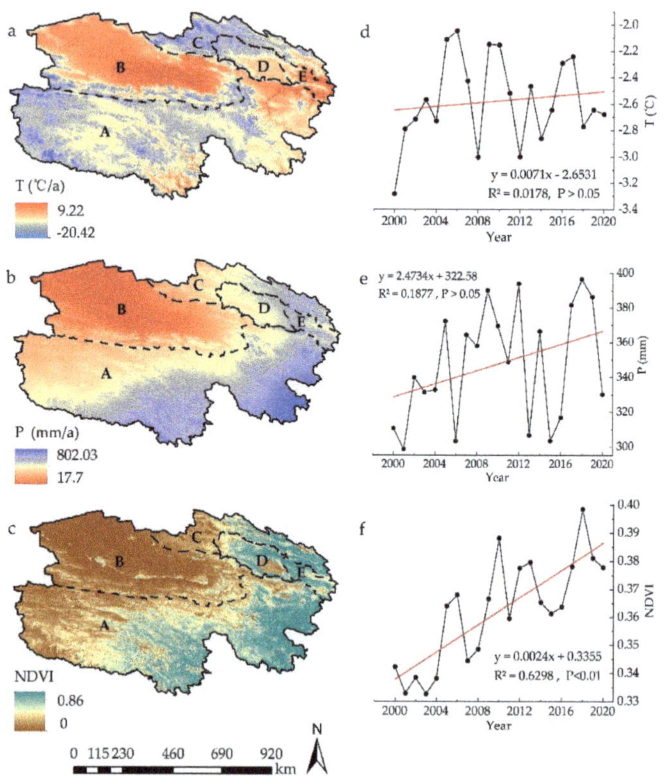

Figure 2. The spatial distribution and temporal changes of annual average T (**a,d**), P (**b,e**), and NDVI (**c,f**) in Qinghai Province during 2000–2020. A–E in the figure represents the Three-River Source region, the Qaidam Basin region, the Qilian Mountain region, the region around Qinghai Lake, and the Hehuang region, respectively.

3.3. Characteristics of Spatial Variations in ET

Figure 5a shows the spatial distribution of ET in Qinghai Province during 2000–2020. It is not difficult to see that ET presented a spatial difference that was high in the southeast and low in the northwest. The annual average ET value in space was between 110.49–684.08 mm. The ET was higher in the Three-River Source region and the middle-eastern part of the Qilian Mountain region, while the ET in the western part of the Three-River Source, Qaidam, and Hehuang region was the lowest.

Figure 5b illustrates the spatial change trend of ET, with slope trend value ranging from −8 to 34 mm/a, and ET gradually increasing from the southwest to the northeast. In general, most regions showed a significant increasing trend, except for the southwestern edge of the Three-River Source region (Figure 4a), which showed a decreasing trend (−8–0 mm/a). The region with the largest increase in ET was recorded in Haidong City in the Hehuang region (9–34 mm/a).

We used box plots to compare the ET per unit region of the five ecological function zones from 2000–2020. (Figure 6). In terms of the ET patterns according to these five divisions, the highest value was 2113.44 mm/km^2 in the Three-River Source region, which was between 1882.36 to 2362.73 mm/km^2, and the lowest value 395.05 mm/km^2 in the Qaidam Basin region, which was between 306.8–448.68 mm/km^2. The unit region ETs of the Qilian Mountain, around Qinghai Lake, and Hehuang region were 395.05 mm/km^2, 1690.29 mm/km^2, 2113.44 mm/km^2, respectively. On the whole, the differences were significant, showing a distribution pattern in which, the south was higher than the north and the east was higher than the west.

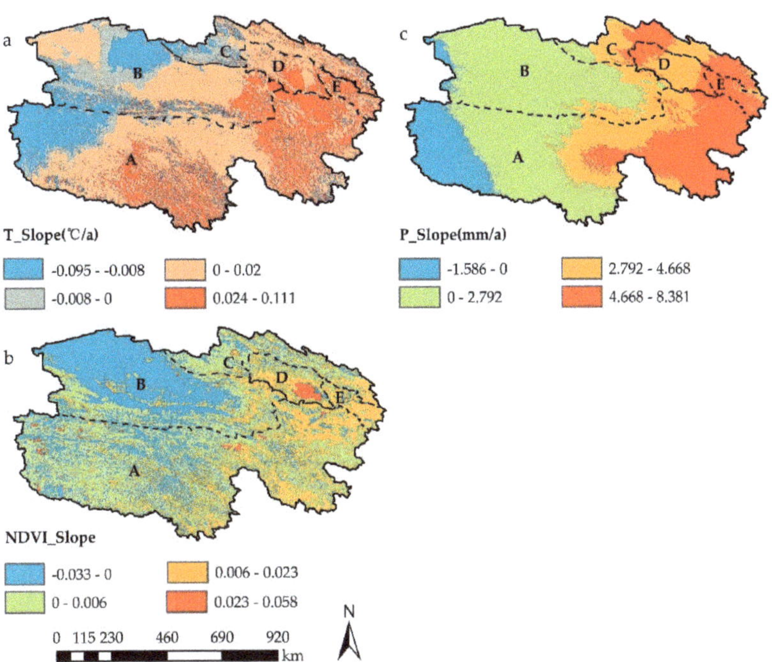

Figure 3. Spatial changes of T (**a**), NDVI (**b**), and P (**c**) in Qinghai Province during 2000–2020. A–E in the figure represents the Three-River Source region, the Qaidam Basin region, the Qilian Mountain region, the region around Qinghai Lake, and the Hehuang region, respectively.

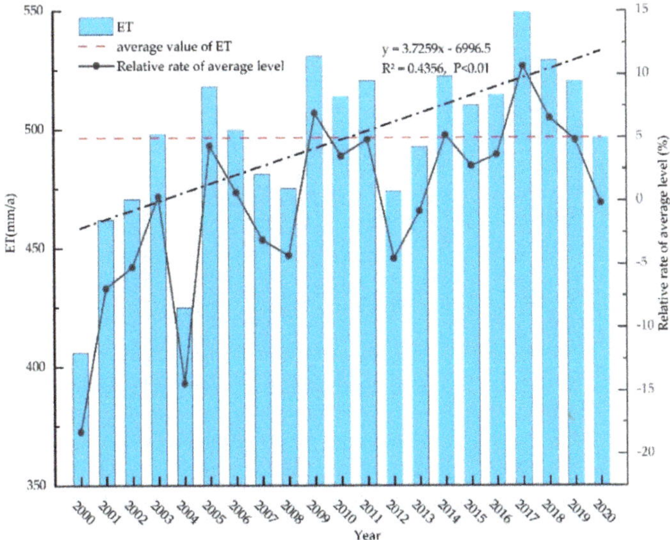

Figure 4. Interannual changes and trends of ET in Qinghai Province during 2000–2020.

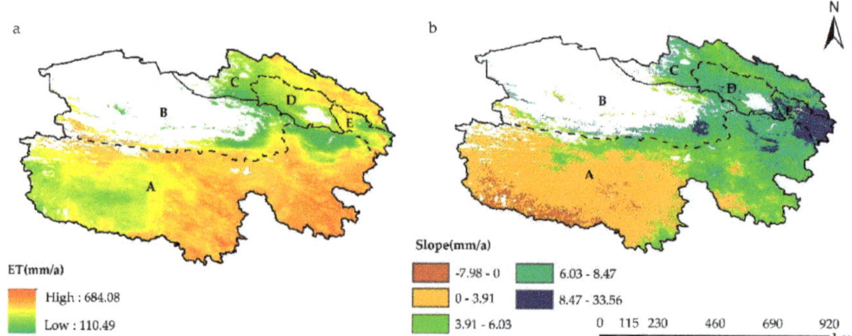

Figure 5. Spatial distribution (**a**) and changes (**b**) of ET in Qinghai Province during 2000–2020. A–E in the figure represents the Three-River Source region, the Qaidam Basin region, the Qilian Mountain region, the region around Qinghai Lake, and the Hehuang region, respectively.

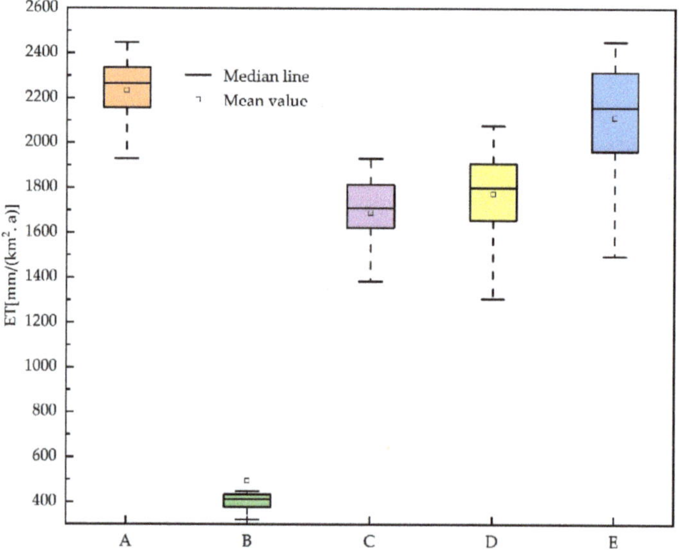

Figure 6. Characteristics of ET per unit region in five ecological function zones in Qinghai Province during 2000–2020. A–E in the figure represents the Three-River Source region, the Qaidam Basin region, the Qilian Mountain region, the region around Qinghai Lake, and the Hehuang region, respectively.

4. Discussion

4.1. Comparison with Existing Research Results on the PLATEAU

This research used MODIS16 data to calculate the ET value of Qinghai Province, and the results showed that there is a significant increasing ET trend with a rate of 3.48 mm/a, which is much higher than that in China [41], and the source of the Yellow River Basin [42], with 1.23 mm/a, and 0.44 mm/a, respectively. It shows that the response of the Plateau to climate change may be higher than that of other regions [18,43].

According to statistics (Table 2), the average annual ET on the Plateau varies between 320–500 mm/a, and the results of this study are within this range. However, due to factors, such as high downward shortwave radiation flux inputs from the Global Modeling and Assimilation Office [44], the estimate based on MODIS16 data (450–500 mm/a) [45] will be higher than the ET combined with the measured and satellite remote sensing data value (350–380 mm/a) [4,44,46,47], with calculated differences between 118–146 mm/a. Song et al. [11] estimated the MODIS16 ET data and obtained a similar gap (111.1 mm/a). In addition, MODIS16 data can reflect the spatial distribution and change pattern of ET on the Plateau; the region where ET decreased was mainly in the Tibet Autonomous Region, while the region where ET increased was mainly in Qinghai Province [44–47]; the spatial distribution of ET decreased from the southeast to the northwest, and the spatial change was just the opposite, decreasing in the southwest and increasing in the northeast. These results show that although the current research methods can show similar spatial patterns, there are significant differences in the magnitude and interannual variation of ET, and different data sources and calculation methods have considerable uncertainty on the ET and time variation of the Qinghai Tibet Plateau. Therefore, an integrated ET product is needed to take advantage of the complementary advantages of each ET data set [48].

Table 2. Comparison of ET in Qinghai Province with the Plateau.

Number	Period	ET (mm/a)	Domain	References
1	2000–2010	320	China	Yao [46]
2	1982–2008	350	TP	Jung [4]
3	1982–2012	378.1	TP	Wang [44]
4	1982–2014	377	TP	Cui [47]
5	1982–2015	373.12	TP	Yang [49]
6	2000–2010	350.3	TP	Song [11]
7	2000–2012	450–500	TP	Zhang [45]
8	1979–2014	364.45–493.32	Yellow River Source	Liu [50]
9	2001–2015	535.01–636.51	Three-River Headwater	Sun [51]
10	2002–2016	66–379	Qaidam	Bibi [52]
11	2001–2011	125.26	Qaidam	Jin [53]
12	2002–2004	341, 407, 426	alpine meadow of Qilian Mountain	Song [54]
13	2000–2015	854.23	Southern Slope of Qilian Mountain	Zhen [55]
14	2014–2015	496.2	Qinghai Lake	Ma [56]
15	2015–2016	633.3–657.2 341–426	two alpine wetland ecosystems of Qinghai Lake	Cao [57]
16	1966–2010	375.54, 378.54	Huangshui River	Zhang [58]

Among the five ecological function regions, the average annual ET of the Three-River Source and around Qinghai Lake regions were 513.59 mm/a and 444.39 mm/a, respectively, which were close to the current research results [50,51,56,57]. The average annual ET of the Qilian Mountain was 460.73 mm/a, which was between the existing research results [54,55]. The average annual ET of the Qaidam region was 428.18 mm/a, which was inconsistent with existing research [52,53]. The dry and hot climate leads to a shallow vegetation index in the Qaidam region, and the MODIS16 data lack many ET values in this region. In addition, the resolution of remote sensing data is too coarse to capture the ET spatial changes in smaller areas [59]. This leads to a significant error with the actual ET value, and more field measurements need to be used for further verification in future studies. Due to the different study areas, the average annual ET in the Hehuang area was 470.32 mm/a, which was higher than the results of Zhang et al. [58].

4.2. Drivers of ET Space Changes

ET results from the complex interaction between climatic factors and environmental factors. The growth rates of T, P, and NDVI in Qinghai Province are 0.07 °C/decade (Figure 2d), 24.73 mm/decade (Figure 2e), and 0.02/decade (Figure 2f), and they spatially present distribution patterns similar to ET (Figure 2a–c). The correlation coefficients of T, P, and NDVI with ET in Qinghai Province are −0.99 to 0.99, −0.93 to 0.99, and −0.99 to 0.99, respectively (Figure 2a–c). Climate factors and environmental factors are positively and negatively correlated with ET, T and ET positively correlated regions are mainly distributed in the northern, eastern, and southwestern margins of Qinghai Province; negatively correlated regions are mainly distributed in the middle and southern parts of Three-River Source (Figure 2a). P and ET positively correlated regions are distributed in most regions of Qinghai Province; negatively correlated regions are mainly distributed in the central and southwest of Qinghai Province with few regions. The relative distribution characteristics of NDVI and ET are essentially the same as P. In general, the distribution patterns of T, P, and NDVI determine the spatial distribution of ET in Qinghai Province.

The results of the significance test showed that T, P, and NDVI had an extremely significant positive or negative correlation ($p < 0.01$) with ET (Figure 7d–f). In the T, extremely significant correlation region between ET and T was 65.27% (Figure 7d), positive correlation regions were mainly distributed in the northern and southwestern of Qinghai Province where the altitude is higher (Figure 1), and the T is lower (Figure 2a). In contrast, in the humid southeast region (Figure 2b), the T was lower than ET showed a very significant negative correlation. P, NDVI, and ET were mainly spatially highly significant positive correlations (Figure 7e,f), with significant correlation regions of 62.52% and 55.41%, respectively, mainly distributed in the southeastern and northern regions of Qinghai Province. This shows that T, P, and NDVI changes were the main driving forces for increasing ET. The T had the most significant impact on ET, which was consistent with previous research results [2,60,61].

Among the five ecological function zones, the driving factors of ET are similar to those of Qinghai Province. The Three-River Source region is in the hinterland of the Plateau. Most regions are high-cold regions above 4600 m, with a unique natural environment and climatic conditions. ET showed a changing pattern during the study period in which the west decreased and increased in the east, mainly due to the "decrease–increase" change pattern of T, P, and NDVI from west to east. In addition, the main reason for the decrease in ET in the southwestern part of the Three-River Source region is the drought caused by the increase in T and the decrease in P [62].

The Qaidam basin region is far from the monsoon climate zone [63], surrounded by mountains, with low P and vegetation coverage (Figure 2b,c), and high T. Although both climatic and environmental factors show an upward trend, the increase in T affects ET. The influence is more vital than other factors, and T is the main reason for the increase in ET [47].

The T in the western part of the Qilian Mountain dropped slightly, P increased, and the climate changed from warm and dry to warm and humid [64], which was the main reason for the increase in ET. The increase in ET in the central and eastern part of the Qilian Mountain region may be due to the combined effects of climate change, vegetation change, and human activities [60]. In addition, the spatial variation of other climatic factors, such as wind speed, is also an essential factor of ET variation in the Qilian Mountain region [65].

The western region around Qinghai Lake has exposed rocks and high altitude, while high mountain regions show strong sensitivity to T [66], and T rise is the main reason for the increase in ET. P and NDVI have increased in the past 20 years, but P can only explain 16.68% of the region (Figure 7b,e). NDVI and ET are mainly negatively correlated, indicating that P has little effect on the growth of ET in the region around Qinghai Lake; the improvement of vegetation conditions restrained the growth of ET to a certain extent.

Although the average annual ET in the Hehuang region is not the largest in Qinghai Province, the average annual growth rate of ET is the largest. There may be several reasons; the Hehuang region is located at a low altitude in the eastern part of Qinghai Province, with better hydrothermal conditions, which are favorable for the evaporation of vegetation and soil. In the past 21 years, T, P, and NDVI in this region have increased the most, reaching 0–0.11 °C/a, 2.79–8.38 mm/a, and 0.006–0.023/a, respectively, and these factors were significantly positively correlated with ET (Figure 7d,e). This may partly explain the rapid growth of ET. The correlation between T, P, NDVI, and ET of more regions in the Hehuang was not significant (Figure 7d,e), accounting for 49.01%, 67.96%, and 49.19% of the region of Hehuang, respectively. ET may also be affected by other factors. Some studies have showed that the contribution rate of human activities to regional hydrological changes in the Hehuang region reached 64.54% [55]. Further in-depth research is needed in the future.

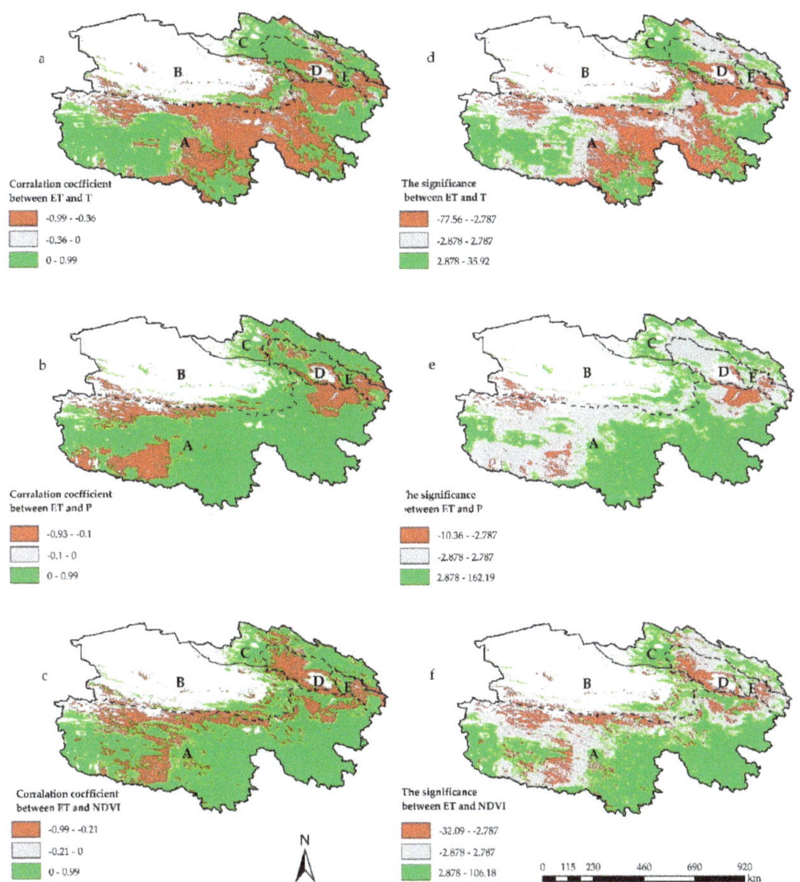

Figure 7. Spatial correlation coefficient distribution map of ET and T, P, and NDVI. (**a–c**) represent the correlation coefficient between ET and T, P, and NDVI; (**d–f**) represent the significance test between ET and T, P, and NDVI. A–E in the figure represents the Three-River Source region, the Qaidam Basin region, the Qilian Mountain region, the region around Qinghai Lake, and the Hehuang region, respectively.

In general, ET in the Three-River Source, Qaidam Basin, and Qilian Mountain regions is driven by T, P, and NDVI, with 54.66–66.51%, 46.89–59.97%, and 51.68–66.34% in the regions, respectively. ET in the Hehuang region and the region around Qinghai Lake is driven by T and NDVI, which are 65–65% and 49.47–56.95% in these regions, respectively. However, the impact of P on ET in these regions is small, only 32% and 17%. In the southwest of Qinghai Province, there is an "evaporation paradox" in the Three-River Source region.

4.3. Altitude Effect of ET Changes

The altitude is one of the critical factors that affects the heterogeneity pattern of ET [67,68]. It affects environmental variables, such as NDVI [69], by affecting climatic conditions, such as T and P, thereby impacting ET. The Plateau was found to have a complex topography, and its average altitude was much higher than that of the surrounding regions at the same latitude. Therefore, it was crucial to analyze the relationship between the altitude gradient and T, P, NDVI, and ET.

Table 3 shows that T was the main factor affecting the change of ET on the elevation gradient in Qinghai Province. T has a significant negative correlation with ET. When the T rises by 1 °C, ET decreased by 5.16 mm (R^2 = 0.53, $p < 0.01$), followed by P, which is significantly positively correlated with ET, with P increasing by 1 mm and ET increasing by 0.23 mm (R^2 = 0.21, $p < 0.01$); the correlation between NDVI and ET is not significant (R^2 = 0.01, $p > 0.05$). The complexity of the overall terrain of Qinghai Province, changes in T, P, and other factors show significant regional differences (Figure 2), resulting in the non-significant dependence of NDVI on altitude.

Table 3. Correlation between ET and T, P, and NDVI on the altitude gradient.

	Fitting Equation	Determination Coefficient (R^2) Test	Correlation Coefficient
T	y = −5.1618x + 447.5	0.5254	$p < 0.01$
P	y = 0.2295x + 370.39	0.2078	$p < 0.01$
NDVI	y = 30.74x + 445.26	0.0091	$p > 0.05$

The distribution characteristics of ET in Qinghai Province were evident on the altitude gradient (Figure 8). The ET value in high-altitude regions was significantly higher than in low-altitude regions. The region with the lowest altitude has the smallest ET, which was 339.51 mm, and the highest ET value appears on the 4300–4400 m altitude gradient, which was 523.29 mm. Meanwhile, ET increased significantly with the increase in altitude; when the altitude increased by 100 m, ET increased by 2.84 mm (R^2 = 0.55, $p < 0.01$). Previous studies on China's Hengduan Mountains and Southwest Mountains also supported the results of other research [70,71], in that ET has a significant positive correlation with altitude. However, there were still differences in the changes of ET on different altitude gradients. Chen et al. found that altitude above 4000 m had a negative and significant correlation with reference ET [68]. This study also found that ET above 4400 m had the same performance (R^2 = 0.95, $p < 0.01$, Table 4). In order to further quantify the influence of altitude on ET, Qinghai Province was divided into five intervals on the altitude gradient (Figure 8) for research.

Table 4. Correlation between ET and T, P, NDVI at different altitude gradients ($p < 0.01$).

Altitudes (m)	T	P	NDVI
1700–2600	y = −11.855x + 482.44 R^2 = 0.5	y = 0.7059x + 102.5 R^2 = 0.78	y = 120.15x + 340.77 R^2 = 0.0212 $p > 0.05$
2600–3800	y = −15.632x + 461.3 R^2 = 0.83	y = 0.3139x + 355.53 R^2 = 0.81	y = 244.98x + 356.99 R^2 = 0.86
3800–4400	y = −16.82x + 451.1 R^2 = 0.98	y = 1.2654x − 30.292 R^2 = 0.89	y = −60.18x + 686.54 R^2 = 0.94
4400–5100	y = 11.598x + 570.71 R^2 = 0.96	y = 0.4916x + 309.31 R^2 = 0.87	Y = 162.61x + 446.56 R^2 = 0.93
5100–6000	y = 2.0736x + 496.2 R^2 = 0.88	y = −0.1975x + 56.05 R^2 = 0.86	y = 96.302x + 458.73 R^2 = 0.84

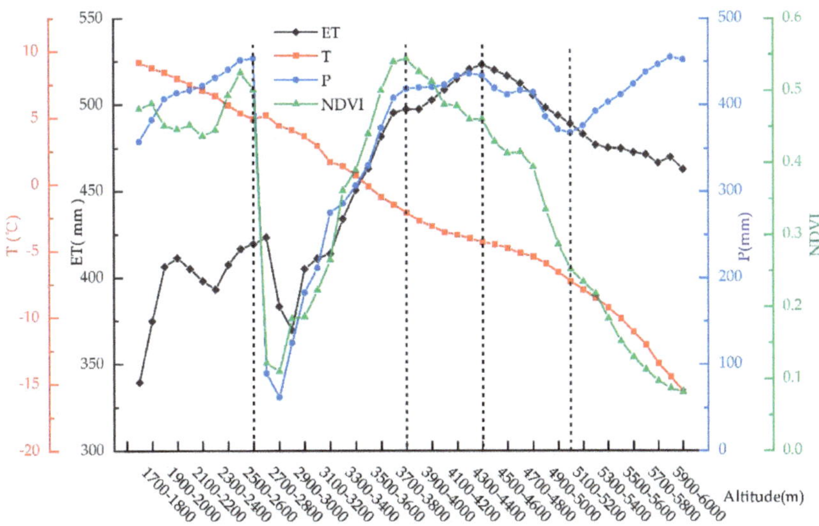

Figure 8. Variations of ET, T, P, and NDVI with altitudes.

Below the altitude of 2600 m, the T kept dropping, but the annual average T was above 4.93 °C, and the P was more and continuously increasing, which was roughly consistent with the changing trend of ET, while the fluctuation of NDVI was very pronounced (Figure 8). It may be because this region was mainly distributed in the eastern agricultural region, and the vegetation types are main crops, and crop ET and the climatic factors affecting it are quite different on the seasonal and annual time scales [72]. Correlation analysis showed that P and ET were significantly positively correlated (R^2 = 0.78, p < 0.01, Table 4), T also had a particular influence on ET (R^2 = 0.5, p < 0.01), and the correlation between NDVI and ET was not significant. Research by Liu et al. showed that the maximum T (Tmax) was the dominant driving factor of ET under the 2800 m altitude gradient of the Plateau [73], which was different from the results of this paper.

On the altitude gradient of 2600–3800 m, ET increased the fastest, which was roughly the same as the changing trend of P and NDVI. Although the T kept decreasing, it stayed above 0 °C, and the P also increased rapidly, which was beneficial to vegetation growth, and NDVI reached its maximum value (Figure 8), becoming the main driving factor of ET (R^2 = 0.86, p < 0.01, Table 4). Consistent with the results of previous studies [73], T and P also had a more significant effect on ET (R^2 > 0.8, p < 0.01). It is worth noting that ET, T, P, and NDVI had abrupt changes on the altitude gradient of 2600–2700 m (Figure 8). According to remote sensing images, this region was in the Qaidam Basin with an extremely arid climate; T and ET had a significant positive correlation (R^2 = 0.99, p < 0.01).

The altitude gradient of 3800–4400 m was a significant response region for the climate transition from warm and dry to warm and humid [74]. There was more P, which slowly increased with the increase in altitude, and the fluctuation of NDVI decreased (Figure 8). The influencing factors of ET are T, P, and NDVI (R^2 > 0.9, Table 4), of which T promotes ET the most (R^2 = 0.98, p < 0.01, Table 4). Previous studies have shown that the high-altitude regions of the Plateau below 5000 m have apparent warming trends [75], leading to regional vegetation greening [76], and climate warming also leads to ET of cryospheric hydrological systems, such as glaciers, frozen soil, snow, and lake ice. The study of Feng's hydrological effects on the glacial permafrost on the Plateau shows that the impact of glacier degradation on ET reaches 14.69 ± 12.82 mm [77]. The effects of these climate changes have promoted the continuous rise of ET on the 3800–4400 m altitude gradient and reached its highest value.

On the altitude gradient of 4400–5100 m, climate and environmental factors (T, P, and NDVI) decreased with the increase in altitude (Figure 8), which limits ET and causes ET to drop continuously ($R^2 > 0.9$, $p < 0.01$, Table 4). The T has the most significant influence on ET changes ($R^2 = 0.96$, $p < 0.01$, Table 4). These results were also verified by measuring the altitude gradient of Qinghai Lake and the scale of the ecosystem [56], indicating that the dominant factor of ET changes with the increase in altitude to the limit of T conditions.

Above an altitude of 5100 m, the ET change was consistent with the changing trend of T and NDVI. The T dropped below −7.89 °C, the vegetation normalization index continued to drop to the lowest level, close to 0.1, and the P increased to the highest value (Figure 8). Due to the high altitude and shallow vegetation coverage, the T was the lowest, although there was more P. P mainly existed in the form of ice and snow. The ability to absorb solar radiation was weakened and soil moisture cannot be effectively converted into evaporation. This may be one of the factors limiting ET [45]. The results showed that the main driving factor of ET was T ($R^2 = 0.88$, $p < 0.01$, Table 4), and P and T also had a certain impact on ET ($R^2 > 0.8$, $p < 0.01$, Table 4).

In summary, the T was the main driving factor for ET on the altitude gradient in Qinghai Province, followed by P, and NDVI had no significant impact on ET. At the same time, the driving variability of ET on the altitude gradient was apparent, changing from P driving to NDVI and T driving with increasing altitude, which was roughly the same as the research results of the Mongolian Plateau and Qinghai Lake [56,78]. Interestingly, ET has prominent distribution characteristics affected by T on the altitude gradient (Figure 9): In regions where the average annual T was less than −4.32 °C and the altitude was less than 4400 m, ET rose significantly ($R^2 = 0.86$, $p < 0.01$). In regions where the annual average T was higher than −4.32 °C, and the altitude was higher than 4400 m, ET decreased significantly ($R^2 = 0.83$, $p < 0.01$). However, P and NDVI had no similar performance. The correlation between P and ET on an altitude gradient was only 0.2 ($p < 0.01$), and the correlation between NDVI and ET was not significant ($p > 0.05$).

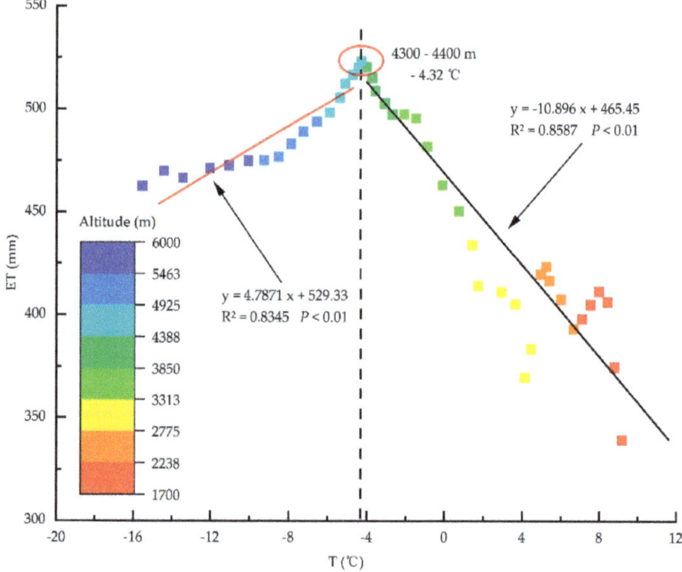

Figure 9. Correlation between T and ET on the altitude gradient.

This paper discussed the spatial distribution and driving factors of ET in Qinghai Province from 2000 to 2020, studied the driving factors of ET in Qinghai Province and five ecological function regions, and quantified the changes and effects of ET on the altitude gradient (Figure 10), and obtained good research results. However, this study did not quantify the spatial driving factors of ET in Qinghai Province. Other natural factors, such as net radiation, wind speed, soil moisture, and the effects of different vegetation types on ET have also not been discussed. At the same time, human activities can promote and inhibit ET [60]. The impact of the large-scale implementation of ecological projects in Qinghai Province by the Chinese government, such as returning farmland to forests, natural forest protection, construction of the "Three Norths" shelter forest system, returning grazing land to grasslands, and land-use changes caused by rapid urbanization in the Hehuang region, are still unknown. By 2016, Qinghai had become the province with the most ecological engineering projects in China, and the accumulated capital investment in the past 21 years has exceeded RMB 60 billion. In addition, the accuracy of MODIS16 data in areas with low vegetation coverage needs to be further verified on seasonal and more minor scales with the measured ET data.

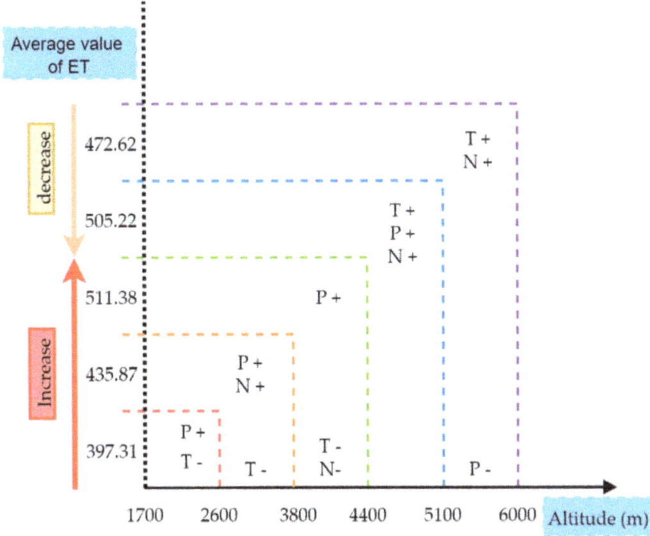

Figure 10. Diagram of influencing factors of ET in Qinghai Province. "+" "−" refers to the positive or negative effects of driving factors varying with altitude gradient on ET.

5. Conclusions

ET is still one of the largest unknowns in the terrestrial ecological cycle. We used MODIS16 ET data to obtain the spatiotemporal changes and driving factors of ET in Qinghai Province. MODIS16 data can accurately reflect the spatial distribution, annual changes, and range of ET. The distribution of ET decreased from southeast to northwest, and the change of ET increased from southwest to northeast during the study period. T is the main factor of the annual average ET change in Qinghai Province, followed by P and NDVI, which account for 65.27%, 62.52%, and 55.41%, respectively. In the five ecological function zones, the driving factors of ET are similar to those of Qinghai Province as a whole. However, in Hehuang and regions around Qinghai Lake, the average annual ET change is mainly affected by T and NDVI, and the driving region of P on ET is only 32% and 17%. The "evaporation paradox" exists in the Three-River Source region in southwestern Qinghai Province. ET is highly dependent on elevation, and there is significant heterogeneity in the driving factors of ET across the altitude gradient: P is the driving factor for 1700–2600 mm, NDVI is the driving factor for 2600–3800 mm, and the driving factor for 3800–6000 m is T. In addition, there is a significant positive correlation between ET and T in regions with an average annual T of less than −4.32 °C and an altitude of less than 4400 m. There was found to be a significant negative correlation between ET and T.

This research can provide a scientific basis and practical support for implementing differentiated water resource management policies and the sustainable development of ecosystems in Qinghai Province and different regions. The high-altitude regions in the west of the Three-River Source, Qilian Mountain, and Qinghai Lake have few human activities, and are mainly affected by climatic factors. We should be alert to the risk of vegetation degradation caused by temperature rise and the risk of increased evapotranspiration and water consumption caused by melting glaciers. By establishing prohibited and restricted development zones, the pilot work in Three-River Source, Qilian Mountain, and Qinghai Lake National Park has controlled the livestock carrying capacity of pastures and reduced the impact of the increase in cultivated land on local water distribution and climate environment. The Hehuang region provides most of the agricultural products in the province, and urbanization and farmland water resource management are the core measures to optimize the regional eco-hydrological system. The Qaidam Basin region is the growth core of industrial development in Qinghai Province. Although the evapotranspiration per unit area is not large, the climate is arid, and the ecosystem is extremely fragile. It is necessary to prevent the impact of industrial production and oasis development on the regional water balance.

Author Contributions: Conceptualization, methodology, code, data curation, investigation, formal analysis, writing—original draft: Z.S.; supervision, writing—review and editing, supervision, funding acquisition: Q.F. and S.C.; writing—review and editing: Z.G. and G.C.; visualization, Z.W. All authors have read and agreed to the published version of the manuscript.

Funding: This study was funded by the Natural Science Foundation of Qinghai Province (2018-ZJ-905).

Institutional Review Board Statement: Not applicable.

Informed Consent Statement: Not applicable.

Data Availability Statement: Publicly available data sets were analyzed in this study. This data can be found here: https://search.earthdata.nasa.gov/ (accessed on 2 June 2021), http://www.gscloud.cn/ (accessed on 20 May 2021), http://www.geodata.cn (accessed on 27 June 2021).

Acknowledgments: Acknowledgement for the data support from "National Earth System Science Data Center, National Science & Technology Infrastructure of China. (http://www.geodata.cn, accessed on 27 June 2021)". The authors would like to thank three anonymous reviewers for their valuable comments that improved the quality of this manuscript.

Conflicts of Interest: The authors declare no conflict of interest.

References

1. Parmesan, C. Ecological and evolutionary responses to recent climate change. *Annu. Rev. Ecol. Evol. Syst.* **2006**, *37*, 637–669. [CrossRef]
2. Yin, Y.; Wu, S.; Zhao, D.; Zheng, D.; Pan, T. Impact of climate change on actual evapotranspiration on the Tibetan Plateau during 1981–2010. *Acta Geogr. Sin.* **2012**, *67*, 1471–1481. [CrossRef]
3. Shukla, J.; Mintz, Y. Influence of Land-Surface Evapotranspiration on the Earth's Climate. *Science* **1982**, *215*, 1498–1501. [CrossRef] [PubMed]
4. Jung, M.; Reichstein, M.; Ciais, P.; Seneviratne, S.I.; Sheffield, J.; Goulden, M.L.; Bonan, G.; Cescatti, A.; Chen, J.; de Jue, R.; et al. Recent decline in the global land evapotranspiration trend due to limited moisture supply. *Nature* **2010**, *467*, 951–954. [CrossRef]
5. Kingston, D.G.; Todd, M.C.; Taylor, R.G.; Thompson, J.R.; Arnell, N.W. Uncertainty in the estimation of potential evapotranspiration under climate change. *Geophys. Res. Lett.* **2009**, *36*, 20403. [CrossRef]
6. Cai, J.; Wang, Y.; Liu, Y. Feature Parameters of Evapotranspiration Estimation Model for Winter Wheat and Summer Maize Based on Lysimeter Monitoring System. *Trans. Chin. Soc. Agric. Mach.* **2021**, *52*, 285–295.
7. Mu, Q.; Zhao, M.; Running, S.W. Improvements to a MODIS global terrestrial evapotranspiration algorithm. *Remote Sens. Environ.* **2011**, *115*, 1781–1800. [CrossRef]
8. Bella, C.M.D.; Rebella, C.M.; Paruelo, J.M. Evapotranspiration estimates using NOAA AVHRR imagery in the Pampa region of Argentina. *Int. J. Remote Sens.* **2000**, *21*, 791–797. [CrossRef]

9.   Sumner, D.M.; Jacobs, J.M. Utility of Penman-Monteith, Priestley-Taylor, reference evapotranspiration, and pan evaporation methods to estimate pasture evapotranspiration. *J. Hydrol.* **2005**, *308*, 81–104. [CrossRef]
10. Su, Z. The Surface Energy Balance System (SEBS) for estimation of turbulent heat fluxes. *Hydrol. Earth Syst. Sci.* **2002**, *6*, 85–99. [CrossRef]
11. Song, L.; Zhuang, Q.; Yin, Y.; Zhu, X.; Wu, S. Spatio-temporal dynamics of evapotranspiration on the Tibetan Plateau from 2000 to 2010. *Environ. Res. Lett.* **2017**, *12*, 014011. [CrossRef]
12. Zhang, K.; Kimball, J.S.; Running, S.W. A review of remote sensing based actual evapotranspiration estimation. *Wiley Interdiscip. Rev. Water.* **2016**, *3*, 834–853. [CrossRef]
13. Liaqat, U.W.; Choi, M. Accuracy comparison of remotely sensed evapotranspiration products and their associated water stress footprints under different land cover types in Korean peninsula. *J. Clean. Prod.* **2017**, *155*, 93–104. [CrossRef]
14. Xu, T.; Guo, Z.; Xia, Y.; Ferreira, V.G.; Liu, S.; Wang, K.; Yao, Y.; Zhang, X.; Zhao, C. Evaluation of twelve evapotranspiration products from machine learning, remote sensing and land surface models over conterminous United States. *J. Hydrol.* **2019**, *578*, 124105. [CrossRef]
15. Ma, Z.; Ray, R.L.; He, Y. Assessing the spatiotemporal distributions of evapotranspiration in the Three Gorges Reservoir Region of China using remote sensing data. *J. Mt. Sci.* **2018**, *15*, 2676–2692. [CrossRef]
16. Sun, Z.; Wang, Q.; Ouyang, Z.; Watanabe, M.; Matsushita, B.; Fukushima, T. Evaluation of MOD16 algorithm using MODIS and ground observational data in winter wheat field in North China Plain. *Hydrol. Process. Int. J.* **2007**, *21*, 1196–1206. [CrossRef]
17. Cheng, L.; Yang, M.; Wang, X.; Wan, G. Spatial and temporal variations of terrestrial evapotranspiration in the upper Taohe River Basin from 2001 to 2018 based on MOD16 ET data. *Adv. Meteorol.* **2020**, *2020*, 3721414. [CrossRef]
18. Wu, G.; Liu, Y.; Zhang, Q.; Duan, A.; Wang, T.; Wan, R.; Liang, X. The influence of mechanical and thermal forcing by the Tibetan Plateau on Asian climate. *J. Hydrometeorol.* **2007**, *8*, 770–789. [CrossRef]
19. Kuang, X.; Jiao, J.J. Review on climate change on the Tibetan Plateau during the last half century. *J. Geophys. Res. Atmos.* **2016**, *121*, 3979–4007. [CrossRef]
20. Chang, S.; Liu, Y.; Hua, S.; Jia, R. Characteristics of Atmospheric Water Vapor over the Qinghai-Tibetan Plateau in Summer with Global Warming. *Plateau Meteorol.* **2019**, *38*, 227–236.
21. Han, C.; Ma, Y.; Wang, B.; Zhong, L.; Ma, W.; Chen, X.; Su, Z. Long-term variations in actual evapotranspiration over the Tibetan Plateau. *Earth Syst. Sci. Data* **2021**, *13*, 3513–3524. [CrossRef]
22. Wang, L.; He, X.; Ding, Y. Characteristics and influence factors of the evapotranspiration from alpine meadow in central Qinghai-Tibet Plateau. *J. Glaciol. Geocryol.* **2019**, *41*, 801–808. [CrossRef]
23. Zhang, Y.; Ma, Y.; Ma, W.; Wang, B.; Wang, Y. Evapotranspiration Variation and Its Correlation with Meteorological Factors on Different Underlying Surfaces of the Tibetan Plateau. *J. Arid Meteorol.* **2021**, *39*, 366–373. [CrossRef]
24. Zhang, F.; Li, H.; Wang, W.; Li, Y.; Lin, L.; Guo, X.; Du, Y.; Li, Q.; Yang, Y.; Cao, G. Net radiation rather than surface moisture limits evapotranspiration over a humid alpine meadow on the northeastern Qinghai–Tibetan Plateau. *Ecohydrology* **2018**, *11*, e1925. [CrossRef]
25. Quan, C.; Zhou, B.; Han, Y.; Zhao, T.; Xiao, J. A study of evapotranspiration on the degraded alpine wetland surface in the Yangtze River source region. *J. Glaciol. Geocryol.* **2016**, *38*, 1249–1257. [CrossRef]
26. Dai, L.; Cao, Y.; Ke, X.; Zhang, F.; Du, Y.; Guo, X.; Cao, G. Response of reference evapotranspiration to meteorological factors in alpine meadows on the Qinghai-Tibet Plateau. *Pratacultural Sci.* **2018**, *35*, 2137–2147. [CrossRef]
27. Li, H.; Zhang, F.; Zhu, J.; Guo, X.; Li, Y.; Lin, L.; Zhang, L.; Yang, Y.; Li, Y.; Cao, G. Precipitation rather than evapotranspiration determines the warm-season water supply in an alpine shrub and an alpine meadow. *Agric. For. Meteorol.* **2021**, *300*, 108318. [CrossRef]
28. Wang, R.; He, M.; Niu, Z.G. Responses of Alpine Wetlands to Climate Changes on the Qinghai-Tibetan Plateau Based on Remote Sensing. *Chin. Geogr. Sci.* **2020**, *30*, 189–201. [CrossRef]
29. Wen, X.; Zhou, J.; Liu, S.; Ma, Y.; Xu, Z.; Ma, J. Spatio-temporal characteristics of surface evapotranspiration in source region of rivers in Southwest China based on multi-source products. *Water Resour. Prot.* **2021**, *37*, 32–42. [CrossRef]
30. Wang, L.; Guo, N.; Wang, W.; Lu, Y.; Sha, S. Actual Evapotranspiration Estimated by TESEBS Model over the Tibetan Plateau. *Remote Sens. Technol. Appl.* **2017**, *32*, 507–513. [CrossRef]
31. Li, H.; Wang, C.; Zhang, F.; He, Y.; Shi, P.; Guo, X.; Wang, J.; Zhang, L.; Li, Y.; Cao, G. Atmospheric water vapor and soil moisture jointly determine the spatiotemporal variations of CO_2 fluxes and evapotranspiration across the Qinghai-Tibetan Plateau grasslands. *Sci. Total Environ.* **2021**, *791*, 148379. [CrossRef] [PubMed]
32. Zhang, H.; Dou, R. Interannual and seasonal variability in evapotranspiration of alpine meadow in the Qinghai-Tibetan Plateau. *Arab. J. Geosci.* **2020**, *13*, 968. [CrossRef]
33. Bonan, G.B. Forests and climate change: Forcings, feedbacks, and the climate benefits of forests. *Science* **2008**, *320*, 1444–1449. [CrossRef] [PubMed]
34. Tian, X.; Zhang, L.; Zhang, X.; Chen, Z.; Zhao, L.; Li, Q.; Gu, S. Evapotranspiration characteristics of degraded meadow and effects of freeze-thaw changes in the Three-River Source Region. *Acta Ecol. Sin.* **2020**, *40*, 5649–5662. [CrossRef]
35. Han, G.; Cao, G.; Cao, S.; Chen, K.; Yang, Y.; Liu, Y. Simulation of Evapotranspiration of Xiaopo Lake and Shaliu River Headwater Wetlands Based on Shuttleworth-Wallace Model. *Wetl. Sci.* **2019**, *17*, 519–526.

36. Liu, X.; Gao, X.; Ma, Y. Spatio-temporal Evolution of Vegetation Coverage in Qinghai Province, China during the Periods from 2002 to 2015. *Arid Zone Res.* **2017**, *34*, 1345–1352. [CrossRef]

37. Zhang, Q.; Zhang, J.; Sun, G.; Di, X. Research on water-vapor distribution in the air over Qilian Mountain. *Acta Meteorol. Sin.* **2008**, *1*, 107–118.

38. Peng, S.; Ding, Y.; Liu, W.; Li, Z. 1 km monthly temperature and precipitation dataset for China from 1901 to 2017. *Earth Syst. Sci. Data* **2019**, *11*, 1931–1946. [CrossRef]

39. Han, H.; Bai, J.; Ma, G.; Yan, J. Vegetation phenological changes in multiple landforms and responses to climate change. *ISPRS Int. J. Geo-Inf.* **2020**, *9*, 111. [CrossRef]

40. Liang, L.; Li, L.; Liu, Q.J.A.; Meteorology, F. Temporal variation of reference evapotranspiration during 1961–2005 in the Taoer River basin of Northeast China. *Agric. For. Meteorol.* **2010**, *150*, 298–306. [CrossRef]

41. Gao, G.; Chen, D.; Xu, C.; Simelton, E. Trend of estimated actual evapotranspiration over China during 1960–2002. *J. Geophys. Res. Atmos.* **2007**, *112*, D11120. [CrossRef]

42. Ye, H.; Zhang, T.; Yi, G.; Li, J.; Bie, X.; Liu, D.; Luo, L. Spatio-temporal characteristics of evapotranspiration and its relationship with climate factors in the source region of the Yellow River from 2000 to 2014. *Acta Geogr. Sin.* **2018**, *73*, 2117–2134. [CrossRef]

43. Zheng, D.; Yao, T. Uplifting of Tibetan Plateau with Its Environmental Effects. *Adv. Earth Sci.* **2006**, *21*, 451–458. [CrossRef]

44. Wang, W.; Li, J.; Yu, Z.; Ding, Y.; Xing, W.; Lu, W. Satellite retrieval of actual evapotranspiration in the Tibetan Plateau: Components partitioning, multidecadal trends and dominated factors identifying. *J. Hydrol.* **2018**, *559*, 471–485. [CrossRef]

45. Zhang, H.; Sun, J.; Xiong, J. Spatial-temporal patterns and controls of evapotranspiration across the Tibetan Plateau (2000–2012). *Adv. Meteorol.* **2017**, *112*, D11120. [CrossRef]

46. Yao, Y.; Liang, S.; Cheng, J.; Liu, S.; Fisher, B.J.; Zhang, X.D.; Jia, K.; Zhao, X.; Qin, Q.; Zhao, B.; et al. MODIS-driven estimation of terrestrial latent heat flux in China based on a modified Priestley–Taylor algorithm. *Agric. For. Meteorol.* **2013**, *171*, 187–202. [CrossRef]

47. Cui, M.; Wang, J.; Wang, S.; Yan, H.; Li, Y. Temporal and spatial distribution of evapotranspiration and its influencing factors on Qinghai-Tibet Plateau from 1982 to 2014. *J. Resour. Ecol.* **2019**, *10*, 213–224. [CrossRef]

48. Khan, M.S.; Baik, J.; Choi, M. Inter-comparison of evapotranspiration datasets over heterogeneous landscapes across Australia. *Adv. Space Res.* **2020**, *66*, 533–545. [CrossRef]

49. Yang, W. A Study on the Dynamic Process of Evapotranspiration and Its Responses to Climate Change on the Tibetan Plateau. Master's Thesis, Lanzhou University, Lanzhou, China, 2020.

50. Liu, R.; Wen, J.; Wang, X. Spatial–Temporal Variation and Abrupt Analysis of Evapotranspiration over the Yellow River Source Region. *Clim. Environ. Res.* **2016**, *21*, 503–511. [CrossRef]

51. Sun, Q. Evapotranspiration Estimation and Its Impact Factors in the Three River Headwater Region. Master's Theisis, China University of Geosciences, Beijing, China, 2018.

52. Bibi, S.; Wang, L.; Li, X.; Zhang, X.; Chen, D. Response of groundwater storage and recharge in the Qaidam Basin (Tibetan Plateau) to climate variations from 2002 to 2016. *J. Geophys. Res. Atmos.* **2019**, *124*, 9918–9934. [CrossRef]

53. Jin, X.; Guo, R.H.; Xia, W. Variation of regional evapotranspiration of Qaidam Basin using MODIS data. *Hydrogeol. Eng. Geol.* **2013**, *40*, 8–13.

54. Gu, S.; Tang, Y.; Cui, X.; Du, M.; Zhao, L.; Li, Y.; Xu, S.; Zhou, H.; Kato, T.; Qi, P. Characterizing evapotranspiration over a meadow ecosystem on the Qinghai–Tibetan Plateau. *J. Geophys. Res. Atmos.* **2008**, *113*. [CrossRef]

55. Zhen, S. Study on Spatial and Temporal Distribution Characteristics of Surface Evapotranspiration and Its Influence Factors in Southern Slope of Qilian Mountain. Master's Thesis, Qinghai Normal University, Xining, China, 2017.

56. Ma, Y.; Li, X.; Liu, L.; Yang, X.; Wu, X.; Wang, P.; Lin, H.; Zhang, G.; Miao, C. Evapotranspiration and its dominant controls along an elevation gradient in the Qinghai Lake watershed, northeast Qinghai-Tibet Plateau. *J. Hydrol.* **2019**, *575*, 257–268. [CrossRef]

57. Cao, S.; Cao, G.; Han, G.; Wu, F.; Lan, Y. Comparison of evapotranspiration between two alpine type wetland ecosystems in Qinghai lake basin of Qinghai-Tibet Plateau. *Ecohydrol. Hydrobiol.* **2020**, *20*, 215–229. [CrossRef]

58. Zhang, T.; Zhu, X.; Wang, Y.; Li, H.; Liu, C. The Impact of Climate Variability and Human Activity on Runoff Changes in the Huangshui River Basin. *Resour. Sci.* **2014**, *36*, 2256–2262.

59. Li, X.; Long, D.; Han, Z.; Scanlon, B.R.; Sun, Z.; Han, P.; Hou, A. Evapotranspiration estimation for Tibetan plateau headwaters using conjoint terrestrial and atmospheric water balances and multisource remote sensing. *Water Resour. Res.* **2019**, *55*, 8608–8630. [CrossRef]

60. Qiu, L.; Zhang, L.; He, Y.; Chen, Y.; Wang, W. Spatiotemporal Variations of Evapotranspiration and Influence Factors in Oilian Mountain from 2000 to 2018. *Res. Soil Water Conserv.* **2020**, *27*, 210–217.

61. Wang, F.; Wang, Z.; Zhang, Y.; Shen, F. Spatio-temporal Variations of Evapotranspiration in Anhui Province Using MOD16 Products. *Resour. Environ. Yangtze Basin* **2018**, *27*, 523–534. [CrossRef]

62. Zhang, F.; Geng, M.; Wu, Q.; Liang, Y. Study on the spatial-temporal variation in evapotranspiration in China from 1948 to 2018. *Sci. Rep.* **2020**, *10*, 17139. [CrossRef]

63. Yao, T.; Thompson, L.; Yang, W.; Yu, W.; Gao, Y.; Guo, X.; Yang, X.; Duan, K.; Zhao, H.; Xu, B.; et al. Different glacier status with atmospheric circulations in Tibetan Plateau and surroundings. *Nat. Clim. Chang.* **2012**, *2*, 663–667. [CrossRef]

64. Shi, Y.; Shen, Y.; Hu, R. Preliminary Study on Signal, Impact and Foreground of Climatic Shift from Warm-Dry to Warm-Humid in Northwest China. *J. Glaciol. Geocryol.* **2002**, *24*, 219–226.

65. Zhang, K.; Pan, S.; Cao, L. Spatial and Temporal Trends of Average Wind Speed in Hexi Region in 1961–2010. *Sci. Geogr. Sin.* **2014**, *34*, 1404–1408. [CrossRef]
66. Seddon, A.W.R.; Macias-Fauria, M.; Long, P.R.; Benz, D.; Willis, K.J. Sensitivity of global terrestrial ecosystems to climate variability. *Nature* **2016**, *531*, 229–232. [CrossRef] [PubMed]
67. Li, Z.; Feng, Q.; Liu, W.; Wang, T.; Gao, Y.; Wang, Y.; Cheng, A.; Li, J.; Liu, L. Spatial and temporal trend of potential evapotranspiration and related driving forces in Southwestern China, during 1961–2009. *Quat. Int.* **2014**, *336*, 127–144. [CrossRef]
68. Chen, S.; Liu, Y.; Axel, T. Climatic change on the Tibetan Plateau: Potential evapotranspiration trends from 1961–2000. *Clim. Chang.* **2006**, *76*, 291–319. [CrossRef]
69. Arturo, G.R.; Fernando, A.Q.J.; Carlos, A.L. Landform instability and land-use dynamics in tropical high mountains, Central Mexico. *J. Mt. Sci.* **2012**, *9*, 414–430. [CrossRef]
70. Thomas, A. Spatial and temporal characteristics of potential evapotranspiration trends over China. *Int. J. Climatol. A J. R. Meteorol. Soc.* **2000**, *20*, 381–396. [CrossRef]
71. Sun, J.; Wang, G.; Sun, X.; Lin, S.; Hu, Z.; Huang, K. Elevation-dependent changes in reference evapotranspiration due to climate change. *Hydrol. Processes.* **2020**, *34*, 5580–5594. [CrossRef]
72. Zhang, C.; Shen, Y.; Liu, F.; Meng, L. Changes in reference evapotranspiration over an agricultural region in the Qinghai-Tibetan plateau, China. *Theor. Appl. Climatol.* **2016**, *123*, 107–115. [CrossRef]
73. Liu, Y.; Yao, X.; Wang, Q.; Yu, J.; Jiang, Q.; Jiang, W.; Li, L. Differences in reference evapotranspiration variation and climate-driven patterns in different altitudes of the Qinghai–Tibet plateau (1961–2017). *Water* **2021**, *13*, 1749. [CrossRef]
74. Jia, W.; He, Y.; Li, Z.; Pang, H.; Yaun, L.; Ning, B.; Song, B.; Zhang, N. Spatio-temporal Distribution Characteristics of Climate Change in Qilian Mountain and Hexi Corridor. *J. Desert Res.* **2008**, *6*, 1151–1155+1215.
75. Qin, J.; Yang, K.; Liang, H.; Guo, X. The altitudinal dependence of recent rapid warming over the Tibetan Plateau. *Clim. Chang.* **2009**, *97*, 321–327. [CrossRef]
76. Shen, M.; Piao, S.; Jeong, S.; Zhou, L.; Zeng, Z.; Ciais, P.; Chen, D.; Huang, M.; Jin, C.; Li, L.Z.X.; et al. Evaporative cooling over the Tibetan Plateau induced by vegetation growth. *Proc. Natl. Acad. Sci. USA* **2015**, *112*, 9299–9304. [CrossRef] [PubMed]
77. Feng, Y. Changes of Glaciers and Permafrost in Qinghai-Tibet Plateau and Their Ecological and Hydrological Effects—Take the Yellow River Source, Brahmaputra River Basin and Permafrost Degradation Turning Zone as Typical Study Region. Ph.D. Thesis, China University of Geosciences, Beijing, China, 2020.
78. Ma, Y.J.; Li, X.; Liu, L.; Huang, Y.; Li, Z.; Hu, X.; Wu, X.; Yang, X.; Wang, P.; Zhao, S. Measurements and modeling of the water budget in semiarid high–altitude Qinghai Lake Basin, Northeast Qinghai–Tibet plateau. *J. Geophys. Res. Atmos.* **2018**, *123*, 10857–10871. [CrossRef]

*water*

Article

Variation Characteristics of Summer Water Vapor Budget and Its Relationship with the Precipitation over the Sichuan Basin

Dongmei Qi [1,2], Yueqing Li [1,2,*] and Changyan Zhou [1,2]

1 Institute of Plateau Meteorology, China Meteorology Administration, Chengdu 610072, China; qidongmei1983@163.com (D.Q.); zcy001124@163.com (C.Z.)
2 Heavy Rain and Drought-Flood Disasters in Plateau and Basin Key Laboratory of Sichuan Province, Chengdu 610072, China
* Correspondence: yueqingli@163.com

Abstract: Based on the daily precipitation data from the meteorological stations in Sichuan and the monthly average ERA-Interim reanalysis data from 1979 to 2016, the variation characteristics of summer water vapor budget in the Sichuan Basin and its relationship with precipitation are discussed in this study. The results show that, in summer, the water vapor in the Sichuan Basin and its four sub-basins flows in from the southern and western boundaries and flows out through the eastern and northern boundaries, and the basin is obviously a water vapor sink. From 1979 to 2016, the water vapor inflow from the southern and western boundaries significantly decreased, as well as the water vapor outflow through the eastern boundary. The summer precipitation in the Sichuan Basin is significantly positively correlated with the water vapor inflow at the southern boundary and net water vapor budget of the basin in the same period, and it is negatively correlated with the water vapor outflow at the northern boundary. The southern and northern boundaries are the two most important boundaries for the summer precipitation in the Sichuan Basin. Additionally, this study reveals that, under the multi-scale topography on the east side of the Tibet Plateau, the spatio-temporal distribution of precipitation in the Sichuan Basin results from the interactions between the unique topography of the Sichuan Basin and the different modes of water-vapor transport from low latitudes. The atmospheric circulation over the key area of air–sea interaction in the tropical region and its accompanying systems, as well as the anomalies of regional circulations and water vapor transport over the eastern China and Sichuan Basin, are the main reasons for the variation in summer precipitation in the Sichuan Basin.

Keywords: Sichuan Basin; water vapor budget; summer precipitation; water resource variation

Citation: Qi, D.; Li, Y.; Zhou, C. Variation Characteristics of Summer Water Vapor Budget and Its Relationship with the Precipitation over the Sichuan Basin. *Water* **2021**, *13*, 2533. https://doi.org/10.3390/w13182533

Academic Editors: Yaoming Ma, Zhongbo Su and Lei Zhong

Received: 12 August 2021
Accepted: 8 September 2021
Published: 15 September 2021

Publisher's Note: MDPI stays neutral with regard to jurisdictional claims in published maps and institutional affiliations.

1. Introduction

Water resources are indispensable natural resources in production and human life as well as other social activities. Factors of climate change, ecological environment evolution and human activities, and climate change and its impact on water resources affect water resources and are important subjects worldwide [1,2]. In the past century, the characteristics of climate and environment have undergone significant changes, such as global warming, and climate change necessarily leads to changes in the hydrological cycle and the spatio-temporal redistribution of water resources [3,4]. Water vapor, as an important component of the atmosphere, absorbs solar radiation. It is an important greenhouse gas, which is closely related to precipitation and climate and plays a key role in the global water cycle and energy cycle [5]. Water vapor transport and budget are important components of the regional water balance, which are directly related to rainfall events and the climate near the ground [6–8]. Many researchers have studied the relationship of water vapor transport and budget with regional drought and flood [9–11]. Xu et al. [12,13] analyzed the water vapor source and sink of the Meiyu belt over the Yangtze River Basin and found that the integrated water vapor transport in drought and wet years in the Yangtze River Basin shows an inverse feature. They have pointed out that the water vapor transport in the "large triangular fan pattern" region composed of the low-latitude activity source areas related to the Tibet Plateau, the South China Sea monsoon, and the Indian monsoon has an important impact on regional drought and flood in China. Zhou et al. [14] illustrated that, when the water vapor source is different from the climatology, the summer precipitation anomaly is induced. The increase in water vapor transport is one of the reasons for the increasing frequency of flood events from 1961 to 2005 [15]. Jiang et al. [16] found that the convergence and divergence of water vapor are closely related to precipitation over the Yangtze River Basin. Zhang et al. [17] studied the characteristics of water vapor transport in East Asia and its impact on drought and flood in North China. Ding et al. [18] referred that the long-lasting drought in North China is the result of prevailing continental warm high and the eastward movement of the water vapor conveyor belt. Simmonds et al. analyzed the summer water vapor transport and budget in China and found that the source of water vapor in southeastern China is from the South China Sea and the Bay of Bengal, while the water vapor in Northeast China and some parts of North China mainly comes from middle-latitude westerlies [19]. Yang et al. [20] compared the characteristics of water vapor transport for four summer rainfall patterns over the monsoon regions in the eastern China. Their study showed that the northern pattern is mainly affected by water vapor transport of Asian monsoon, the central pattern and Yangtze River pattern are affected by water vapor transport from the Pacific, and the South China pattern is impacted by water vapor transport from the Indian Ocean, the Pacific, and the South China Sea. Li et al. [21] studied two main water vapor conveyor belts that influence the summer drought and flood in eastern Southwest China. Chu et al. and Dong et al. [22–24] employed different models to analyze the water vapor source of the first rainy season in South China and the water vapor sources of midsummer precipitation in different regions of eastern China. They also revealed the relationship between water vapor transport over the Yangtze River basin in the Meiyu period and the key region in the Tibet Plateau. All the above studies have shown that water vapor transport and budget significantly influence regional drought and flood. Studies on the air water resource and its utilization are directly related to regional economic construction, human life, and economic development, so it is of great significance.

The Sichuan Basin is located in the east of the Tibet Plateau. Due to the special geographical location and complex topography, it is a sensitive, vulnerable, and key area for climate change. The water resource distribution and regional climate change there have distinct particularity and diversity. In summer, affected by the combined influence of various monsoon circulations, the cyclone made by the southern branch flow around the Tibet Plateau and the water vapor transport from the Pacific Ocean, the Bay of Bengal, and the Arabian Sea make the region rich in water vapor and precipitation. The atmospheric water vapor source, the track of water vapor transport, and the water vapor budget are the key links of regional water cycle in this region, especially the air water resources. They play a key role in the regional water balance and are closely related to the atmospheric circulation evolution and the variation of regional drought and flood [25–27]. Summer precipitation in the Sichuan Basin shows a decreasing trend, especially after the 1990s [28–30]. Zhou et al. [31] pointed out that the water vapor transported from the southern boundary of eastern Tibet Plateau and its surrounding regions decreases. Moreover, the decreasing trend is especially significant after the 1990s, and is not prone to precipitation in the Sichuan Basin. In the Sichuan Basin, the atmospheric forcing and circulation factors causing precipitation anomalies are complex, resulting in the great variations of inter-annual precipitation and its spatial distribution and the alternating occurrence of drought and flood [32]. Summer precipitation in the Sichuan Basin is affected by the simultaneity of East Asian summer monsoon, Tibetan Plateau summer monsoon, and southwest monsoon [33–36], as well as the anomalous variations of the location and intensity of the western Pacific subtropical high, South Asian high, and the westerly jet [37–41]. Rainfall events and climate in the basin are also closely related to variation in the atmospheric heat source in the Tibet Plateau and its surrounding regions [42–45]. In addition, as an important external forcing source for the atmosphere ocean has a great impact on the anomalies of atmospheric circulation, and the intensity of western Pacific warm pool, and the sea surface temperature (SST) in the equatorial central-eastern Pacific and Indian Ocean is also closely related to the precipitation in the Sichuan Basin [46–50].

In summary, the anomalies of water vapor transport and budget are direct reasons for summer precipitation anomalies in China. Many researchers have studied the spatio-temporal distribution of summer precipitation in the Sichuan Basin. The analysis of the precipitation anomaly mainly focuses on the atmospheric circulation and external forcing signals such as SST, but few studies have systematically and deeply analyzed the water vapor transport and budget closely related to precipitation. Climate change in the Sichuan Basin, especially precipitation change, has significant regional characteristics. Consequently, it is necessary to investigate the precipitation characteristics and the specific response to climate change on a regional scale. It is enormously helpful to understand the characteristics and impacts of climate change and its regional response. The Sichuan Basin is a sensitive area for responding to climate change. Under global warming, the water vapor budget/transport in this region will change significantly, which will further result in abnormal changes of precipitation. In the past, research on the variation characteristics of water vapor budget in Sichuan mainly focused on a large spatial scale, and there were no in-depth discussions on each region within the basin. Then, what are the spatio-temporal variation characteristics of water vapor budget over the Sichuan Basin on a regional scale? In particular, what are the similarities and differences of water vapor budget changes among different regions in the basin? Is there is a close relationship between the water vapor budget/transport and regional droughts/floods, and how does the water vapor budget/transport over the Sichuan Basin influence summer precipitation there? What are the main factors for the water vapor variations and the mechanisms? At present, the possible answers to these questions are not clear. Focusing on these problems, this study employs the ERA-Interim reanalysis data to study the variation characteristics of the summer water vapor budget in the Sichuan Basin and uses the observed precipitation data at meteorological stations to reveal the impact of regional water vapor budget on the summer precipitation in the Sichuan Basin. It can improve the understanding of water vapor budget anomalies and its impact on the summer precipitation, and can support for the forecast of climatic disasters.

2. Material and Methods

The daily rain gauge data from the latest version (V3) of surface climatological data is compiled by the China National Meteorological Information Center. Out of the daily precipitation data at 114 stations in Sichuan Province from 1979 to 2016, 74 stations are selected to represent the Sichuan Basin. In addition, we use the monthly mean ERA-Interim reanalysis data with a horizontal resolution of $1.5° \times 1.5°$ from 1979 to 2016, including the specific humidity from 1000 hPa to 300 hPa, horizontal zonal wind, horizontal meridional wind, geo-potential height, and the corresponding surface pressure.

As shown in Figure 1, the boundary shows the Sichuan province, and the rectangles outline the Sichuan Basin. The study area is the rectangular area of the Sichuan Basin (Figure 1), in which the Sichuan Basin has six boundaries. Of these, 1 and 5 are the southern boundaries; 2 is the western boundary; 3 is the northern boundary; and 4 and 6 are the eastern boundaries (Figure 1a). The water vapor transport through the southern boundary of the basin is the sum of the water vapor transport at boundaries 1 and 5, and the water vapor transport through the eastern boundary is the sum of the water vapor transport at boundaries 4 and 6. Bounded by 105° E, the Sichuan Basin is divided into the eastern basin and the western basin (Figure 1b). Additionally, boundaries 1, 2, 3, and 4 are defined as the southern, western, northern, and eastern boundaries of the western basin, respectively; boundaries 9 and 7 are the southern boundaries of the eastern basin; boundary 4 is the western boundary of the eastern basin; boundary 5 is the northern boundary of the eastern basin; and boundaries 6 and 8 are defined as the eastern boundaries of the eastern basin.

The water vapor transport through the southern boundary of the eastern basin is the sum of the water vapor transport at boundary 9 and 7. The water vapor transport through the eastern boundary of the eastern basin is the sum of the water vapor transport at boundary 6 and 8. Along 30° N, the Sichuan Basin is divided into the northern basin and the southern basin (Figure 1c), and both regions have four refined boundaries. Boundaries 1, 2, 3, and 4 are the southern, western, northern, and eastern boundaries of the southern basin; and boundaries 5, 6, 7, and 8 are the southern, western, northern, and eastern boundaries of the northern basin. In this study, the total inflow and outflow over the Sichuan Basin (the eastern, western, northern, and southern basin) are the sum of the inflow and outflow at the southern, western, northern, and eastern boundaries, and the difference between the total inflow and outflow is the net budget. The net water vapor budget over the Sichuan Basin is defined as B_T, including the water vapor budgets at the southern (B_S), western (B_W), northern (B_N), and eastern (B_E) boundaries. Here, S, W, N, and E, respectively, stands for the southern, western, northern, and eastern boundary. The positive (negative) values of B_S, B_W, B_N, and B_E indicate water vapor inflow (outflow) on the boundary of the region. The variations in summer water vapor inflow and outflow over the Sichuan Basin and its eastern, western, northern and southern sub-basins, and their relationship with precipitation are analyzed in the following.

The Equations to calculate the vertically integrated water vapor flux are as follows.

The zonal and meridional water vapor fluxes are calculated as follows:

$$Q_\lambda = -\frac{1}{g}\int_{P_s}^{P_t} qu\,dp \tag{1}$$

$$Q_\varphi = -\frac{1}{g}\int_{P_s}^{P_t} qv\,dp \tag{2}$$

where P_s is the surface pressure, and P_t is set as 300 hPa; q is the specific humidity, u is the meridional wind, v is the zonal wind, λ is longitude, φ is latitude, and g is the gravity acceleration. The water vapor transport calculated by the boundary integral is as follows:

$$F_u = \int Q_\lambda a\,d\varphi \tag{3}$$

$$F_v = \int Q_\varphi a \cos\varphi\,d\lambda \tag{4}$$

In this Equation, F_u is the east–west water vapor transport, and F_v is the north–south water vapor transport.

The regional total water vapor budget is

$$D_s = \sum (F_u, F_v) = F_i - F_a \tag{5}$$

In this Equation, F_i is the total water vapor inflow, and F_a is the total water vapor outflow.

Correlation and composite analyses are utilized in detecting the relationships between pairs of variables. The change trend is analyzed by linear trend estimation method and Student's t test is used to assess the statistical significance. The summer is the averages of June, July, and August.

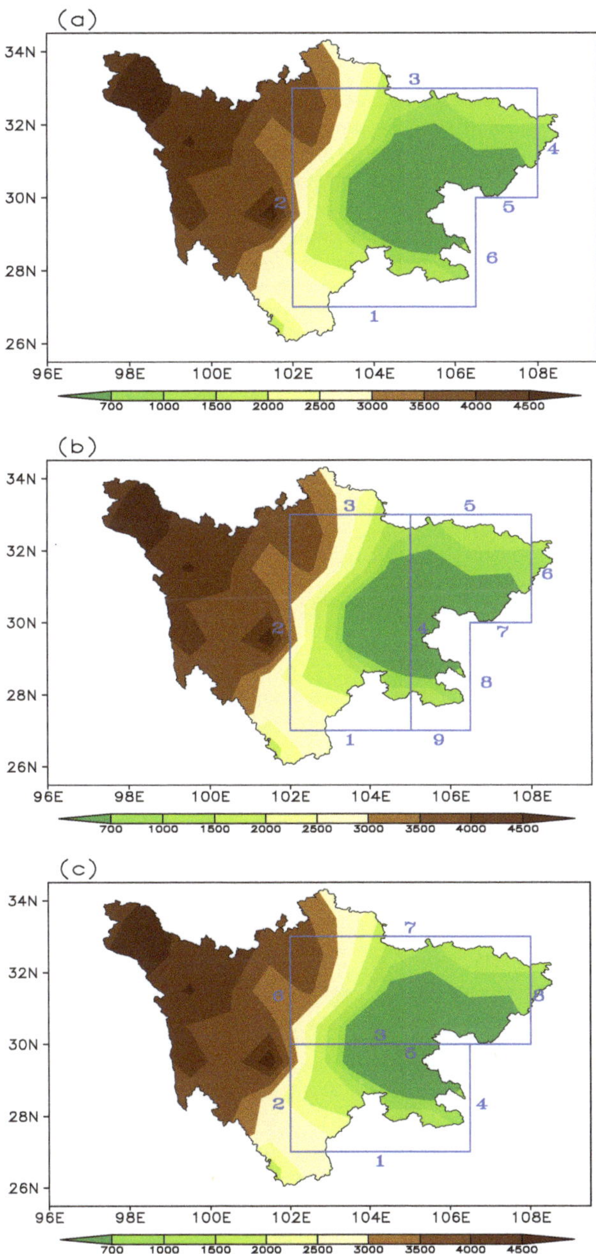

Figure 1. The boundaries of the Sichuan Basin and the sub-basins. (Map shows the topography of Sichuan Province, different colors indicate terrain height, units: m. The boundary shows the Sichuan province, and the rectangles outline the Sichuan Basin. (**a**) The Sichuan Basin; (**b**) eastern and western basin; (**c**) northern and southern basin).

3. Results

3.1. Variation Characteristics of the Summer Water Vapor Budget in the Sichuan Basin

As can be seen in Table 1, the summer water vapor flows in from the southern and western boundaries and flows out through the eastern and northern boundaries of the Sichuan Basin and the four sub-basins. The amount of outflow through the eastern boundary is much greater than that through the northern boundary, and thus the eastern boundary is the main boundary of the water vapor outflow over the Sichuan Basin. The amount of water vapor inflow from the southern boundary is much greater than that from the western boundary for all the sub-basins in the Sichuan Basin, except for the western basin. The total inflow is much greater than the total outflow at the boundaries of the Sichuan Basin and the four sub-basins, and the net water vapor budget is positive. Therefore, an obvious water vapor sink is observed in the Sichuan Basin and the four sub-basins in summer. Comparing the eastern and western basin, except for the inflow from the western boundary, the water vapor inflow and outflow at each boundary, the total inflow, the total outflow, and the net budget in the eastern basin are larger than those in the western basin, indicating that the variation in water vapor budget over the eastern basin is more significant than that over the western basin. This is closely related to the unique topography of high and steep mountain ranges on the boundary of the western basin and the relatively low and gentle mountain ranges on the edge of eastern basin, and their different interactions with the water vapor from the peripheral circulation. In comparing the northern and southern basin, the water vapor inflow at the western and southern boundaries, and the water vapor outflow at the eastern and northern boundaries, the total inflow and the total outflow are all greater in the southern basin than those in the northern basin, while the net water vapor budget over the two sub-basins is quite close.

Table 1. The net water vapor transport, total inflow, total outflow, and the net budget at all boundaries in the Sichuan Basin (the eastern, western, northern and southern basin) in summer (unit: 10^6 kg s^{-1}).

Region	Water Vapor Budget						
	B_S	B_W	B_N	B_E	Inflow	Outflow	B_T
Basin	106.76	59.79	−24.09	−67.63	166.55	−91.72	74.83
eastern	69.06	55.55	−20.32	−67.63	124.61	−87.95	36.66
western	33.88	59.79	−3.32	−55.55	93.67	−58.87	34.8
northern	68.22	28.22	−24.09	−31.01	96.44	−55.1	41.34
southern	70.74	30.6	−30.85	−36.62	101.34	−67.47	33.87

From the inter-annual variation in the water vapor budget at all boundaries of the Sichuan Basin in summer (Figure 2a), it is found that the water vapor inflow from the southern and western boundaries of the Sichuan Basin decreases significantly from 1979 to 2016 (significant at the 90% and 95% confidence level, respectively), and the water vapor outflow through the northern boundary increases (but not significant at the 90% confidence level), while the water vapor outflow through the eastern boundary decreases significantly (significant at the 95% confidence level). Moreover, the net budget of the water vapor over the whole basin also decreases in summer (but not significant at the 90% confidence level). The decrease in water vapor inflow at the western and southern boundaries is mostly balanced with the decrease in water vapor outflow at the eastern boundary, while the increase in water vapor outflow at the northern boundary is the main reason for the decrease in net water vapor budget in the Sichuan Basin. From 1979 to 2016, the variation trends of water vapor at all the boundaries of the eastern and western Sichuan Basin are the same (Figure 2b,c), and the water vapor inflow at their southern and western boundaries shows a decreasing trend (significant at the 90% and 95% confidence level, respectively). In both regions, the decline at the western boundary is greater than that at the southern boundary. The water vapor outflow at the northern boundary shows a weak increasing trend (but not significant at the 90% confidence level), while the water vapor outflow at the eastern boundary obviously decreases (significant at the 95% confidence level). The net water vapor budget in the eastern and western basin shows a decreasing trend (but not significant at the 90% confidence level), and the decrease in net water vapor budget in the western basin is more obvious than that in the eastern basin in summer. In comparison, the decline in water vapor inflow (outflow) at each boundary in the western basin is weaker than that in the eastern basin, but the decline in net water vapor budget in the western basin is greater than that in the eastern basin. In fact, the decrease in water vapor inflow at the western and southern boundaries is largely balanced with the decrease in water vapor outflow at the eastern boundary, and the increase in water vapor outflow at the northern boundary is the main reason for the decrease in net water vapor budget in the eastern and western basin. In the summers of 1979–2016, the water vapor inflow at the southern boundary exhibits a decreasing trend (but not significant at the 90% confidence level). Meanwhile, the water vapor inflow at the western boundary exhibits a significant decreasing trend in the northern basin (significant at the 99% confidence level), and the decrease in water vapor inflow at the western boundary is larger than that at the southern boundary. The water vapor outflow shows an increasing trend at the northern boundary (but not significant at the 90% confidence level) and an obviously decreasing trend at the eastern boundary (significant at the 99% confidence level) (Figure 2d). For the southern basin (Figure 2e), the water vapor inflow at the southern boundary shows a decreasing trend (significant at the 90% confidence level), and the inflow at the western boundary also shows a decreasing trend (but not significant at the 90% confidence level). The decrease in water vapor inflow at the southern boundary is larger than that at the western boundary. The water vapor outflow at the northern and eastern boundaries shows a decreasing trend (but not significant at the 90% confidence level), and the water vapor outflow at the eastern boundary decreases more strongly than that at the northern boundary. It is worth noting that the decline in water vapor inflow at the western boundary of the northern basin is greater than that in the southern basin; the decline in water vapor inflow at the southern boundary is weaker than that in the southern basin, and the decline in water vapor outflow at the eastern boundary of the northern basin is significantly greater than that in the southern basin. In summer, the net water vapor budget increases slightly in the northern basin (but not significant at the 90% confidence level) but decreases in the southern basin (significant at the 90% confidence level).

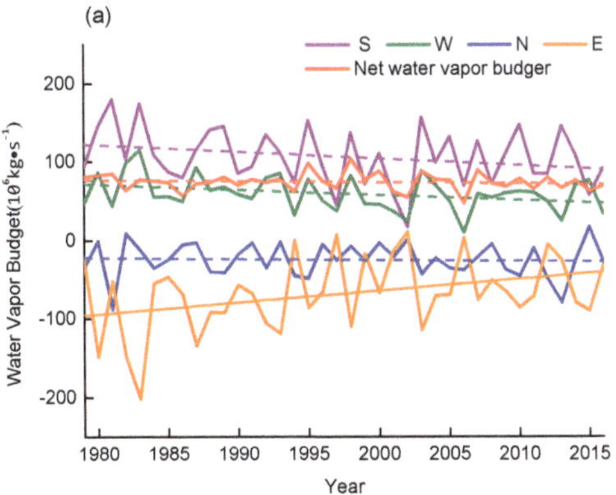

Figure 2. *Cont.*

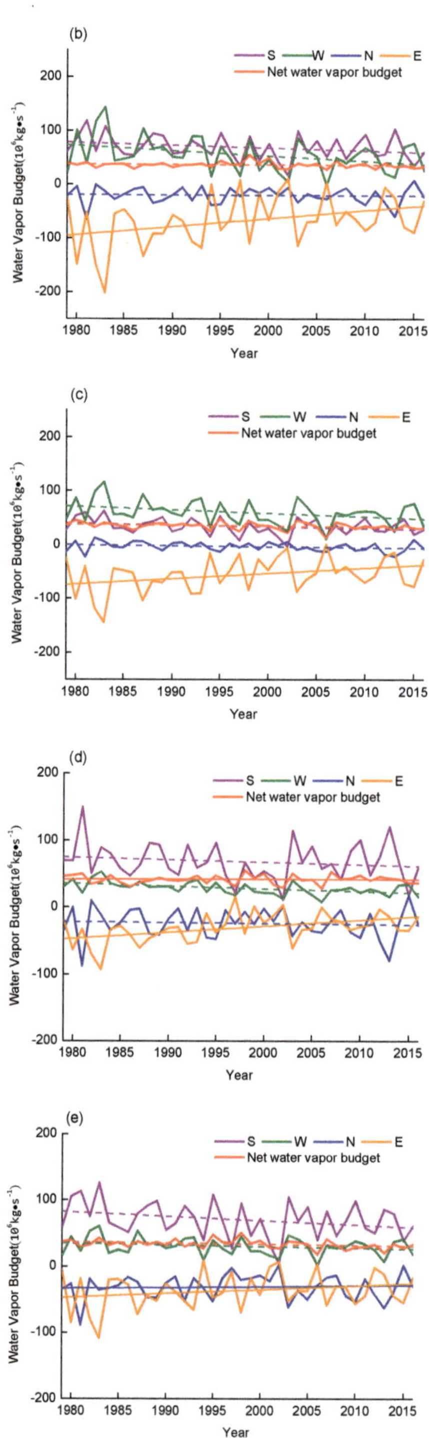

Figure 2. Inter-annual variation of water vapor inflow and outflow and net water vapor budget at

each boundary of the Sichuan Basin in summer ((**a**) the Sichuan Basin, (**b**) eastern basin, (**c**) western basin, (**d**) northern basin, and (**e**) southern basin; S, W, N, and E indicate the southern, western, northern, and eastern boundaries, respectively; unit: 10^6 kg s^{-1}. The thin lines indicate linear trends, the solid and dashed lines show positive and negative trends, respectively).

3.2. Spatio-Temporal Variation of Summer Precipitation in the Sichuan Basin

Sichuan is located in the subtropical climate zone. It is mainly composed of the Western Sichuan Plateau in the west and the Sichuan Basin in the east. The meteorological and geological conditions are complex and diverse, with high incidence of natural disasters, resulting in serious losses and wide impact. Due to the special geographical location and significantly different topography, as well as the alternating influence of different monsoon circulations, a unique regional climate type is formed, with significantly different and various variations of precipitation, and the causes are complex. As shown in Figure 3, the summer precipitation in Sichuan province is mainly over the Sichuan Basin, and there is more precipitation in the western basin than in the eastern basin, while there is less precipitation in the Western Sichuan Plateau. Moreover, there are three centers of precipitation maxima in the Sichuan Basin, which are located in the northwest of the basin (centered around Beichuan) with the summer precipitation of 765 mm, in the northeast of the basin (centered around Wanyuan) with the summer precipitation of 672 mm and in the southwest of the basin (centered around Ya'an) with the summer precipitation of 960 mm.

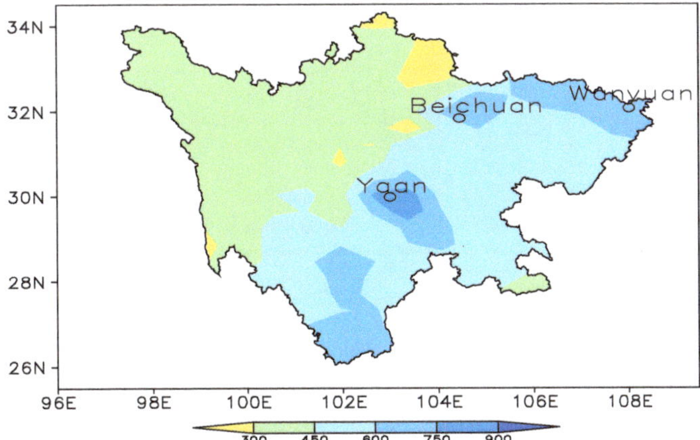

Figure 3. Spatial distribution of summer mean precipitation in the Sichuan from 1979 to 2016 (shadow; unit, mm; black circles are representative stations of the three precipitation centers in the basin).

In addition, the southern margin of the Western Sichuan Plateau is also a relative center of precipitation maxima. The locations of the three precipitation centers are closely related to the topography. Ya'an, in the southwestern basin, has the largest precipitation, known as the "sky leakage", followed by Beichuan in the northwestern basin and Wanyuan in the northeastern basin. The large precipitation is mainly distributed along the southwest, northwest, and northeast transition areas at the junction of the Sichuan Basin and its surrounding mountains in the western and northern basin.

Figure 4 shows the temporal variation of precipitation at the representative stations of the three precipitation centers in the Sichuan Basin from 1979 to 2016. The summer precipitation in the Sichuan Basin shows a decreasing trend from 1979 to 2016, with a decrease of 18.87 mm 10a^{-1} (significant at the 90% confidence level); the summer precipitation in Beichuan in the northwestern basin shows a decreasing trend, with a decrease of 50.25 mm·10a^{-1}; the summer precipitation in Wanyuan in the northeastern basin also displays a decreasing trend, with a decrease of 36.57 mm 10a^{-1}. The summer precipitation in Ya'an in the southwestern basin presents a decreasing trend, with a decrease of 21.42 mm 10a^{-1} (but not significant at the 90% confidence level). In comparison, the decreasing trend of summer precipitation is the most obvious in Beichuan, followed by Wanyuan, and it is the weakest in Ya'an.

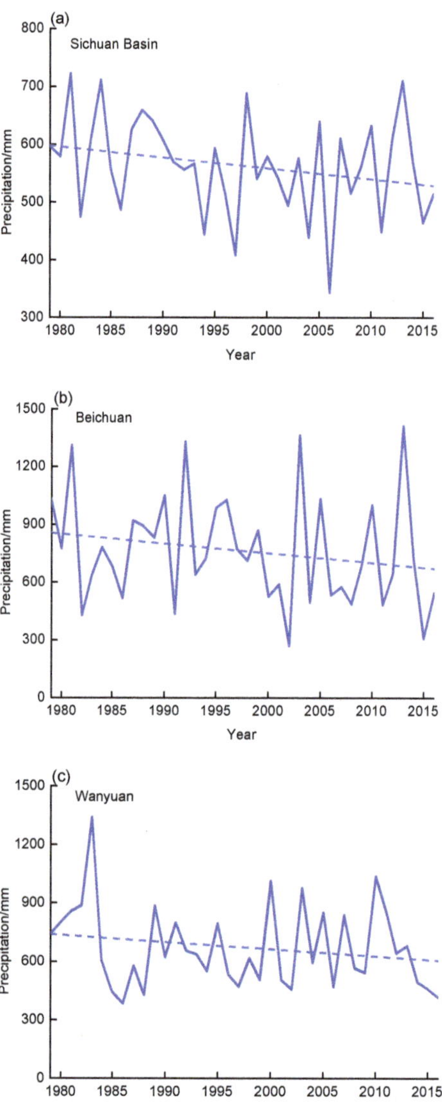

Figure 4. *Cont.*

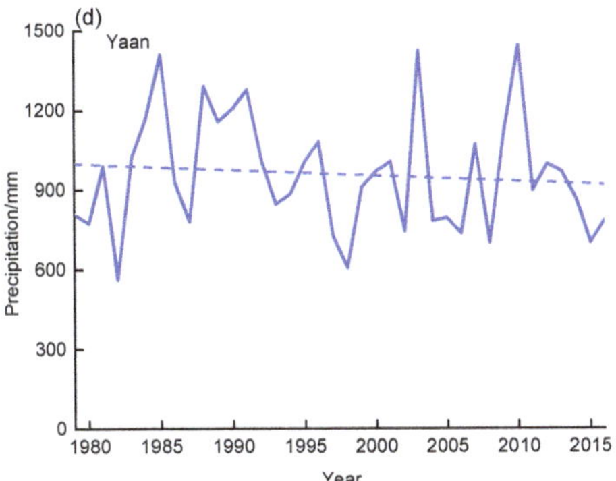

Figure 4. Inter-annual variation of summer precipitation at the three representative stations with large precipitation in the Sichuan Basin from 1979 to 2016. ((**a**) the Sichuan Basin, (**b**) Beichuan, (**c**) Wanyuan, and (**d**) Ya'an. The thin lines indicate linear trends.).

3.3. Relationship between the Summer Precipitation and Water Vapor Budget in the Sichuan Basin

3.3.1. Correlation Analysis

As shown in Table 2, the summer precipitation in the Sichuan Basin is significantly positively correlated with the water vapor inflow at the southern boundary of the basin (eastern, western, northern, and southern basin) and the net water vapor budget in the same period, and negatively correlated with the water vapor outflow at the northern and eastern boundaries of the basin (the eastern and northern basin) in the same period. In addition, the water vapor inflow from the southern boundary plays a key role for the summer precipitation in the whole basin and its sub-basins, followed by the water vapor outflow at the northern boundary. The water vapor outflow through the northern boundary makes a more important impact in the eastern basin than that in the western basin. The net water vapor budget in the whole basin as well as the eastern, western, and northern basin has a great impact on the summer precipitation over the basin. The summer precipitation in Beichuan has a significant positive correlation with the water vapor budget at the southern boundary of the basin (four sub-basins) in the same period and a significant negative correlation with the water vapor budget at the northern boundary. Except for the northern basin, the influence of water vapor budget at the northern boundary is greater than that at the southern boundary. Moreover, it is also positively correlated with the net water vapor budget of the basin (the eastern, western, and northern basin) in the same period. The summer precipitation in Ya'an has a positive correlation with the water vapor budget at the southern boundary of the basin (four sub-basins) and a negative correlation with the water vapor budget at the northern boundary (the eastern, northern, and southern basin). The influence of the water vapor budget over the eastern basin on the precipitation in Ya'an is greater than that over the western basin, the influence of the water vapor budget at the southern boundary of the northern basin is greater than that at the northern boundary, and the influence of water vapor budget at the northern boundary of the southern basin is greater than that at the southern boundary. The summer precipitation in Wanyuan has a significant positive correlation with the water vapor budget and net water vapor budget at the southern and western boundaries of the basin (four sub-basins) in the same period, a significant negative correlation with the water vapor budget at the eastern boundary, and a negative correlation with the water vapor budget at the northern boundary of the southern basin. In comparison, the influence of the net water vapor budget in the western basin is greater than that in the eastern basin, and the influence of the net water vapor budget in the northern basin is greater than that in the southern basin. The influence of the water vapor budget at each boundary is greater in the southern basin than that in the northern basin.

Table 2. Correlation Coefficients of the summer precipitation in the Sichuan Basin and at the three precipitation centers with the water vapor budget in the same period from 1979 to 2016. (**** represents passing the 99.9% confidence level, *** represents passing the 99% confidence level, ** represents passing the 95% confidence level, and * represents passing the 90% confidence level).

Water Vapor Budget		$R_{\text{Sichuan Basin}}$	R_{Beichuan}	R_{Yaan}	R_{Wanyuan}
Sichuan basin	B_S	0.686 ****	0.601 ****	0.349 **	0.654 ****
	B_W	0.235	0.056	0.037	0.471 ***
	B_N	−0.423 ***	−0.684 ****	−0.369 **	−0.111
	B_E	−0.291 *	−0.081	−0.057	−0.56 ****
	B_T	0.614 ****	0.34 **	0.22	0.468 ***
eastern basin	B_S	0.715 ****	0.635 ****	0.38 **	0.629 ****
	B_W	0.227	0.055	0.034	0.511 ****
	B_N	−0.412 ***	−0.68 ****	−0.395 **	−0.12
	B_E	−0.291 *	−0.081	−0.057	−0.56 ****
	B_T	0.566 ****	0.295 *	0.143	0.377 **
western basin	B_S	0.599 ****	0.518 ****	0.29 *	0.63 ****
	B_W	0.235	0.056	0.037	0.471 ***
	B_N	−0.311 *	−0.582 ****	−0.246	0.002
	B_E	−0.227	−0.055	−0.034	−0.511 ****
	B_T	0.587 ****	0.3 *	0.27	0.488 ***
northern basin	B_S	0.666 ****	0.747 ****	0.435 ***	0.468 ***
	B_W	0.232	0.063	0.053	0.377 **
	B_N	−0.423 ***	−0.684 ****	−0.369 **	−0.111
	B_E	−0.313 *	−0.16	−0.122	−0.503 ***
	B_T	0.672 ****	0.391 **	0.257	0.469 ***
southern basin	B_S	0.631 ****	0.537 ****	0.307 *	0.671 ****
	B_W	0.205	0.027	0.012	0.49 ***
	B_N	−0.556 ****	−0.752 ****	−0.421 ***	−0.349 **
	B_E	−0.253	−0.012	−0.002	−0.567 ****
	B_T	0.302 *	−0.01	0.034	0.346 **

From the above demonstration, the impact of the water vapor budget at the southern boundary on the summer precipitation in the Sichuan Basin is the greatest, followed by the northern boundary, and it is also closely related to the regional net water vapor budget. For the precipitation in Beichuan, in the northern basin, the water vapor budget at the southern boundary makes influence on it most significantly, followed by the northern boundary, while the influence of the water vapor budget in other sub-basins are all the most significant at the northern boundaries. For the precipitation in Ya'an, the influence of water vapor budget is the greatest at the northern boundaries of the eastern, southern, and whole basin and at the southern boundaries of the western and northern basin. For the precipitation in Wanyuan, the impact of water vapor budget at the eastern boundary is the greatest in the northern basin, while in the whole basin and the other parts of the basin, the impact is the greatest at the southern boundaries.

3.3.2. Composite Analysis

Five wet summers (summer precipitation is greater than 650 mm; 1981, 1984, 1988, 1998, and 2013) and five drought summers (summer precipitation is less than 450 mm; 1994, 1997, 2004, 2006, and 2011) in the Sichuan Basin are selected for composite analysis to discuss the differences in summer water vapor inflow, outflow and budget in the Sichuan Basin between the wet and drought years.

As seen in Table 3, the water vapor inflow at the southern and western boundaries, the water vapor outflow at the northern and eastern boundaries, and the total inflow, total outflow, and net water vapor budget in the wet summers in the Sichuan Basin are much greater than those in the drought summers. Between the wet summers and drought summers, for the whole basin, the difference in the water vapor inflow is the largest at the southern boundary, followed by the water vapor outflow at the eastern and northern boundaries, and the difference in the water vapor inflow at the western boundary is the least; for the four sub-basins, the difference in the water vapor inflow is the greatest at the southern boundary. The difference in the water vapor outflow is the least at the northern boundaries of the eastern and western basin and at the western boundaries of the northern and southern basin. Out of the four sub-basins, the difference in water vapor inflow between the wet and drought summers at the southern boundary is the greatest in the northern basin, followed by that in the southern and eastern basin, and that in the western basin is the least.

Table 3. The water vapor flux, total inflow, total outflow, and net budget at each boundary of the Sichuan Basin in the wet and drought summers (unit: 10^6 kg s^{-1}).

Region	Summers	Water Vapor Budget Region						
		B_S	B_W	B_N	B_E	Inflow	Outflow	B_T
Sichuan	wet	713.4	272.7	−248.4	−331.5	986.1	−579.9	406.2
basin	drought	374.4	213.3	−138.3	−129	587.7	−267.3	320.4
eastern	wet	480.9	248.4	−194.7	−331.5	729.3	−526.2	203.1
basin	drought	254.7	152.4	−116.7	−129	407.1	−245.7	161.4
western	wet	211.5	272.7	−47.4	−248.4	484.2	−295.8	188.4
basin	drought	99.9	213.3	−25.5	−152.4	313.2	−177.9	135.3
northern	wet	513.3	136.2	−248.4	−167.1	649.5	−415.5	234
basin	drought	262.2	103.8	−138.3	−53.1	366	−191.4	174.6
southern	wet	452.4	129.6	−251.7	−164.4	582	−416.1	165.9
basin	drought	234	106.2	−117.3	−75.9	340.2	−193.2	147

The water vapor transport, especially the water vapor inflow at the southern boundary, plays a dominant role in the summer precipitation over the Sichuan Basin. To a great extent, the amount of water vapor flux determines whether it is a wet year or a drought year in the Sichuan Basin, especially the water vapor inflow at the southern boundary of the northern basin.

By analyzing the water vapor flux, the source and path of water vapor over the Sichuan Basin can be learned. For the wet summers in the Sichuan Basin (Figure 5a), over the sea east of Philippines and north of East Indonesia–Papua New Guinea (120–180° E, 0–10° N), there are consistent easterly airflows, the area (120–170° E, 10–20° N) is dominated by an anticyclone, the area (130–160° E, 20–35° N) is dominated by an obvious cyclone, and the area (60–85° E, 5°–30° N) is dominated by a weak cyclone. Over southern and eastern China, there is an anomalous divergence of water vapor. The vast region east of the Tibet Plateau, including the Sichuan Basin, is dominated by southerly airflows. Due to the terrain of the nearly north–south steep slope east of the Tibet Plateau and the block of nearly east–west mountains in the north of the basin, the anomalous water vapor convergence extends from the whole Sichuan Basin to the south of the Western Sichuan Plateau, which is conducive to precipitation in the Sichuan Basin. On the contrary, in the drought summers (Figure 5b), over the sea (120–180° E, 0–10° N), there are consistent westerly airflows, the area (120–160° E, 10–25° N) is dominated by a cyclone, and the area (60–90° E, 5° S–20° N) is dominated by an anticyclone. There are convergence anomalies of water vapor in southern and eastern China. The vast region east of the Tibet Plateau, including the Sichuan Basin, is dominated by easterly airflows. There is an anomalous water vapor divergence over the Sichuan Basin and its surroundings, leading to less precipitation in the basin. It shows that the anomalous southerly (easterly) airflow and water vapor convergence (divergence) persist in wet (drought) summers in the Sichuan Basin, and the anomalous water vapor divergence (convergence) is maintained in southern and eastern China.

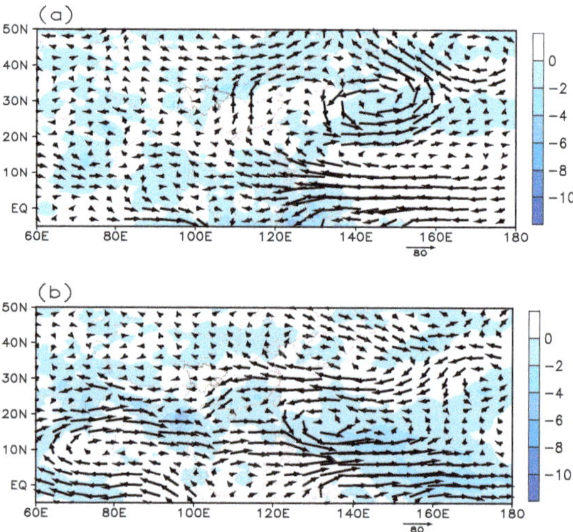

Figure 5. Anomalies of integrated water vapor flux in the whole layer (vector; unit: kg m^{-1} s^{-1}) and the water vapor divergence (shaded; unit: 10^{-5} kg m^{-2} s^{-1}) in (**a**) wet summers and (**b**) drought summers in the Sichuan Basin.

Moreover, in the drought summers, the water vapor transport and convergence in eastern and southern China are strong and link to the strong water vapor convergence over the South China Sea at low latitudes. In the wet summers, the water vapor divergence is located south of the basin at low latitudes, which is broken from the water vapor convergence over the South China Sea. Therefore, the amount of summer precipitation in the Sichuan Basin is closely related to the circulation anomalies over the key sea area east of Philippines and north of East Indonesia–Papua New Guinea (120–180° E, 0–10° N), and the circulation anomalies to the north of the key area. When there is a consistent easterly airflow in the key sea area and an anticyclonic circulation to its north (120–170° E, 10–20° N), the corresponding descending motion over it is enhanced. The significant easterly wind anomaly over the equatorial western Pacific extends westward, which helps the warm and humid airflow at low latitudes travel to the Sichuan Basin and increases the water vapor inflow at the southern boundary of the Sichuan Basin. As the Sichuan Basin is located in the convergence zone of water vapor, it favors summer precipitation, and vice versa.

The summer precipitation in the Sichuan Basin is closely related to the anomalies of the regional water vapor budget and convergence. The atmospheric circulation anomaly is the direct reason for the anomalies of water vapor and precipitation in summer over the Sichuan Basin. What is the circulation anomaly that causes the anomalies of water vapor and precipitation in summer over the Sichuan Basin? We analyze the characteristics of the atmospheric circulation in the wet and drought summers in the following. From the difference in composite circulation fields between the wet summers and drought summers in the Sichuan Basin, it is shown that there is more water vapor transported by the easterly wind (Figure 6a) near the equator at 700 hPa, which causes the water vapor to continuously travel westward from the South China Sea and the western Pacific regions. Part of the warm and humid air moves along the western edge of subtropical high and turns into southwesterly wind to transport the water vapor to the Sichuan Basin. Another part transports to the west and converges with the warm and humid flow in the Bay of Bengal. Influenced by the anomalous anticyclone above the area from the south of the plateau to the Bay of Bengal, part of the water vapor turns around and follows the anticyclonic circulation to transports along the southern edge of the plateau into the Sichuan Basin. As for the 850 hPa geo-potential height (Figure 6b), the low pressure trough east of Baikal Lake is obviously strong. Over the sea east of Philippines and north of East Indonesia–Papua New Guinea (135–175° E, 10–20° N), there is a significantly positive anomaly, which indicates that, when the anticyclonic circulation over the ocean east of Philippines is anomalously strong, the corresponding descending motion over the region is enhanced. A significant easterly wind anomaly exists over the equatorial western Pacific region and extends westward, which helps the humid and warm flow at low latitudes transport to the Sichuan Basin, leading to the increase in water vapor inflow at the southern boundary and the increase in summer precipitation in the Sichuan Basin. It is consistent with the water vapor flux shown in Figure 5. The configuration of 700 hPa wind field and 850 hPa geo-potential height field is the main reason for the variations in water vapor budget and precipitation in summer in the Sichuan Basin.

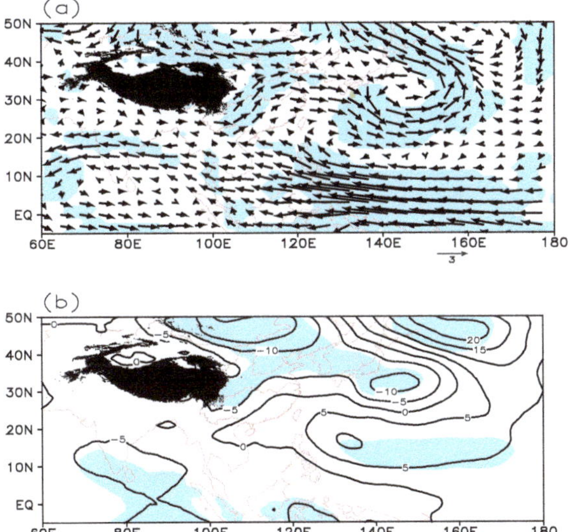

Figure 6. The difference of the composites of (**a**) 700 hPa wind (unit: m s^{-1}) and (**b**) 850 hPa geopotential height (unit: gpm) between wet summers and drought summers in the Sichuan Basin. The blue shadow indicates passing the 95% confidence level, and the black shadow indicates the Tibet Plateau.

4. Conclusions

Using many statistical and diagnostic methods, this study employed the daily precipitation data from the meteorological stations in Sichuan, and the monthly averaged ERA-Interim reanalysis data to study the variation characteristics of summer water vapor budget and its relationship with the precipitation in the Sichuan Basin. The main conclusions are as follows.

The summer water vapor flows in from the southern and western boundaries and flows out at the eastern and northern boundaries in the Sichuan Basin and the four sub-basins. As the main boundary of water vapor outflow, the amount of water vapor outflow at the eastern boundary is much larger than that at the northern boundary. Except for the western basin, where the water vapor inflow at the western boundary is the largest, the water vapor inflow is the largest at the southern boundaries in the other sub-basins. From 1979 to 2016, the water vapor inflow from the southern and the western boundaries of the Sichuan Basin and its sub-basins to Sichuan Province decreased significantly, as well as the water vapor outflow through the eastern boundary. Except for the southern basin, all the outflows through the northern boundaries increased; except for the northern basin, the net water vapor budget showed a decreasing trend in summer.

There are three large-value centers of the summer precipitation over the Sichuan Basin located in Beihuan, Wanyuan, and Ya'an, respectively, and Ya'an has the largest precipitation. The summer precipitation shows a significantly positive correlation with both the water vapor inflow from the southern boundary and the net budget in the Sichuan Basin and a negative correlation with water vapor outflow at the northern boundary in the same period. For the whole Sichuan Basin and its four sub-basins, the southern and northern boundaries are the most important boundaries for summer precipitation. Relatively, the influence of water vapor budget at all the boundaries in the eastern basin on the precipitation is greater than that in the western basin, while the influence is greater in the northern basin than that in the southern basin. By comparing the three centers with large precipitation values, it is found that the influence of outflow at the northern boundaries on the summer precipitation in Beichuan and Ya'an is the greatest, and the inflow at the southern boundary impacts the precipitation in Wanyuan the most greatly.

In the wet summers of the Sichuan Basin, the water vapor inflow from the southern and western boundaries, the total water vapor inflow, the outflow at the northern and eastern boundaries, and the total water vapor outflow are significantly larger than those in the drought summers, as well as the net budget. The difference in water vapor inflow between the wet summers and drought summers are greatest at the southern boundary of the northern basin. The water vapor flux in summer, especially the water vapor inflow at the southern boundary, plays a dominant role in summer precipitation over the basin. To some extent, the summer water vapor in the Sichuan Basin, especially the water vapor inflow from the southern boundaries of the northern basin, decides whether the summer is wet or dry. The impact of water vapor inflow from the southern boundary of the northern basin plays a particularly significant role.

The summer precipitation in the Sichuan Basin is not only directly related to the regional circulation and water vapor transport in eastern China and the Sichuan Basin, but also closely related to the atmospheric circulation over the key area of air–sea interaction in the tropics. It reveals that, under the complex topography on the east side of the Tibet Plateau, the spatio-temporal distribution of precipitation in the Sichuan Basin is the result of the new mechanism of multi-scale interactions between the unique topography of the Sichuan Basin and the different water vapor transports at low latitudes. When there are easterly (westerly) airflows east of the Philippines and north of East Indonesia–Papua New Guinea (120–180° E, 0–10° N) and an anticyclonic (cyclonic) circulation in the north, the anomalous southerly (easterly) airflow and water vapor divergence (convergence) maintain in the key region of the eastern and southern China. Meanwhile, due to the complex topography of the plateau and basin, the anomalies of regional water vapor budget and divergence in the Sichuan Basin is caused, and the anomalous southerly (easterly) airflow and water vapor convergence (divergence) maintain in the basin, which is prone (not prone) to the low-latitude warm and humid air transporting to the basin, resulting in the increase (decrease) of water vapor inflow at the southern boundary of the basin and more (less) summer precipitation in the basin.

In generally, in the context of global climate change, the water vapor inflow at the western and southern boundaries of Sichuan Basin has decreased significantly in the past 38 years, and the summer precipitation in the basin also has a decreasing trend. The decrease in net water vapor budget in the Sichuan basin has led to a shortage of water resources, which has affected all aspects of social economy and people's lives in Sichuan. Under the influence of global warming, accelerated urbanization, and abnormal atmospheric circulation, the reduction in water vapor budget in the basin may be more obvious in the future, and so the water resources problem in the Sichuan Basin deserves more attention.

Author Contributions: Conceptualization, Y.L.; Methodology, C.Z.; writing—original draft preparation, D.Q. All authors have read and agreed to the published version of the manuscript.

Funding: This research was jointly funded by the Second Tibetan Plateau Scientific Expedition and Research (STEP) program (Grant No.2019QZKK0103), the National Natural Science Foundation of China (Grant No.91937301), the Scientific and Technological Research Program of China Railway Eryuan Engineering Group CO. LTD (Grant No. KYY2020066(20–22)), the co-sponsored project by Sichuan Meteorological Bureau and Nanjing University of Information Science & Technology (SCJXHZ04), and Heavy Rain and Drought-Flood Disasters in Plateau and Basin Key Laboratory of Sichuan Province (SCQXKJQN2019012).

Institutional Review Board Statement: Not applicable.

Informed Consent Statement: Not applicable.

Data Availability Statement: The daily precipitation data used in the study were obtained from the China Meteorological Administration Information Center. The monthly average ERA-Interim reanalysis data are available through the ECMWF, https://apps.ecmwf.int/datasets/ (accessed on 15 September 2018).

Acknowledgments: We thank the two anonymous reviewers for their time and support in reshaping the manuscript. Their comments helped us to improve the final version of the manuscript. Special appreciation to the China National Meteorological Information Center and ECMWF, for the provision of the datasets used in the study.

Conflicts of Interest: The authors declare no conflict of interest.

References

1. Michael, D.; Dettinger, F.; Martin, R.; Tapash, D.; Paul, J. Atmospheric Rivers, Floods and the Water Resources of California. *Water* **2011**, *3*, 445–478.
2. Mostafa, S.M.; Wahed, O.; El-Nashar, W.Y.; El-Marsafawy, S.M.; Zelenáková, M.; Abd-Elhamid, H.F. Potential Climate Change Impacts on Water Resources in Egypt. *Water* **2021**, *13*, 1715. [CrossRef]
3. James, J.M.; Osvaldo, F.; Canziani, N.A. *IPCC Climate Change 2001, Impacts, Adaptation, and Vulnerability*; Cambridge University Press: Cambridge, UK, 2001.
4. Antunes, S.; Pires, O.; Rocha, A. Detecting spatiotemporal precipitation variability in Portugal using multichannel singular spectral analysis. *Int. J. Climatol.* **2010**, *26*, 1358–1373.
5. Trenberth, K.E.; Smith, L.; Qian, T. Estimates of the Global Water Budget and Its Annual Cycle Using Observational and Model Data. *J. Hydrometeorol.* **2007**, *8*, 758–769. [CrossRef]
6. Trenberth, K.E. Atmospheric moisture residence times and cycling: Implications for rainfall rates and climate change. *Clim. Chang.* **1998**, *39*, 667–694. [CrossRef]
7. Starr, V.P.; Peixoto, J.P. On the global balance of water vapor and the hydrology of deserts. *Tellus* **1958**, *10*, 189–194. [CrossRef]
8. Xie, C.Y.; Li, M.J.; Zhang, X.Q.; Guan, X.F. Moisture transport features in summer and its rainfall effects over key region in Southern margin of Qinghai-Xizang Plateau. *Plateau Meteorol.* **2015**, *34*, 327–337. (In Chinese)
9. Ding, Y.H.; Hu, G.Q. A study on water vapor budget over China during the 1998 severe flood periods. *Acta Meteor. Sin.* **2003**, *61*, 129–145. (In Chinese)
10. Zhou, Y.S.; Gao, S.Y.; Deng, G. A diagnostic study of water vapor transport and budget during Heavy precipitation over the Changjiang river and the Huaihe river basins in 2003. *Chin. J. Atmos. Sci.* **2005**, *29*, 195–204. (In Chinese)
11. Miao, Q.J.; Xu, X.D.; Zhang, S.J. Whole layer water vapor budget of Yangtze river valley and moisture flux components transform in the key areas of the Plateau. *Acta Meteor. Sin.* **2005**, *63*, 93–99. (In Chinese)
12. Xu, X.D.; Chen, L.S.; Wang, X.R. Source-sink structure of water vapor transport in Meiyu belt of Yangtze River basin. *Chin. Sci. Bull.* **2003**, *48*, 2288–2294. (In Chinese)
13. Xu, X.D.; Tao, S.Y.; Wang, J.Z.; Chen, L.S.; Zhou, L.; Wang, X.R. The relationship between water vapor transport features of Tibetan Plateau-Monsoon "Large triangle" affecting region and drought-flood abnormality of China. *Acta Meteorol. Sin.* **2002**, *60*, 258–264. (In Chinese)
14. Zhou, T.J.; Yu, R.C. Atmospheric water vapor transport associated with typical anomalous summer rainfall patterns in China. *J. Geophys. Res. Atmos.* **2005**, *110*, D08104. [CrossRef]
15. Qiang, Z.; Xu, C.Y.; Zhang, Z. Changes of atmospheric water vapor budget in the Pearl River basin and possible implications for hydrological cycle. *Theor. Appl. Climatol.* **2010**, *102*, 185–195.
16. Jiang, X.W.; Li, Y.Q.; Wang, X. Water vapor transportation over China and its relationship with draught and flood in the Yangtze river Basin. *Acta Geogr. Sin.* **2008**, *63*, 482–490. (In Chinese)
17. Zhang, Z.; Zhang, Q.; Chen, X.; Zhang, J.; Zhou, J. Statistical properties of moisture transport in East Asia and their impacts on wetness/dryness variations in North China. *Theor. Appl. Climatol.* **2011**, *104*, 337–347. [CrossRef]
18. Ding, Y.H.; Sun, Y.; Li, Y.F. *Large Scale Conditional Analysis of Severe Drought and Flood Events in East Asia in the 1990s*; Meteorological Press: Beijing, China, 2003; pp. 260–275.

19. Simmonds, I.; Bi, D.H.; Hope, P. Atmospheric water vapor flux and its association with rainfall over China in summer. *J. Clim.* **1999**, *12*, 1353–1367. [CrossRef]
20. Yang, L.; Zhao, J.H.; Feng, G.L. Characteristics and differences of summertime moisture transport associated with four rainfall patterns over eastern China monsoon region. *Chin. J. Atmos. Sci.* **2018**, *42*, 81–95. (In Chinese)
21. Li, Y.H.; Xu, H.M.; Gao, Y.H.; Li, Q. The characteristics of moisture transport associated with drought/flood in summer over the east of the southwestern China. *Acta Meteor. Sin.* **2010**, *68*, 932–943. (In Chinese)
22. Chu, Q.C.; Wang, Q.G.; Feng, G.L. The roles of moisture transports in intraseasonal precipitation during the preflood season over South China. *Int. J. Climatol.* **2020**, *40*, 2239–2252. [CrossRef]
23. Chu, Q.C.; Zhi, R.; Wang, Q.G.; Feng, G.L. Roles of moisture sources and transport in precipitation variabilities during boreal summer over East China. *Clim. Dyn.* **2019**, *53*, 5437–5457. [CrossRef]
24. Dong, L.; Xu, X.; Zhao, T.; Ren, H. Linkage between moisture transport over the Yangtze River Basin and a critical area of the Tibetan Plateau during the Meiyu. *Clim. Dyn.* **2019**, *53*, 2643–2662. [CrossRef]
25. Zhang, R.H. Relations of water vapor transport from Indian monsoon with that over East Asia and the summer rainfall in China. *Adv. Atmos. Sci.* **2001**, *18*, 1005–1017.
26. Huang, Y.; Cui, X. Moisture sources of torrential rainfall events in the Sichuan basin of China during summers of 2009–13. *J. Hydrometeorol.* **2015**, *16*, 1906–1917. [CrossRef]
27. Wang, N.; Gu, W.Z.; Qiu, C.; Meng, X.X.; Zhou, F. Characteristics of atmospheric water vapor distribution and transport during summer over Shandong province. *Plateau Meteorol.* **2021**, *40*, 159–168. (In Chinese) [CrossRef]
28. Zhou, C.Y.; Li, Y.Q.; Peng, J. The characteristics and variation of precipitation and water resource of Sichuan and Chongqing basin on the eastern side of the plateau. *Chin. J. Atmos. Sci.* **2006**, *30*, 1217–1226. (In Chinese)
29. Chen, C.; Pang, Y.M.; Zhang, Y.F. On the characteristics of climate change in Sichuan basin in the recent 50 years. *J. Southwest Univ.* **2010**, *32*, 115–120. (In Chinese)
30. Bai, Y.Y.; Zhang, Y.; Li, Q.; Li, Y.H.; Lei, T. Preliminary study on regional difference of summer rainfall in Sichuan Basin and their connections with summer monsoons. *Meteorol. Mon.* **2014**, *40*, 440–449. (In Chinese)
31. Zhou, C.Y.; Jiang, X.W.; Li, Y.Q.; Wei, G.C. Features of climate change of water vapor resource over eastern region of the Tibetan plateau and its surroundings. *Plateau Meteorol.* **2009**, *28*, 55–63. (In Chinese)
32. Chen, Q.L.; Ni, C.J.; Wan, W.L. Features of summer precipitation change of the Chuanyu basin and its relationship with large-scale circulation. *Plateau Meteorol.* **2010**, *29*, 587–594. (In Chinese)
33. Jie, C.; Hu, J.; Yun, T. An index for the interface between the Indian summer monsoon and the East Asian summer monsoon. *J. Geophys. Res. Atmos.* **2012**, *117*, D18108. [CrossRef]
34. Wang, Y.; Li, D.L. Variation of the Tibetan Plateau summer monsoon under the background of global warming and its impact on the climate in southwestern China. *Acta Meteor. Sin.* **2015**, *73*, 910–924. (In Chinese)
35. Lv, J.M.; Ren, J.Z.; Ju, J.H. The interdecadal variability of East Asia monsoon and its effect on the rainfall over China. *J. Trop. Meteorol.* **2004**, *20*, 73–80. (In Chinese)
36. Huan, Y.; Li, Y.Q. The synergy between the East Asian summer monsoon and the South Asian summer monsoon and its relations with anomalous rainfall in southern China. *Plateau Meteorol.* **2018**, *37*, 1563–1577. (In Chinese)
37. Yan, H.M.; Zi, Y.C. Characteristics of the Subseasonal-Scale Zonal Movement of Subtropical High in Summer and Its Relationship with Precipitation in Southwest China. *Chin. J. Atmos. Sci.* **2021**, *45*, 1–20. (In Chinese)
38. Chen, Y.R.; Li, Y.Q.; Qi, D.M. Variations of South Asia High and west pacific subtropical high and their relationships with precipitation. *Plateau Meteorol.* **2011**, *30*, 1148–1157. (In Chinese)
39. Hu, D.Q.; Lu, R.Y.; Su, Q.; Fan, G.Z. Interannual variation in the mid-summer rainfall over the western Sichuan basin and the associated circulation anomalies. *Chin. J. Atmos. Sci.* **2014**, *38*, 13–20. (In Chinese)
40. Yang, X.B.; Yang, S.Q.; Ma, Z.F. Influence of the position of east Asian subtropical westerly jet in summer on precipitation over Sichuan-Chongqing region. *Plateau Meteorol.* **2014**, *33*, 384–393. (In Chinese)
41. Wei, W.; Zhang, R.H.; Wen, M.; Kim, B.-J.; Nam, J.-C. Interannual variation of the South Asian high and its relation with Indian and East Asian summer monsoon rainfall. *J. Clim.* **2015**, *28*, 2623–2634. [CrossRef]
42. Li, Y.Q.; Li, D.J.; Yang, S.; Liu, C.; Zhong, A.H.; Li, Y. Characteristics of the precipitation over the eastern edge of the Tibetan Plateau. *Meteorol. Atmos. Phys.* **2010**, *106*, 49–56. [CrossRef]
43. Li, Y.H.; Lu, C.H.; Xu, H.M.; Cheng, B.Y.; Wang, Y. Contemporaneous relationships between summer atmospheric heat source over the Tibetan plateau and drought/flood in eastern southwest China. *Chin. J. Atmos. Sci.* **2011**, *35*, 422–434. (In Chinese)
44. Cen, S.X.; Gong, Y.F.; Lai, X. Impact of heat source over Qinghai-Xizang plateau and its surrounding areas on rainfall in Sichuan-Chongqing basin in summer. *Plateau Meteorol.* **2014**, *33*, 1182–1189. (In Chinese)
45. Chen, D.; Zhou, C.Y.; Qi, D.M. Relationship between atmospheric heat source over Qinghai-Tibetan plateau and its surrounding area and rainstorm in Sichuan basin during summer. *Plateau Meteorol.* **2019**, *38*, 1149–1157. (In Chinese)
46. Li, Y.H.; Lu, C.H.; Xu, H.M.; Chen, B.Y. Anomalies of sea surface temperature in Pacific-Indian Ocean and effects on drought/flood in summer over Eastern of Southwest China. *J. Trop. Meteorol.* **2012**, *28*, 145–156. (In Chinese)
47. Chen, B.Y.; Guo, Q.; Sun, W.G. Relationship between SST in Nino3 and the precipitation over Chongqing and Sichuan and its stability. *Meteorol. Mon.* **2010**, *36*, 27–33. (In Chinese)

48. Pang, Y.S.; Qin, N.S.; Wang, C.X.; Luo, Y. Analysis on the impact of ENSO events seasonal evolution on summer rainfall anomalies in Southwest China. *Plateau Meteorol.* **2020**, *39*, 581–593. (In Chinese)
49. Wang, L.; Huang, G.; Chen, W.; Zhou, W.; Wang, W. Wet-to-dry shift over Southwest China in 1994 tied to the warming of tropical warm pool. *Clim. Dyn.* **2018**, *51*, 3111–3123.
50. Liu, J.P.; Ren, H.L.; Li, W.J.; Zuo, J.Q. Diagnosing the leading mode of interdecadal covariability between the Indian Ocean sea surface temperature and summer precipitation in southern China. *Theor. Appl. Climatol.* **2019**, *135*, 1295–1306. [CrossRef]

Article

Characteristics of the Water Vapor Transport in the Canyon Area of the Southeastern Tibetan Plateau

Maoshan Li [1,*], Lingzhi Wang [1], Na Chang [1], Ming Gong [1], Yaoming Ma [2,3,4], Yaoxian Yang [5], Xuelong Chen [2,3], Cunbo Han [2,3] and Fanglin Sun [5]

[1] Plateau Atmosphere and Environment Key Laboratory, School of Atmospheric Sciences, Chengdu University of Information Technology, Chengdu 610225, China; 18428333633@163.com (L.W.); changna0816@163.com (N.C.); gongming9799@163.com (M.G.)

[2] Land-Atmosphere Interaction and Its Climatic Effects Group, State Key Laboratory of Tibetan Plateau Earth System, Resources and Environment (TPESRE), Institute of Tibetan Plateau Research, Chinese Academy of Sciences, Beijing 100101, China; ymma@itpcas.ac.cn (Y.M.); x.chen@itpcas.ac.cn (X.C.); cunbo.han@itpcas.ac.cn (C.H.)

[3] College of Earth and Planetary Sciences, University of Chinese Academy of Sciences, Beijing 100049, China

[4] College of Atmospheric Science, Lanzhou University, Lanzhou 730000, China

[5] Key Laboratory of Land Surface Process and Climate Change in Cold and Arid Regions, Chinese Academy of Sciences, Lanzhou 730000, China; yangyaoxian@nieer.ac.cn (Y.Y.); fanglin.sun@gmail.com (F.S.)

* Correspondence: lims@cuit.edu.cn

Citation: Li, M.; Wang, L.; Chang, N.; Gong, M.; Ma, Y.; Yang, Y.; Chen, X.; Han, C.; Sun, F. Characteristics of the Water Vapor Transport in the Canyon Area of the Southeastern Tibetan Plateau. *Water* **2021**, *13*, 3620. https://doi.org/10.3390/w13243620

Academic Editor: Maria Mimikou

Received: 9 November 2021
Accepted: 12 December 2021
Published: 16 December 2021

Publisher's Note: MDPI stays neutral with regard to jurisdictional claims in published maps and institutional affiliations.

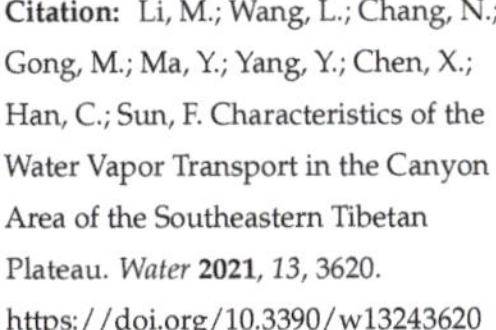

Abstract: Changes in the surface fluxes cause changes in the annular flow field over a region, and they affect the transport of water vapor. To study the influence of the changes in the surface flux on the water vapor transport in the upper layer in the canyon area of southeastern Tibet, in this study, the water vapor transport characteristics were analyzed using the HYSPLIT_v4 backward trajectory model at Danka and Motuo stations in the canyons in the southeastern Tibetan Plateau from November 2018 to October 2019. Then, using ERA-5 reanalysis data from 1989 to 2019 and the characteristics of the high-altitude water vapor transportation, the impact of the surface flux changes on the water vapor transportation was analyzed using singular value decomposition (SVD). The results show that the main sources of the water vapor in the study area were from the west and southwest during the non-Asian monsoon (non-AMS), while there was mainly southwest air flow and a small amount of southeast air flow in the lower layer during the Asian monsoon (AMS) at the stations in southeastern Tibet. The water vapor transmission channel of the westward airflow is higher than 3000 m, and the water vapor transmission channel of the southwestward and southeastward airflow is about 2000 m. The sensible heat and latent heat are negatively correlated with water vapor flux divergence. The southwest boundary of southeastern Tibet is a key area affecting water vapor flux divergence. When the sensible heat and latent heat exhibit downward trends during the non-Asian monsoon season, the eastward water vapor flux exhibits an upward trend. During the Asian monsoon season, when the sensible heat and latent heat in southeastern Tibet increase as a whole, the eastward water vapor flux in the total-column of southeastern Tibet increases.

Keywords: surface fluxes; HYSPLIT_v4 model; water vapor transport; singular value decomposition

1. Introduction

The Tibetan Plateau (TP) is located in Southwestern China. It is the highest plateau in the world. It is known as the third pole of the Earth and is also the source area of many rivers [1–4]. For the same troposphere height in summer, the moisture content over the plateau is much higher than that in the other surrounding areas. The sensible heating of the plateau is an important reason for the abrupt change in the East Asian circulation, which plays an important role in modulating the East Asian monsoon [5–10]. Ye [11] pointed out that the southeastern part of the plateau is an exceedingly high humidity center in summer compared with the surrounding areas. The plateau serves the function of transferring the water vapor from the south to the east, and the strength of this effect directly affects the drought and flood conditions in the middle and lower reaches of the Yangtze River [11–14]. Previous studies have also shown that the southeastern part of the plateau is a high value center of the total water vapor in summer [15–19]. Qu and Zhang [19] studied the distribution of the summer water vapor flux field in East Asia and concluded that there are three water vapor transport channels in East Asia in July. The first is from the Bay of Bengal and the east coast of India to China; the second is from Southern and Southeastern China to Eastern China; and the third channel trends east-west, from East Asia to China [19]. Therefore, it is crucial to study the surface flux and its influence on the water vapor transport in the southeastern part of the plateau to gain a better understanding of the land–atmosphere interactions and their influence on the high-increasing water vapor transport on the plateau.

In recent years, the Lagrangian method has been gradually applied to the study of water vapor transport. Massacand et al. inferred the mesospheric humidity source of heavy precipitation on the southern side of the Alpine area by examining the specific humidity along the back trajectories [20]. Bertò et al. used the Lagrangian trajectory model (i.e., the HYSPLIT model) to analyze the water vapor source during a heavy precipitation event in Trentino, Italy, in 2002 [21]. They found that the main water vapor channel was transported from subtropical Africa to Trentino through the Mediterranean. James et al. investigated the change in the net water along a large number of backward trajectories to identify the water source in the flooded areas of the Elbe River in August 2002 [22]. Sodemann and Stohl [23] employed the recently developed Lagrangian moisture source diagnostic of Sodemann et al. [24] to determine the seasonality of moisture sources for all of Antarctica over a 5-year period. Previous studies indicate that the moisture source and transport path can change rapidly during a precipitation event [25,26]. Using the Lagrangian method, Jiang et al. (2013) studied the characteristics of the moisture contributions during the boreal summer over the Yangtze River valley (YRV) [27]. Chu et al. (2021) focused on the effect of water vapor transport processes on the variations in the seasonal mean rainfall over East China [28]. Chen and Luo (2018) used the Lagrangian model to explore the paths and sources of the water vapor carried to Southern China (SC) during the pre-flood season [29]. Moreover, based on a Lagrangian model, Sun and Wang (2014) quantitatively calculated the water vapor transport from every water vapor source to Eastern China during 2000–2009 [30].

The progress in meteorology research on the TP depends to a large extent on the development of various data about the plateau. With the launch of the three Field Observation Experiments of Atmospheric Science on the Tibetan Plateau, the research data about the plateau have been gradually improved [31–34]. In addition, it has been reported that even along the same latitudinal belt, the atmospheric circulation patterns [35] and the surface heat fluxes [36–38] regulating the moisture transport to the western TP are different from those of the eastern TP. The canyon area in southeastern Tibet is an important channel for water vapor transport from the Bay of Bengal to the south of the plateau to mainland China. The heat flux anomaly over the plateau affects the vertical movement and convergence and divergenceover the plateau, which leads to anomalies in the height field and wind field in East Asia [18]. The changes in the surface flux cause the changes in the annular flow field over the region, and they affect the water vapor transport. To study the influence of the changes in the surface flux on the water vapor transport in the upper layer, the singular value decomposition (SVD) method was used to analyze the correlation between the water vapor flux divergence field and the surface heat fluxes fields and to separate multiple coupling modes from the two element fields to the greatest extent possible to reveal the temporal and spatial relationships between the water vapor flux divergence field and the surface heat flux fields.

In this study, the water vapor transport characteristics were analyzed using the Hybrid Single-Particle Lagrangian Integrated Trajectory (HYSPLIT)_v4 backward trajectory model at Danka and Motuo stations in the canyons in the southeastern TP from November 2018 to

October 2019. The contribution rates of the different water vapor paths were quantitatively analyzed to further deepen our scientific understanding of the water vapor transport paths on the TP. Then, using ERA-5 reanalysis data and the characteristics of the high-altitude water vapor transport, the impact of the changes in the surface fluxes on the water vapor transport was analyzed using SVD. The results help reveal the source of the water vapor and the mechanism by which the Earth-atmosphere interactions influence the water vapor transport on the TP.

2. Data and Methods

2.1. Study Area

The study area was the canyon area in the southeastern TP, and the research stations were Danka Station (29.89° N, 95.68° E, a.s.l. (above sea level) 2701 m) and Motuo Station (29.31° N, 95.32° E, a.s.l. 1154 m). Danka station is located in the northwestern part of Bomi County, Nyingchi City, Tibet Autonomous Region, on the southern bank of the Palongzangbo River Valley, and the underlying surface is mainly grassland. Motuo Station is located in the lower reaches of the Yarlung Zangbo River, Motuo County, and the underlying surface is mainly grassland. The topography of the observation area and the distribution of observation sites are shown in Figure 1.

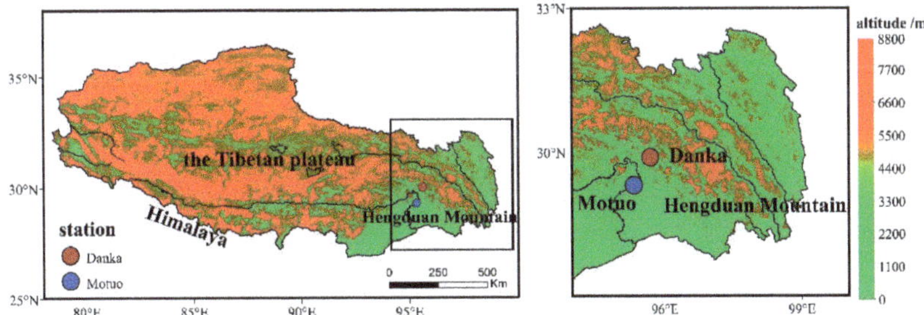

Figure 1. Topography of the study area and locations of the observation sites, Solid dots indicate the site location.

2.2. Data

(1)   The global data assessment system (GDAS) data were obtained from the National Centers for Environmental Prediction (NCEP) of the United States (https://www.ready.noaa.gov/gdas1.php, accessed on 8 September 2020). The data were stored as 5 weeks for each month. The horizontal resolution of the data is $1° \times 1°$.

(2)   The NCEP/NCAR reanalysis data were obtained from the NCEP of the United States. We used the air temperature (air), u wind speed (uwnd), v wind speed (vwnd), and relative humidity (Rhum) data as the input for HYSPLIT_v4, and the horizontal resolution of the data is $2.5° \times 2.5°$ (https://psl.noaa.gov/data/gridded/data.ncep.reanalysis.html, accessed on 10 May 2020).

(3)   The variables used in the reanalysis of the ERA-5 data (https://www.ecmwf.int/en/about/media-centre/science-blog/2017/era5-new-reanalysis-weather-and-climate-data, accessed on 18 October 2020) from the European Center for Medium Range Weather Forecasts (ECMWF) are the sensible heat flux (SSHF), the latent heat flux(SLHF), water vapor flux divergence integral (VIMDF, the vertical integral of the moisture flux is the horizontal rate of flow of moisture, per meter across the flow, for a column of air extending from the surface of the Earth to the top of the atmosphere), and the monthly mean water vapor flux from north to east. The horizontal resolution of the data is $0.1° \times 0.1°$. The time range of the data used in this study is from 1989 to 2019.

2.3. Water Vapor Transport Trajectory Model (HYSPLIT_v4)

The HYSPLIT_v4 model is a professional model for calculating and analyzing the transport and diffusion trajectories of air pollutants. It was jointly developed by the Air Resources Laboratory of the National Oceanic and Atmospheric Administration (NOAA) of the United States and the Australian Meteorological Service in the past 20 years [39,40]. The HYSPLIT_v4 model is a mixed Eulerian-Lagrangian model, in which the Lagrangian method is used for the advection and diffusion. The model is usually used to track the movement direction of the particles or gas carried by the airflow, and it can be used to forecast the wind situation in real time, to analyze precipitation, and to study the trajectory. In this study, the HYSPLIT_v4 model was used to simulate the water vapor transport from November 2018 to October 2019 at Danka and Motuo stations in the canyon area of southeastern Tibet, and the water vapor transport trajectory was studied. The model uses 11-day backward water vapor transport trajectory simulation, and the top height of the model is 10 km. The characteristics of the water vapor trajectories in different seasons were obtained.

2.4. Singular Value Decomposition (SVD) Method

The SVD method is based on the maximum covariance between two element fields. It is an effective diagnostic analysis method for studying the correlation structure of two fields. It is suitable for climate diagnostic analysis and for studying the teleconnections between large-scale meteorological fields. Compared with typical correlation analysis methods, the singular value decomposition method can maximally separate the independent coupling distribution patterns of the two meteorological fields. These distribution patterns reveal the spatial relationship and the temporal correlation between the two meteorological fields in order to determine the real teleconnection pattern and the key area of influence. From the spatial distribution pattern of the modal anisotropy correlation, we found the large value region of anisotropy correlation coefficient in the left and right fields, which represents the key region of the interactions between the left and right fields. The anisotropy correlation coefficients between the two meteorological element fields can be used to analyze their relationship. The Monte Carlo method was used to test the significance of the SVD results.

3. Results Analysis

3.1. Water Vapor Transport Trajectories in the Canyon Area in Southeastern Tibet

Figure 2 shows the water vapor transport track at Danka Station and Motuo Station during the non-Asian monsoon season (non-AMS) (i.e., spring) for 11 days (the map projection is the Lambert projection). The northwest air flow is the main water vapor source at the two stations in the non-AMS. Seven percent of the water vapor at Danka Station came from the northeastern part of the Pacific Ocean, passed over Canada, passed over the Arctic Ocean, turned to the northwest over the Norwegian Sea, and reached Danka Station. Similarly, Motuo Station received a small amount of water vapor from the northern Pacific, 25% from India, 40% from the northwest path, and 32% from the nearby area to the west. Compared with the spring atmospheric circulation pattern on the TP, strong westerlies occurred to the south of the plateau during spring and were the main water vapor transport channels in the southeastern gorge area of Tibet. The high terrain of the plateau affected the wind flow around this latitude, forming a pattern characterized by a southern trough and a northern Ridge. The northwestward air flow of the northern ridge delivered about 40% of the water vapor to southeastern Tibet during the non-AMS. Based on the height of the water vapor transmission channels, the water vapor transmission channel in the southwest was below 3000 m, while the water vapor transmission channel in the west and the northwest water vapor transport height reached 4500 m.

Compared with the sources of water vapor transport during the non-AMS (spring), and the main water vapor source path of the two stations came from the southwest air flow (Figure 3). During the Asian monsoon season (AMS) (i.e., summer), 20% of the water vapor at Danka Station came from the northwest path and 80% came from the southwest, mainly from the Arabian Sea and the Bay of Bengal, whereas almost all of the water vapor at Motuo station came from the Arabian Sea. The height of the water vapor transmission path between the two stations through the Bay of Bengal was less than 1500 m. During this period, the South Asian Monsoon prevailed on the plateau, and the southwest winds at this time made a major contribution to the water vapor transport from the Bay of Bengal to this area, forming a main water vapor transport channel over southeastern Tibet and into inland China.

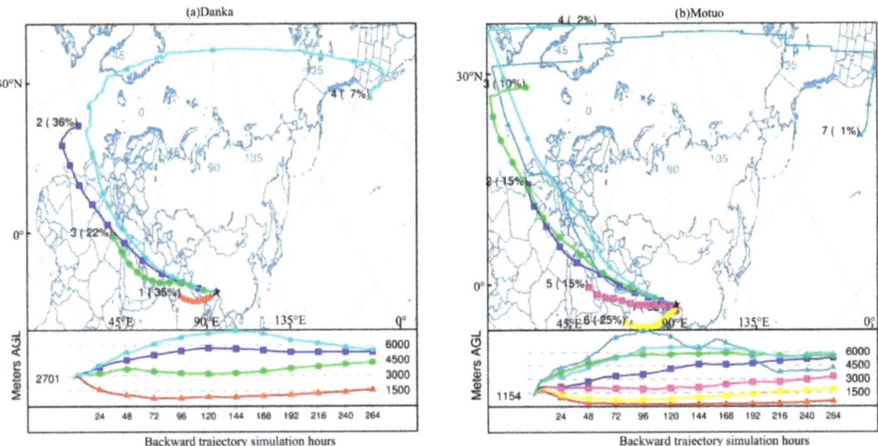

Figure 2. Eleven-day backward water vapor transport trajectories for Danka station and Motuo station during the non-AMS.

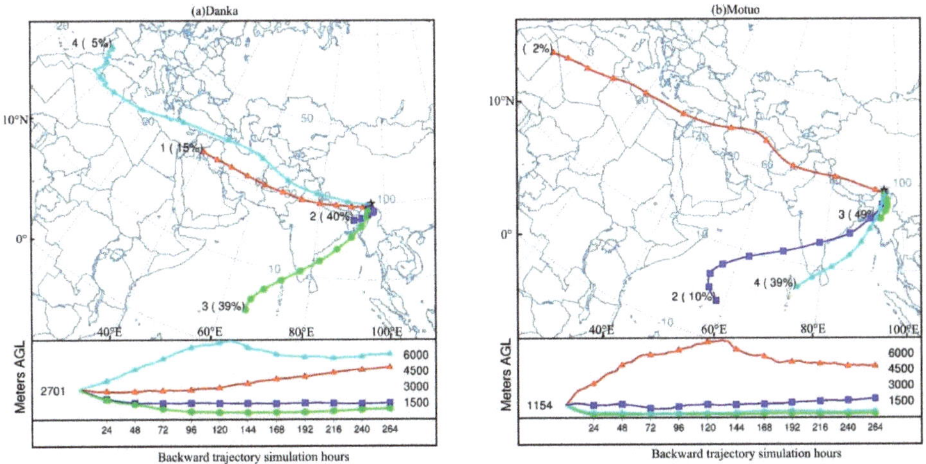

Figure 3. Eleven-day backward water vapor transport trajectories for Danka station and Motuo station during the AMS.

The water vapor transport in the canyon area of the southeastern TP is closely related to the Asian monsoon climate. Correlation analysis was conducted between the water vapor transport path and the circulation field, ground heating, and water vapor flux during the non-AMS and AMS in order to determine the mechanism by which the monsoon climate affects the water vapor transport path in the canyon area of the southeastern TP.

3.2. Effect of Surface Fluxes on Water Vapor Transport

3.2.1. Relationship between Sensible Heat Flux and Water Vapor Flux Divergence in the Canyon Area of Southeastern Tibet

Based on the results of the water vapor transport simulation obtained using the HYSPLIT_v4 backward trajectory model (see Section 3.1), the water vapor transport in the canyon area of the southeastern TP exhibited significant climate characteristics during the AMS. In this section, we analyze the spatial relationship of the temporal correlation between the sensible heat and water vapor flux divergence from the perspective of the variations in the AMS. The water vapor flux divergence is a very important meteorological factor, which characterizes the water vapor transport. The SVD method was used to calculate the correlation coefficient (R) between the first four pairs of singular vector covariance contributions (SCFs), the cumulative covariance contributions (CSCFs), and the expansion coefficient. The time span was from 1989 to 2019 (Table 1).

Table 1. SVD results of sensible heat field and water vapor flux divergence field.

Singular Vector	First Mode		Second Mode		Third Mode		Fourth Mode		CSCF (%)
	SCF (%)	R	SCF (%)	R	SCF (%)	R	SCF (%)	R	
spring	33.25	0.93	16.74	0.91	13.19	0.83	8.39	0.86	80.17
summer	75.35	0.94	11.43	0.93	6.08	0.92	1.64	0.90	94.49
autumn	48.79	0.91	19.57	0.93	8.93	0.84	5.12	0.87	82.41
winter	38.03	0.89	20.66	0.85	12.22	0.82	7.15	0.87	78.06
all year	44.57	0.85	20.58	0.86	10.19	0.80	5.80	0.84	81.13

As can be seen from Table 1, the cumulative covariance contributions of the first four pairs of the coupling modes of the sensible heat and the water vapor flux divergence in each season and the whole year were greater than 78%, the covariance contribution rate of the first two modes was greater than 58%, and it reached 94.49% during the AMS, indicating that the first two pairs of coupling modes represent most of the characteristics of the sensible heat field and the water vapor flux divergence. In addition, the first two modes passed the 95% Monte Carlo test, so in this section, the first two coupling modes of the sensible heat field and the water vapor flux divergence field in southeastern Tibet are analyzed.

Figure 4 shows the results of the SVD anisotropy correlation analysis between the standardized anomaly of the sensible heat field and the standardized anomaly of the water vapor flux divergence field during the non-AMS. The first mode (Figure 4a,b) shows that the sensible heat in the southwestern and southwestern parts of southeastern Tibet was mainly positive. There was a significant negative value on the southern boundary line, and the values in the central and eastern regions were negative. In the corresponding water vapor flux divergence field, there was a positive center of the water vapor flux divergence on the southern boundary of the plateau and a negative region in the west. Except for the sporadic negative anomaly regions in the eastern, southern, and central parts of southeastern Tibet, the other regions were positive. Figure 4e,f show that the correlation coefficient between the sensible heat and water vapor flux divergence of the first mode is 0.93. There is a significant positive correlation between the sensible heat flux field and the water vapor flux divergence field. This indicates that during the non-AMS, when the sensible heat of the western and southwestern parts of the southeastern Tibet increased (decreased) and the sensible heat of the eastern and central parts of the southeastern Tibet decreased (increased), the water vapor flux divergence in the eastern part of Tibet decreased (increased), while it increased (decreased) in the central and eastern regions. From the second type of modal space, we can see that the sensible heat field was significantly negative in the Hengduan Mountains on the southeastern part of the TP, and there was a positive area in the northeast. The negative anomaly area in the northeastern part of southeastern Tibet corresponded to the water vapor flux divergence, and the rest of the areas were in the same positive region. The two large value areas corresponded to each other. The correlation coefficient between the sensible heat field and the whole water vapor flux divergence field of the second mode reached 0.91, indicating that when the sensible heat of the northeast area increased (decreased) during the non-AMS, the sensible heat in the rest of the areas decreased (increased), the water vapor flux divergence in the northeastern region decreased (increased), and it increased (decreased) in the rest of the areas.

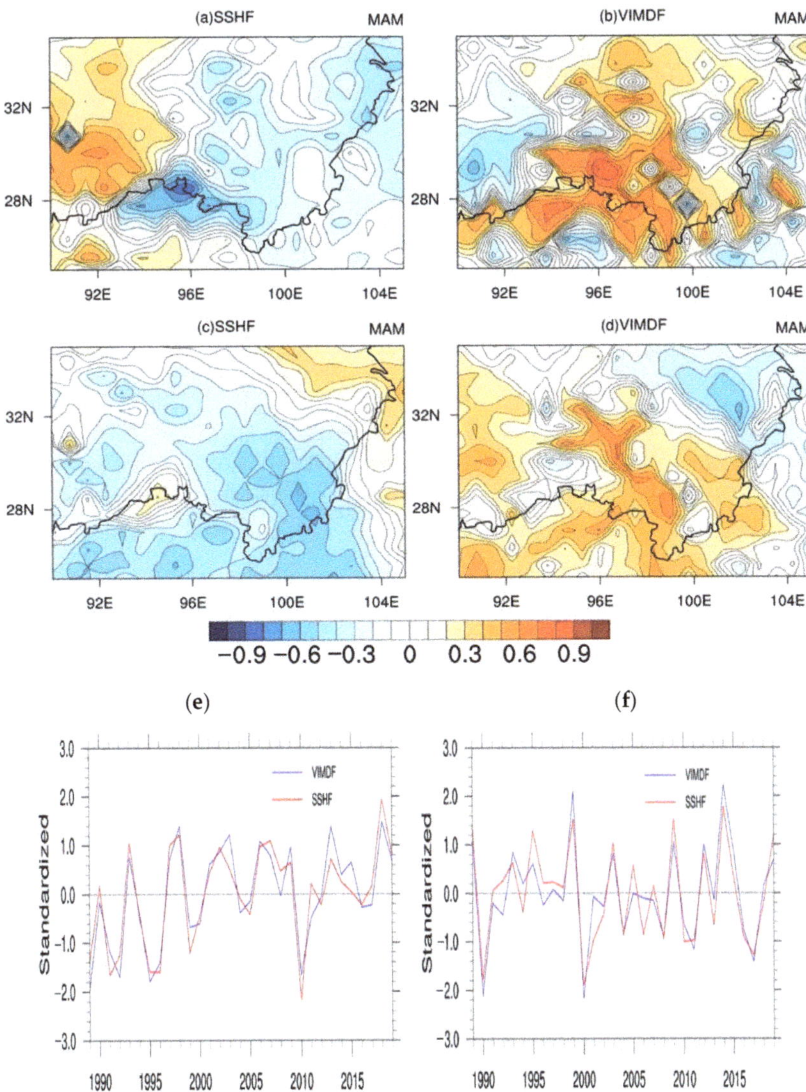

Figure 4. (**a–d**) SVD analysis and (**e,f**) corresponding time series of the spring sensible heat field and the water vapor flux divergence field in southeastern Tibet during the non-AMS. (**a**) the first mode of the thermal field; (**b**) the first mode of water vapor flux divergence field; (**c**) the second mode of the sensible heat field; (**d**) the second mode of the water vapor flux divergence field; (**e**) the first mode; and (**f**) the second mode. The solid black line is the boundary of the Tibetan Plateau in (**a–d**). The solid red line is SSHF time series and the solid blue line is VIMDF time series in (**e,f**).

The AMS is the main season for water vapor transport. The results of the SVD analysis of the sensible heat and the water vapor flux divergence fields during the AMS show that the cumulative variance contribution of the first four modes of the two fields reached 94.49%, which represents most of the characteristics of the sensible heat and the moisture flux divergence fields during the AMS (Table 1). The first mode of the SVD analysis of the summer surface flux field and the moisture flux divergence field shows that the sensible heat field was a negative region with a uniform type, and there was a significant negative value region on the southern and eastern boundaries of the plateau (Figure 5a–d). The whole water vapor flux divergence field was clearly positive in southeastern Tibet. The correlation coefficient between the sensible heat and the moisture flux divergence of the first mode was 0.94 during the AMS, indicating that there was a clear positive correlation between the sensible heat and the water vapor flux divergence fields. That is, when the sensible heat decreased (increased) during the AMS, the water vapor flux divergence increased (decreased), and the large negative area on the southern boundary and the southeast boundary was the key area and affected the whole water vapor flux divergence during the AMS. The contribution of the second mode's covariance was 11.43%, and the significant positive area of the sensible heat field was located in the northern part of southeastern Tibet. The negative value area was located in the southwest, while the right field was opposite to this distribution. The correlation coefficient for the sensible heat and water vapor flux divergence fields was 0.93, exhibiting a positive correlation, indicating that the sensible heat field in the canyon area of southeastern Tibet increased in the northeast during the AMS. In addition, when it decreased in the southwest, the water vapor flux divergence field decreased in the northeast and increased in the southwest. The time series changes in the sensible heat and the water vapor flux divergence fields of the first two modes were basically synchronous (Figure 5e,f), and the fluctuation amplitude was large.

3.2.2. Relationship between the Latent Heat and the Water Vapor Flux Divergence in the Canyon Area of Southeastern Tibet

In this section, the SVD method is used to investigate the spatial relationship between the latent heat field and the water vapor flux divergence field. The latent heat field is the left field, and the water vapor flux divergence field is the right field. The covariance contribution (SCF), cumulative covariance contribution (CSCF), and correlation coefficient (R) of the first four pairs of singular vectors in each season were calculated (Table 2).

Table 2. SVD results of latent heat field and water vapor flux divergence field.

Singular Vector	First Mode		Second Mode		Third Mode		Fourth Mode		CSCF (%)
	SCF (%)	R	SCF (%)	R	SCF (%)	R	SCF (%)	R	
spring	36.98	0.95	20.20	0.86	12.87	0.82	8.85	0.83	78.10
summer	77.46	0.94	9.88	0.93	4.05	0.87	2.25	0.85	93.64
autumn	52.31	0.90	21.15	0.95	8.88	0.91	4.34	0.92	86.68
winter	43.20	0.93	17.11	0.89	12.91	0.93	5.94	0.89	79.15
all year	47.67	0.87	21.13	0.89	8.98	0.82	5.93	0.85	83.71

The cumulative covariance contributions of the first four pairs of coupling modes of the latent heat flux and the water vapor flux divergence in each season and the whole year were greater than 78%. The covariance contributions of the first two modes were greater than 57% and 93.64% during the AMS, respectively. It indicated that the first two pairs of coupling modes represent most of the characteristics of the latent heat flux and the water vapor flux divergence field (Table 2). The first two modes passed the 95% Monte Carlo test, so in this section, the first two coupling modes of the latent heat and water vapor flux divergence field in southeastern Tibet are analyzed.

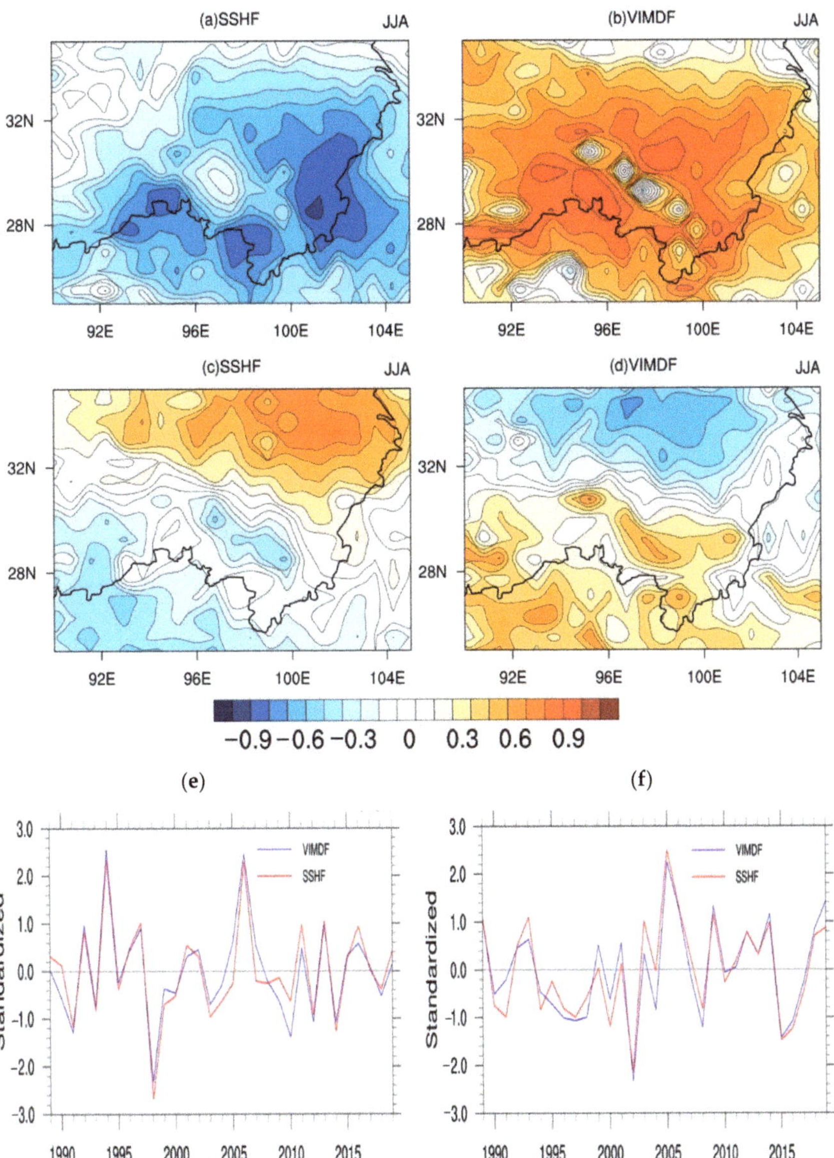

Figure 5. (**a–d**) SVD analysis and (**e,f**) corresponding time series of the sensible heat field and the water vapor flux divergence field in southeastern Tibet during the AMS. (**a**) the first mode of the thermal field; (**b**) the first mode of water vapor flux divergence field; (**c**) the second modal of the sensible heat field; (**d**) the second mode of water vapor flux divergence field; (**e**) the first mode; and (**f**) the second mode. The solid black line is the boundary of the Tibetan Plateau in (**a–d**). The solid red line is SSHF time series and the solid blue line is VIMDF time series in (**e,f**).

Figure 6a–d shows the results of the SVD analysis of the normalized anomalies of the latent heat field and the water vapor flux divergence field during the non-AMS. From spatial distribution of the first mode, we can see that there was a positive region in the latent heat field, except for in the western region. The remaining areas were negative.

There were remarkably positive values in the eastern and southern parts of the plateau. There was a negative value in the western part of the whole water vapor flux divergence field, and there were sporadic negative anomaly centers in the other areas. The correlation coefficient between the latent heat field and the water vapor flux divergence field of the first mode was 0.95, indicating that the two fields had a good positive correlation. That is, when the latent heat in southeastern Tibet decreased (increased) in the southwestern and eastern areas and the latent heat increased (decreased) in the west, the water vapor flux divergence increased (decreased) in the southwest and east and decreased (increased) in the west. In the second modal spatial distribution, the latent heat field was negative in the northwest and positive in the southeast, and the negative value of the whole water vapor flux divergence field was positive in the northeast. That is, when the sensible heat in southeastern Tibet was weak (strong) in northwest-southeast direction, the water vapor flux divergence field was strong (weak) in southwest-northeast direction. Based on the time series curve of the first mode and the second mode of the two fields, the correlations were 0.95 and 0.86, respectively (Figure 6e,f). The fluctuation trend was basically consistent, and the first mode time series exhibited an upward trend.

Figure 7a–d show the first mode of the spatial distribution of the latent heat field and the water vapor flux divergence field in southeastern Tibet during the AMS. It can be concluded that the latent heat field and the water vapor flux divergence field exhibited good consistent distributions, and the latent heat field during the AMS was negative. The significant negative centers were located on the plateau and the eastern and southern boundaries. The whole water vapor flux divergence field was a positive region. The positive centers were located in the southern and central parts of southeastern Tibet. The correlation coefficient for the first mode of the latent heat field and the water vapor flux divergence field was 0.94 during the AMS. There was a significant positive correlation. This shows that the water vapor flux divergence increased (decreased) when latent heat decreased in southeastern Tibet during the AMS. The second modal spatial distribution pattern was drawn as follows. When the latent heat field in southwestern Tibet increased in the southwest and decreased in the northeast, the water vapor flux divergence field in the total-column was reversed. Based on the corresponding time series of the first mode and the second mode, the change trend was more consistent, and the correlation coefficients reached 0.94 and 0.93, respectively, exhibiting the same phase of change (Figure 7e,f).

3.2.3. Relationships between the Surface Sensible Heat, Latent Heat, and Northward Water Vapor Flux

According to the trajectory simulation results of the HYSPLIT_v4 11-day backward trajectory, the source of the water vapor transport was closely related to the changes in the atmospheric circulation situation. The source of the water vapor transport was quite different during the non-AMS and the AMS. During the non-AMS, the water vapor in southeastern Tibet mainly came from a westward path, while during the AMS, the water vapor mainly came from a southwestward path, accounting for a relatively small number of westward paths. Therefore, the surface fluxes and the water vapor flux to the north and east in winter and summer are discussed in this section. Similarly, the surface sensible heat flux and the latent heat flux were taken as the left field in the SVD, the water vapor flux to the north and the water vapor flux to the east were taken as the right field, and the time period was from 1989 to 2019. The covariance contribution (SCF), cumulative covariance contribution (CSCF), and correlation coefficient (R) between the expansion coefficients of the first four pairs of singular vectors are shown in Table 3. The north (east) water vapor flux refers to the horizontal water vapor velocity of a column of air extending from the Earth's surface to the top of the atmosphere. A positive value represents the flux from the west (south) to the east (north). The mean cumulative variance contribution of the first mode to the SVD analysis of the surface flux and the northward and eastward vapor fluxes exceeded 61%. This represents the main characteristics of the relationship between the two physical quantity fields. The first mode passed the 95% level, so the first mode was analyzed in this section. The first four pairs of modal results are shown in Table 3.

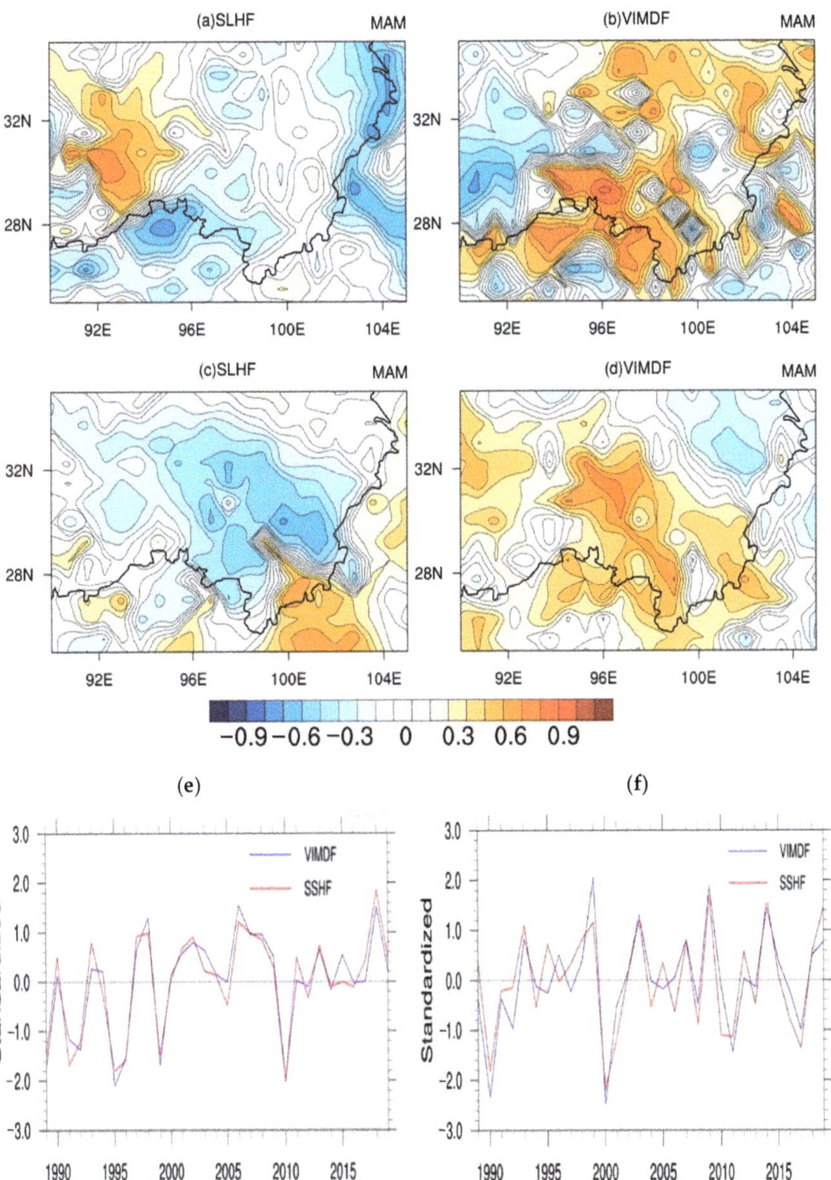

Figure 6. SVD analysis and corresponding time coefficients of the surface latent heat field and the water vapor flux divergence field in southeastern Tibet during the non-AMS (**a**) the first mode of the latent heat field; (**b**) the first mode of water vapor flux divergence field; (**c**) the second mode of the latent heat field; (**d**) the second mode of water vapor flux divergence field; (**e**) the first mode; and (**f**) the second mode. The black solid line is the boundary of the Tibetan Plateau in (**a**–**d**). The red solid line is SSHF time series and the blue solid line is VIMDF time series in (**e**,**f**).

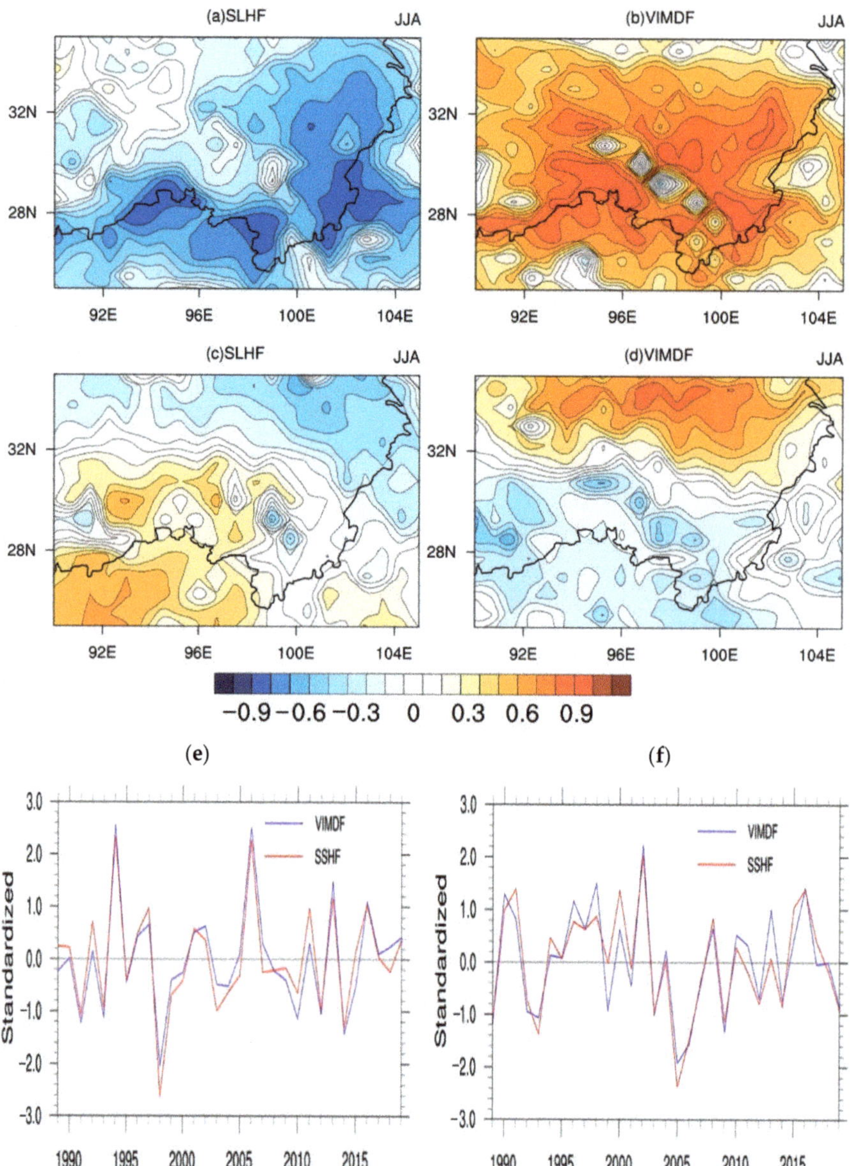

Figure 7. (**a–d**) SVD analysis and (**e,f**) corresponding time series of the latent heat field and the water vapor flux divergence field in southeastern Tibet in summer. (**a**) the first mode of the latent heat field; (**b**) the first mode of water vapor flux divergence field; (**c**) the second mode of the latent heat field; (**d**) the second mode of water vapor flux divergence field; (**e**) the first mode; and (**f**) the second mode. The solid black line is the boundary of the Tibetan Plateau in (**a–d**). The solid red line is SSHF time series and the solid blue line is VIMDF time series in (**e,f**).

Table 3. SVD Analysis of surface flux and northward water vapor flux in Southeast Tibet.

Singular Vector	SSHF				SLHF			
	Summer		Winter		Summer		Winter	
	SCF (%)	R	SCF (%)	R	SCF (%)	R	SCF (%)	R
1	68.15	0.83	76.41	0.78	61.89	0.85	78.52	0.80
2	20.82	0.74	13.31	0.78	24.91	0.77	11.70	0.81
3	5.86	0.76	4.66	0.82	6.40	0.79	3.96	0.83
4	2.08	0.84	1.94	0.87	3.03	0.81	1.81	0.82
CSCF (%)	96.90	/	96.32	/	96.23	/	95.98	/

For the first mode of the SVD analysis of the sensible heat and moisture during the non-AMS (Figure 8a,b), the results show that the sensible heat field was negative in both the southeast boundary and the southern part of the plateau, and the northern region was a positive anomaly, which was the key area affecting the northward water vapor flux during the same period. The northward water vapor flux field was positive, except for in the western region of the plateau, and the center of the positive anomaly was located in the central region. The correlation coefficient of the first mode was 0.78, indicating that there was a positive correlation between the sensible heat and the northward vapor flux. When the sensible heat increased in northeastern Tibet and decreased in southeastern Tibet and the southern part of the plateau, the water vapor flux in most parts of the plateau increased to the north and decreased to the west, and vice versa. From the corresponding time series (Figure 8e), it can be seen that the overall change trends of the surface flux and the northward water vapor flux were roughly the same. The fitting degree before 2008 was worse than that after 2008, and there were large fluctuations in 2009 and 2013.

There was also a significant coupling mode between the latent heat and northward water vapor flux during the non-AMS (Figure 8c,d). In the first mode, the contribution of the variance was 78.52%, and the correlation coefficient of the latent heat and the northward water vapor flux fields reached 0.8. The distribution characteristics of the time series were similar to those of the sensible heat field. The first mode was similar to the first mode of the sensible heat and the plateau's northward moisture flux. The difference is that in the sensible heat field, the values were negative in northern and northwestern Tibet, while the northward water vapor flux field moved eastward. However, the whole plateau was still a positive anomaly, and the western part was negative. The correlation coefficient of the first mode was 0.8. This shows that when the surface latent heat increased in the western, central, southern, and eastern parts of the plateau in winter, the northward water vapor flux increased in most parts of the plateau and decreased in the western part, and vice versa (Figure 8f).

During the AMS, the relationship between the surface flux field and the northward water vapor flux field was more pronounced. There were significant coupling modes between the sensible and latent heat fluxes and the northward water vapor transport flux during the AMS, and the correlation coefficients were 0.83 and 0.85, respectively (Figure 9e,f). The sensible and latent heat flux fields had consistent positive values (Figure 9a,c). The large values were located in the eastern part of southeastern Tibet and the southern boundary of the plateau, which were the key areas in which the summer climate influenced the northward water vapor flux. The northward water vapor flux field was basically positive on the main part of the plateau, and the area with the largest positive value was in the southern part of southeastern Tibet. There was a good positive correlation between the surface sensible and latent heat fluxes and the northward water vapor flux, that is, when the sensible and latent heat fluxes increased (decreased) in the eastern part of southeast Tibet and the southern boundary of the plateau during the AMS, the northward water vapor flux in the main part of the plateau increased, especially in southeastern Tibet, and decreased in the Himalayas in southern Tibet. It can be seen from the time series diagram that the change trend of the surface flux field in southeastern Tibet was consistent with that of the northward water vapor flux field during the AMS (Figure 9e,f).

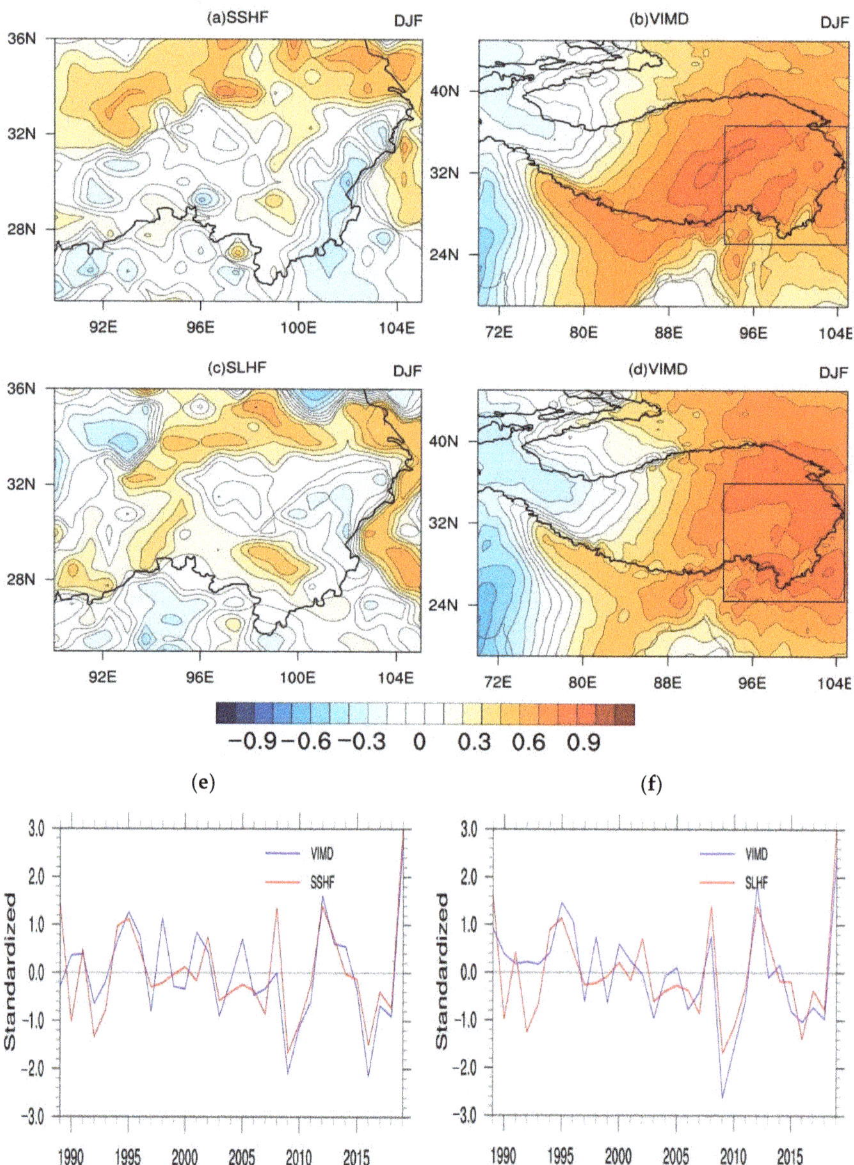

Figure 8. (**a–d**) SVD analysis and (**e,f**) corresponding time series of the surface flux field and northward water vapor flux field in southeastern Tibet during the non-AMS. (**a**) the first mode of the thermal field; (**b**) the first mode of the northward water vapor flux field; (**c**) the first mode of the latent heat field; (**d**) the first mode of the northward water vapor flux field; (**e**) the sensible heat flux and northward water vapor flux field; and (**f**) the latent heat flux and northward water vapor flux field. The solid black line is the boundary of the Tibetan Plateau in (**a–d**). The solid red line is SSHF time series and the solid blue line is VIMD time series in (**e,f**).

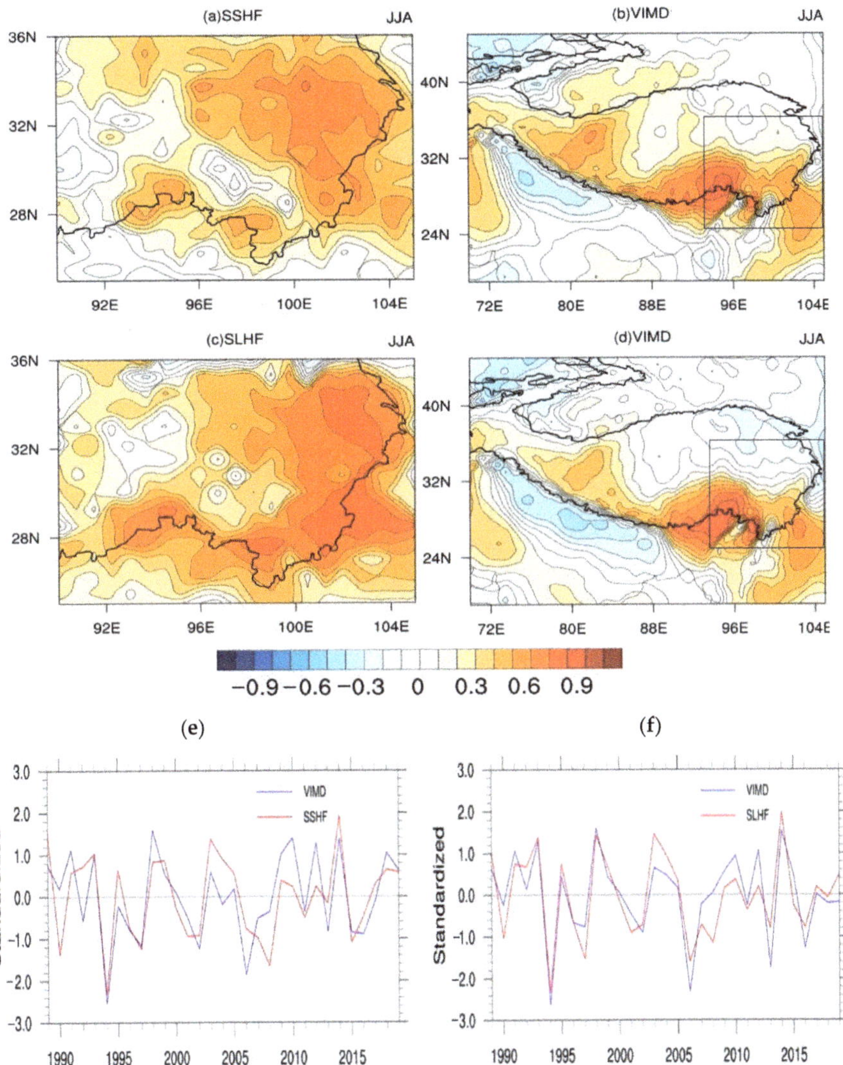

Figure 9. (**a–d**) SVD analysis and (**e,f**) corresponding time series of the surface flux field and northward water vapor flux field in southeastern Tibet during the AMS. (**a**) the first mode of the thermal field; (**b**) the first mode of the northward water vapor flux field; (**c**) the first mode of the latent heat field; (**d**) the first mode of the northward water vapor flux field; (**e**) the sensible heat flux and northward water vapor flux field; and (**f**) the latent heat flux and northward water vapor flux field. The solid black line is the boundary of the Tibetan Plateau in (**a–d**). The solid red line is SSHF time series and the solid blue line is VIMDF time series in (**e,f**).

3.2.4. Influences of the Surface Sensible Heat and Latent Heat on the Water Eastward Vapor Flux

Table 4 shows the contribution rates and correlation coefficients of the first four pairs of principal component modes of the variance for the SVD analysis of the surface fluxes and eastward water vapor flux in southeast Tibet. It can be seen that the contribution rate of the cumulative variance of the first four pairs was greater than 97%, and there was a significant coupling mode. The analysis period was from 1989 to 2019. Similarly, the first mode of the SVD analysis passed the 95% level Monte Carlo test, so the first mode was mainly analyzed.

Table 4. SVD Analysis of surface flux and eastward water vapor flux in Southeast Tibet.

Singular Vector	SSHF				SLHF			
	Summer		Winter		Summer		Winter	
	SCF (%)	R	SCF (%)	R	SCF (%)	R	SCF (%)	R
1	83.37	0.77	73.02	0.80	87.85	0.85	72.01	0.79
2	11.22	0.63	19.55	0.71	7.88	0.74	19.51	0.71
3	2.29	0.79	2.83	0.74	2.12	0.71	3.64	0.89
4	1.68	0.80	2.08	0.79	0.87	0.80	2.31	0.84
CSCF (%)	98.55	/	97.48	/	98.72	/	97.46	/

It can be seen from the time series diagram of the surface fluxes and the eastward water vapor flux during the non-AMS that the trends of the two-time series were quite consistent (Figure 10). The sensible heat and eastward water vapor flux exhibited fluctuating upward trends during the non-AMS, while the latent heat and eastward water vapor flux exhibited fluctuating downward trends. Figure 10e,f show the first mode of the corresponding time series of the SVD anisotropy between the surface fluxes and the eastward vapor flux during the non-AMS. Figure 10a,b show that the surface sensible heat field was negative in the northwest and positive in the southeast. The negative region was located at the boundary of the southeast plateau, and the eastward water vapor flux field was consistent as a whole. The main body of the plateau and the northern part of the plateau were positive. The high value area was located in the eastern part of Qinghai Province on northern part of the TP, and the southern part of the Himalayas was a consistent negative value area. The correlation coefficient between the sensible heat and the first mode of the eastward water vapor flux was 0.80, indicating that there was a positive correlation between the surface fluxes and the eastward water vapor flux fields. That is, when the sensible heat increased in the southeast and decreased in the northwest in the winter, the eastward water vapor flux increased over the plateau and decreased in the Himalayas, and vice versa.

It can be seen from Figure 10c,d that the latent heat had a good consistency. The positive high value area was located on the southeastern boundary of the plateau, and there was a small negative value area in the southeastern part. In the eastward water vapor flux field, the main body of the plateau and the area to the north were negative value areas, the high value area was in the Qinghai area on the northeastern part of the TP, and the positive area was in the southern part of the Himalayas, with a distribution opposite that of sensible heat field. The correlation coefficient of the first mode of the latent heat flux and the eastward vapor flux fields reached 0.79, which indicates that when the southeast latent heat accumulation increased in the southeast, the overall water vapor flux to the east decreased and that in the Himalaya Mountains increased, and vice versa.

The spatial distribution of the first mode of the SVD analysis of the sensible heat and latent heat fluxes during the AMS was consistent. In the first mode, the sensible and latent heat fields in southeast Tibet were positive, and the two large positive areas were located in southeastern Tibet (Figure 11a,c). This shows that this area was the key area in which the surface flux affected the eastward water vapor flux during the AMS. The eastward water vapor flux field was positive in the main part of the plateau and the Himalayas, and there were negative areas in the northwest and southwest (Figure 11b,d). The maximum positive area was located in southeastern Tibet, which corresponded well with the maximum positive area of the surface fluxes. There were significant coupling modes between the eastward water vapor flux and the sensible heat flux and latent heat flux during the AMS, and the correlation coefficients were 0.77 and 0.85, respectively. That is, when the sensible heat and latent heat fluxes increased in southeastern Tibet during the AMS, the water vapor flux from the main part of the plateau and the Himalayas to the east increased, and vice versa. It can be seen from the time series diagram that the change trend corresponded well and the fluctuation range was large (Figure 11e,f).

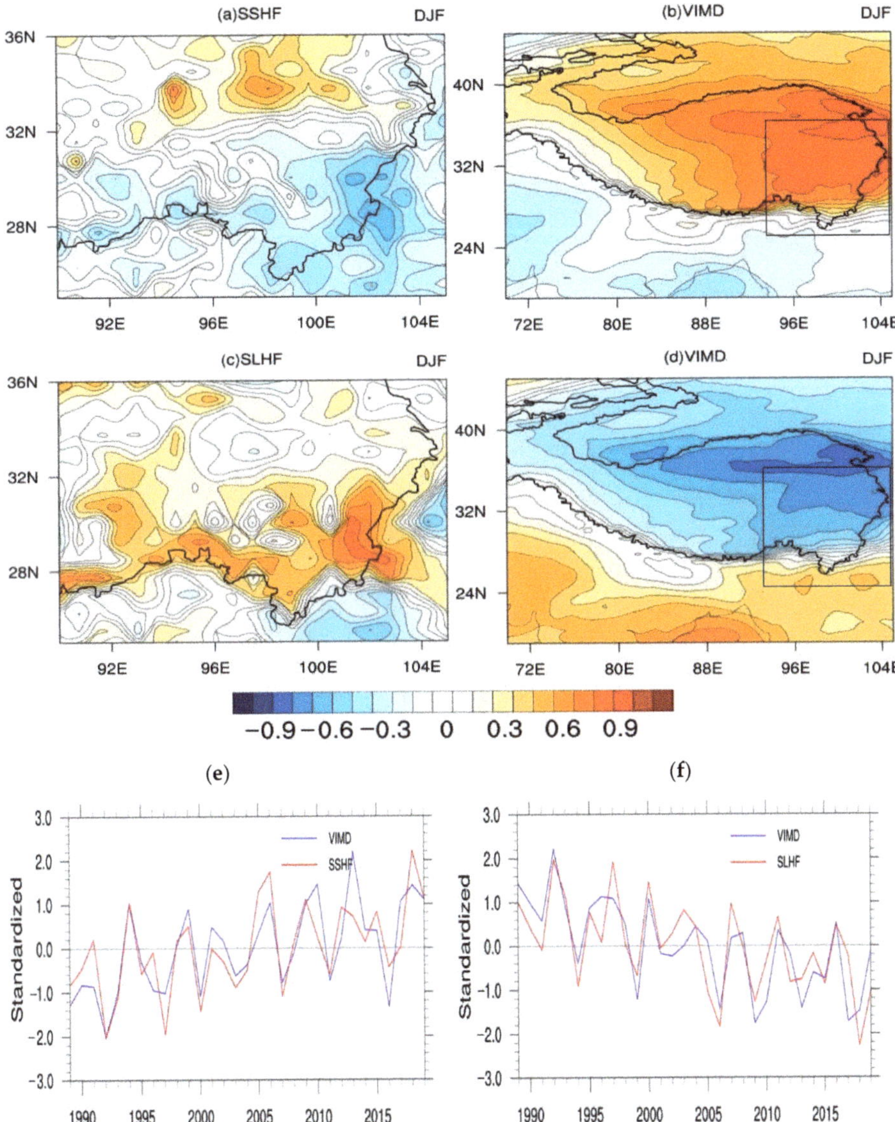

Figure 10. (**a–d**) SVD analysis and (**e,f**) corresponding time series of the surface flux field and eastward water vapor flux field in southeastern Tibet during the non-AMS. (**a**) the first mode of the thermal field; (**b**) the first mode of the eastward water vapor flux field; (**c**) the first mode of the latent heat field; (**d**) the first mode of the eastward water vapor flux field; (**e**) the sensible heat flux and eastward water vapor flux field; and (**f**) the latent heat flux and eastward water vapor flux field. The solid black line is the boundary of the Tibetan Plateau in (**a–d**). The solid red line is SSHF time series and the solid blue line is VIMD time series in (**e,f**).

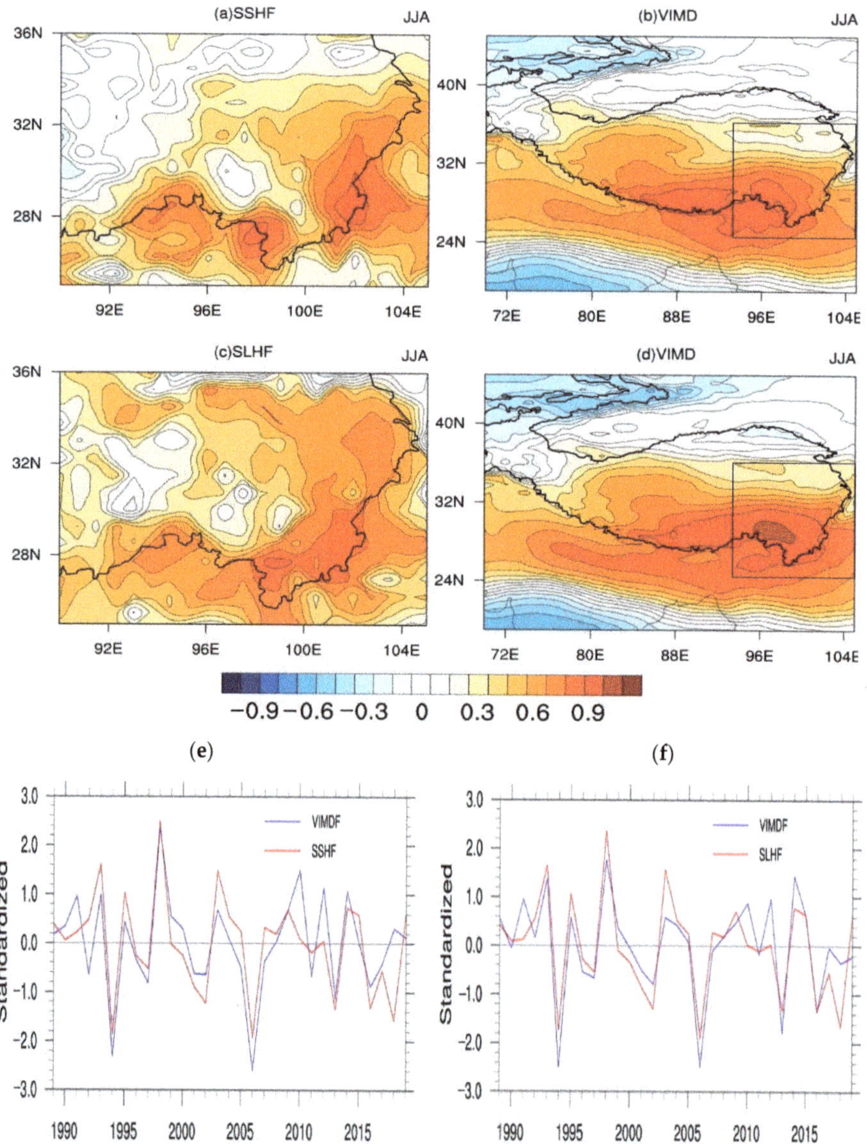

Figure 11. (**a–d**) SVD analysis and (**e,f**) corresponding time series of the surface flux field and eastward water vapor flux field in southeastern Tibet during the AMS. (**a**) the first mode of the thermal field; (**b**) the first mode of the eastward water vapor flux field; (**c**) the first mode of the latent heat field; (**d**) the first mode of the eastward water vapor flux field; (**e**) the sensible heat flux and eastward water vapor flux field; and (**f**) the latent heat flux and eastward water vapor flux field. The solid black line is the boundary of the Tibetan Plateau in (**a–d**). The red solid line is SSHF time series and the blue solid line is VIMD time series in (**e,f**).

4. Summary and Conclusions

In this study, the 11-day backward trajectories of two observation stations located in the southeastern Tibetan Canyon from November 2018 to October 2019 were analyzed using the HYSPLIT_v4 backward trajectory model. Then the SVD method was used to analyze the relationships between the sensible heat and latent heat and the water vapor flux divergence in the southeastern Tibet gorge region. Due to the effects of the atmospheric circulation patterns and seasonal heat fluxes, the patterns of the moisture sources for southeastern Tibet exhibited significant seasonal differences. The main conclusions of this study are as follows.

(1)   The sources of the water vapor were different during the Asian monsoon and non-Asian monsoon seasons, and the main sources of the water vapor in the study area during the non-AMS were from the west and southwest. During the AMS, there was mainly southwest air flow and a small amount of southeast air flow in the lower layer. The westerly flow and northwesterly flow were the main sources of water vapor in winter. During the AMS, the southwestern water vapor transport accounted for more than half of the total. There was a certain correlation between the transportation height of each station and the source of water vapor. The height of the water vapor transportation channel of the western air flow was higher than 3000 m, and the height of the water vapor transportation channel of the southwestern and southeastern air flows was about 2000 m.

(2)   The sensible heat and latent heat in the northern part of the southeastern Tibet Canyon during the non-AMS were directly proportional to the change in the northward water vapor flux in the central and eastern parts of the plateau. During the AMS, the sensible heat and latent heat were directly proportional to the northward water vapor flux. When the sensible heat and latent heat decreased during the non-AMS, the eastward water vapor flux increased. The sensible heat and latent heat were negatively correlated with the eastward water vapor flux, while the sensible heat and latent heat were positively correlated with the eastward water vapor flux during the AMS.

(3)   There was a negative correlation between the surface fluxes and the water vapor flux divergence in this area. The southwest boundary of southeast Tibet was the key area affecting the water vapor flux divergence. When the surface sensible and latent heat fluxes increased in southeastern Tibet, the divergence of the water vapor flux decreased, that is, the water vapor transport to the region was weakened. When the sensible and latent heat fluxes decreased, the divergence of water vapor flux increased and the water vapor transport increased.

(4)   During the non-AMS, when the sensible heat in the canyon area of southeastern Tibet decreased, the eastward water vapor flux increased. Additionally, the northward water vapor fluxes on the plateau east of 75° E increased, while they decreased on the western part. When the latent heat increased, the eastward water vapor flux decreased, and the sensible heat and latent heat were negatively correlated with the northward water vapor flux during the non-AMS. That is, when the surface flux increased in southeastern Tibet, the water vapor transport from west to east increased. During the AMS, when the sensible heat and latent heat in southeastern Tibet increased as a whole, the eastward water vapor flux in the total-column in southeastern Tibet increased. This indicates that when the surface flux in southeastern Tibet increased during the AMS, the water vapor transport increased from west to east.

Our results show that the source of water vapor in the study area is different in different seasons, which will provide a certain theoretical basis for further research on the extreme precipitation of the Tibetan Plateau in the future. It is of great value to further study the different source areas and density of water vapor sources to improve extreme precipitation forecasts. In addition, some questions remain to be addressed. For example, this article only analyzes the seasonal characteristics of water vapor sources in southeastern Tibet from 2018 to 2019, only the data of the past 30 years is selected for analysis. Why were there large fluctuations in 2009 and 2013, whether the VIMDF is mainly influenced by the surface thermal effect on the TP, and whether the relationship is regulated by other external forcing factors, such as sea surface temperature (SST). Cui et al. (2015) pointed out that during the positive phase of the North Atlantic Oscillation (NAO) in winter, it can inspire a stable downstream Rossby wave train, inducing the Asian subtropical westerly jet to intensify and the India-Burma trough to deepen, and it also increases the snow depth on the TP in winter, followed by a positive SSHF anomaly in spring in most areas of the TP [41]. What are the synergetic effect and contribution rates of the NAO and the SSHF on the TP? These issues need further study.

Author Contributions: M.L., L.W., N.C. and Y.Y. mainly wrote the manuscript and were responsible for the research design, data preparation and analysis. Y.M. and M.L. supervised the research, including methodology development, as well as manuscript structure, writing and revision. ML drafted the manuscript. F.S., M.G., X.C. and C.H. prepared the data and wrote this paper. All authors have read and agreed to the published version of the manuscript.

Funding: This work was financially supported by the Second Tibetan Plateau Scientific Expedition and Research (STEP) program (Grant No. 2019QZKK0103).

Institutional Review Board Statement: Not applicable.

Informed Consent Statement: Not applicable.

Data Availability Statement: The global data assessment system (GDAS) data were obtained from the National Centers for Environmental Prediction (NCEP) of the United States (https://www.ready.noaa.gov/gdas1.php, accessed on 8 September 2020). The NCEP/NCAR reanalysis data were obtained from the NCEP of the United States. We used the air temperature (air), u wind speed (uwnd), v wind speed (vwnd), and relative humidity (Rhum) data as the input for HYSPLIT_v4, and the horizontal resolution of the data is $2.5° \times 2.5°$ (https://psl.noaa.gov/data/gridded/data.ncep.reanalysis.html, accessed on 10 May 2020). The variables used in the reanalysis of the ERA-5 data (https://www.ecmwf.int/en/about/media-centre/science-blog/2017/era5-new-reanalysis-weather-and-climate-data, accessed on 18 October 2020).

Acknowledgments: This work was financially supported by the Second Tibetan Plateau Scientific Expedition and Research (STEP) program (Grant No. 2019QZKK0103), the National Natural Science Foundation of China (Grant No. 41675106, 41805009), National key research and development program of China (2017YFC1505702) and Scientific Research Project of Chengdu University of Information Technology (KYTZ201721).

Conflicts of Interest: The authors declare no conflict of interest.

References

1. Li, G.P. *Dynamic Meteorology of Qinghai-Tibet Plateau*; Meteorological Press: Beijing, China, 2002.
2. Xu, X.; Lu, C.; Shi, X.; Gao, S. World water tower: An atmospheric perspective. *Geophys. Res. Lett.* **2008**, *35*. [CrossRef]
3. Jane, Q. The third pole. *Nature* **2008**, *454*, 393–396.
4. Yao, T.; Masson-Delmotte, V.; Gao, J.; Yu, W.; Yang, X.; Risi, C.; Sturm, C.; Werner, M.; Zhao, H.; He, Y.; et al. A review of climatic controls on δ18O in precipitation over the Tibetan Plateau: Observations and simulations. *Rev. Geophys.* **2013**, *51*, 525–548. [CrossRef]
5. Wu, G.; Liu, Y.; Zhang, Q.; Duan, A.; Wang, T.; Wan, R.; Liu, X.; Li, W.; Wang, Z.; Liang, X. The Influence of Mechanical and Thermal Forcing by the Tibetan Plateau on Asian Climate. *J. Hydrometeorol.* **2007**, *8*, 770–789. [CrossRef]
6. Meng, X.N.; Liu, H.Z.; Du, Q.; Liu, Y.; Xu, L.J. Factors controlling the latent and sensible heat fluxes over erhai lake under different atmospheric surface layer stability conditions. *Atmos. Ocean. Sci. Lett.* **2020**, *13*, 400–406. [CrossRef]
7. Ren, Q.; Zhou, C.Y.; Xia, Y.; Cen, S.X.; Long, Y. Interannual relationship between spring sensible heat flux over the eastern Tibetan Plateau and temperature in eastern China. *J. Glaciol. Geocryol.* **2019**, *41*, 783–792.
8. Wang, J.; Tong, J.L.; Xiao, Y.Q.; Wu, X.Y.; Zhang, W.Y. The interannual variation of summer sensible heat flux in typical arid and semi-arid regions of East Asia. *Arid Meteorol.* **2018**, *36*, 203–211.
9. Odhiambo, G.O.; Savage, M.J. Sensible heat flux by surface layer scintillometry and eddy covariance over a mixed grassland community as affected by bowen ratio and most formulations for unstable conditions. *J. Hydrometeorol.* **2009**, *10*, 479–492. [CrossRef]
10. Zou, H.; Zhou, L.B.; Ma, S.P.; Wang, P.; Li, W.; Jia, A.G.; Gao, J.J.; Deng, Y. Local wind system in the Rongbuk valley on the northern slope of Mt. Everest. *Geophys. Res. Lett.* **2008**, *35*, 344–349. [CrossRef]

11. Ye, D.Z. *Meteorology of Qinghai Tibet Plateau*; Science Press: Beijing, China, 1979.
12. Zhang, Q.; Wang, R.; Yue, P.; Zhao, Y.D. Discussion on some scientific problems in the field of land air interaction under complex conditions. *Acta Meteorol. Sin.* **2017**, *75*, 39–56.
13. Xu, X.; Zhao, T.; Lu, C.; Guo, Y.; Chen, B.; Liu, R.; Li, Y.; Shi, X. An important mechanism sustaining the atmospheric "water tower" over the Tibetan Plateau. *Atmos. Chem. Phys.* **2014**, *14*, 11287–11295. [CrossRef]
14. Tao, S.; Ding, Y. Observational evidence of the influence of the Qinghai-Xizang (Tibet) Plateau on the occurrence of heavy rain and severe convective storms in China. *Bull. Am. Meteor. Soc.* **1981**, *62*, 23–30. [CrossRef]
15. Liang, H. Distribution and Variation of Atmospheric Water Vapor over the Tibetan Plateau and Its Surrounding Areas. Ph.D. Thesis, Chinese Academy of Meteorological Sciences, Beijing, China, 2005.
16. Liang, H.; Liu, J.M.; Zhang, J.C.; Bi, Y.M.; Wang, K.C. Research on Retrieval of the Amount of Atmospheric Water Vapor over Qinghai-Xizang Plateau. *Plateau Meteorol.* **2006**, *025*, 1055–1063.
17. Jiang, J.X.; Fan, M.Z. A preliminary study on the relationship between TBB field and water vapor distribution over the Tibetan Plateau in summer. *Plateau Meteorol.* **2002**, *21*, 20–24.
18. Ding, Y.G.; Jiang, Z.H. Universality of SVD method in meteorological field diagnosis and analysis. *Acta Meteorol. Sin.* **1996**, *54*, 365–372.
19. Qu, Y.L.; Zhang, C.D. The distribution and structure of water vapour flux field over East Asia in summer. *Plateau Meteorol.* **1985**, *4*, 121–128.
20. Massacand, A.C.; Wernli, H.; Davies, H.C. Heavy precipitation on the Alpine southside: An upper-level precursor. *Geophys. Res. Lett.* **1998**, *25*, 1435–1438. [CrossRef]
21. Bertò, A.; Buzzi, A.; Zardi, D. Back—tracking water vapour contributing to a precipitation event over Trentino: A case study. *Meteorol. Z.* **2004**, *13*, 189–200. [CrossRef]
22. James, P.; Stohl, A.; Spichtinger, N.; Eckhardt, S.; Forster, C. Climatological aspects of the extreme European rainfall of August 2002 and a trajectory method for estimating the associated evaporative source regions. *Nat. Hazards Earth Syst. Sci.* **2004**, *4*, 733–746. [CrossRef]
23. Sodemann, H.; Stohl, A. Asymmetries in the moisture origin of Antarctic precipitation. *Geophys. Res. Lett.* **2009**, *36*, 273–289. [CrossRef]
24. Sodemann, H.; Schwierz, C.; Wernli, H. Interannual variability of Greenland winter precipitation sources: Lagrangian moisture diagnostic and North Atlantic Oscillation influence. *J. Geophys. Res.* **2008**, *113*, D03107. [CrossRef]
25. Li, J.; Tao, T.; Pang, Z.; Tan, M.; Kong, Y.; Duan, W.; Zhang, Y. Identification of different moisture sources through isotopic monitoring during a storm event. *J. Hydrometeorl.* **2015**, *16*, 1918–1927. [CrossRef]
26. Han, T.; Zhang, M.; Wang, S.; Qu, D.; Du, Q. Sub-hourly variability of stable isotopes in precipitation in the marginal zone of East Asian monsoon. *Water* **2020**, *12*, 2145. [CrossRef]
27. Jiang, Z.; Ren, W.; Liu, Z.; Yang, H. Analysis of water vapor transportcharacteristics during the Vleiyu over the Yangtze-Huaihe River valley using the Lagrangian method. *Acta Meteor Sin.* **2013**, *71*, 295–304.
28. Chu, Q.; Wang, Q.; Feng, G.; Jia, Z.; Liu, G. Roles of water vapor sources and transport in the intraseasonal and interannual variation in the peak monsoon rainfall over East China. *Clim. Dyn.* **2021**, *57*, 2153–2170. [CrossRef]
29. Chen, Y.; Luo, Y. Analysis of Paths and Sources of Moisture for the South China Rainfall during the Presummer Rainy Season of 1979–2014. *Acta Meteorol. Sin.* **2018**, *32*, 744–757. [CrossRef]
30. Sun, B.; Wang, H. Moisture Sources of Semiarid Grassland in China489 Using the Lagrangian Particle Model FLEXPART. *J. Clim.* **2014**, *27*, 2457–2474. [CrossRef]
31. Tao, S.; Luo, S.; Zhang, H. The Qinghai-Xizang Plateau Meteorological Experiment (Qxpmex) May–August 1979. In Proceedings of the International Symposium on the Qinghai-Xizang Plateau and Mountain Meteorology, Beijing, China, 20–24 March 1986; pp. 3–13.
32. Wang, J. Land surface process experiments and interaction study in China-From HEIFE to IMGRASS and GAME-TIBET/TIPEX. *Plateau Meteorol.* **1999**, *18*, 280–294.
33. Ma, Y.; Yao, T.; Wang, J. Experimental study of energy and water cycle in Tibetan plateau—The progress introduction on the study of GAME/Tibet and CAMP/Tibet. *Plateau Meteorol.* **2006**, *25*, 344–351.
34. Ma, Y.; Hu, Z.; Xie, Z.; Ma, W.; Wang, B.; Chen, X.; Li, M.; Zhong, L.; Sun, F.; Gu, L.; et al. A long-term (2005–2016) dataset of hourly integrated land-atmosphere interaction observations on the Tibetan Plateau. *Earth Syst. Sci. Data (ESSD)* **2020**, *12*, 2937–2957. [CrossRef]
35. Dong, W.; Lin, Y.; Wright, J.S.; Ming, Y.; Xie, Y.; Wang, B.; Huang, W.; Huang, J.; Wang, L.; Tian, L.; et al. Summer rainfall over the southwestern Tibetan Plateau controlled by deep convection over the Indian subcontinent. *Nat. Commun.* **2016**, *7*, 10925. [CrossRef] [PubMed]
36. Ma, Y.; Wang, Y.; Han, C. Regionalization of land surface heat fluxes over the heterogeneous landscape: From the Tibetan Plateau to the Third Pole region. *Int. J. Remote. Sens.* **2018**, *39*, 5872–5890. [CrossRef]
37. Li, M.; Babel, W.; Tanaka, K.; Foken, T. Note on the application of planar-fit rotation for non-omnidirectional sonic anemometers. *Atmos. Meas. Tech.* **2013**, *6*, 221–229. [CrossRef]
38. Zhong, L.; Ma, Y.; Hu, Z.; Fu, Y.; Hu, Y.; Wang, X.; Cheng, M.; Ge, N. Estimation of hourly land surface heat fluxes over the Tibetan Plateau by the combined use of geostationary and polar-orbiting satellites. *Atmos. Chem. Phys.* **2019**, *19*, 5529–5541. [CrossRef]

39. Wernli, H. A Lagrangian-based analysis of extratropical cyclones II: A detailed case-study. *Q. J. R. Meteorol. Soc.* **1997**, *123*, 1677–1706. [CrossRef]
40. Nieto, R.; Gimeno, L.; Trigo, R.M. A Lagrangian identificantion of major sources of Sahel moisture. *Geophy. Res. Lett.* **2006**, *33*, L18707. [CrossRef]
41. Cui, Y.F.; Duan, A.M.; Liu, Y.M.; Wu, G.X. Interannual variability of the spring atmospheric heat source over the Tibetan Plateau forced by the North Atlantic SSTA. *Clim. Dyn.* **2015**, *45*, 1617–1634. [CrossRef]

 *water* 

Article

Vertical Motion of Air over the Indian Ocean and the Climate in East Asia

Rongxiang Tian [1,2,*], Yaoming Ma [3,4,5] and Weiqiang Ma [3,4,5]

1   School of Earth Sciences, Zhejiang University, Hangzhou 310027, China
2   Key Laboratory of Geoscience Big Data and Deep Resource of Zhejiang, Hangzhou 310027, China
3   Institute of Tibetan Plateau Research, Chinese Academy of Sciences, Beijing 100101, China;
    ymma@itpcas.ac.cn (Y.M.); wqma@itpcas.ac.cn (W.M.)
4   CAS Center for Excellence in Tibetan Plateau Earth Sciences, Chinese Academy of Sciences,
    Beijing 100101, China
5   College of Earth and Planetary Sciences, University of Chinese Academy of Sciences, Beijing 100049, China
*   Correspondence: trx@zju.edu.cn

Abstract: The Indian Ocean and East Asia are the most famous monsoonal regions, and the climate of East Asia is affected by the change in wind direction due to monsoons. The vertical motion of the atmosphere is closely related to the amount of precipitation in whichever particular region. Climate diagnosis and statistical analysis were used to study the vertical motion of air over the Indian Ocean and its relationship with the climate in East Asia. The vertical motion of air over the Indian Ocean had a significant correlation with the climate in China—especially with precipitation in the Tibetan Plateau and the Yangtze River Basin—as a result of the interaction of the vertical motion of air from the Indian Ocean, the Tibetan Plateau and the subpolar region in the Northern Hemisphere. The vertical motion over the Indian Ocean was weakened from 1981 to 2010, except at a height of 500 hPa in winter. The vertical motion of air over the Indian Ocean had a period of 7–9 years in summer and 9–12 years in winter. An ascending motion was dominant over most of the Indian Ocean throughout the year and the central axis of the ascending motion changed from a clockwise rotation from winter to summer to a counterclockwise rotation from summer to winter as a result of the monsoonal circulation over the Indian Ocean. These results will provide a theoretical reference for a comprehensive understanding of the climate in Asia and contribute to work on climate prediction in these regions.

Keywords: Indian Ocean; East Asia climate; vertical motion of air; Tibetan Plateau

Citation: Tian, R.; Ma, Y.; Ma, W. Vertical Motion of Air over the Indian Ocean and the Climate in East Asia. *Water* **2021**, *13*, 2641. https://doi.org/10.3390/w13192641

Academic Editor: Aizhong Ye

Received: 15 August 2021
Accepted: 19 September 2021
Published: 25 September 2021

Publisher's Note: MDPI stays neutral with regard to jurisdictional claims in published maps and institutional affiliations.

1. Introduction

The Indian Ocean and East Asia have a monsoonal climate [1]. The Indian Ocean is surrounded by continental regions. There are different radiation and energy balances between the land and ocean as a result of the different natures of the underlying land and ocean surfaces (e.g., the heat capacity and albedo, etc.), and these balances change with the seasons. Different pressure fields are produced in different seasons, which generate a seasonal change in the wind field, namely, monsoons [1,2]. The summer monsoon brings abundant precipitation, while the winter monsoon brings drought and little rain. In terms of the influence of the change in wind direction due to the monsoon on the climate, there has been much research concerning the influence of change in the horizontal wind field on climate, while the relationship between climate and change in the vertical wind field has not received enough attention. The vertical motion of air is the result of both thermal and dynamic action [3], and can be directly linked to precipitation. Heating of the underlying surface can be expressed directly as the vertical motion of air [4]. The characteristics and intensity of vertical motion in the atmosphere are closely related to precipitation. Precipitation is associated with updrafts and drought is associated with downdrafts [5]. For example, on a temporal scale, short-term vertical motion is associated with small-scale precipitation—such as typhoons (hurricanes) and rainstorms—whereas long-term vertical motion is associated with large-scale rainy weather (mostly low-pressure areas) and droughts (mostly high-pressure areas) [6]. On a spatial scale, areas of ascent in the Hadley circulation correspond to low pressure and a rainy zone in the meridional direction, such as the region of ascending air near the equator, whereas areas of descending air correspond to high pressure and arid zones (e.g., the subtropical arid region) [7,8]. The area of ascent related to the Walker and anti-Walker circulations [9] corresponds to the rainy zone around the equator, whereas the area of descending air corresponds to regions with only small amounts of rain in the zonal direction [10,11].

The upward motion of air over the Indian Ocean and the downward motion of air over the surrounding continents form many vertical circulations [12] which affect the amount of precipitation. Bjerknes [13] studied the relationship between the vertical motion of the atmosphere and precipitation over India as early as 1910 and showed that the vertical motion of air over India is closely related to the amount of precipitation in the northern side of the Indian Ocean. On the western side of the Indian Ocean, the vertical motion of air over Africa is also closely related to the local precipitation, and has led to droughts in Africa during the last century [14,15]. Since China is on the path of the Indian monsoon, there has been much research on the influence of change in the horizontal wind field on climate in China [16,17], while the research on the relationship between the vertical movement of air from the Indian Ocean and the climate of faraway China is rare.

We investigated the effects of the vertical motion of air over the Indian Ocean on the climate in East Asia. We aimed to find out whether the vertical motion of air associated with the Indian Ocean monsoonal circulation plays a role in the formation of and change in the climate of East Asia.

The research results are of significance to understand comprehensively the climate of East Asia and make more accurate climate forecasts. The paper is organized as follows: Section 2 describes the data and methods. The results and discussion are presented in Sections 3–5. The study concludes with a brief summary in Section 6.

2. Data and Methods

2.1. Data

Data for the monthly mean vertical wind speed were obtained from the National Centers for Environmental Prediction/National Center for Atmospheric Research (NCEP/NCAR) reanalysis dataset [18] with a resolution of $2.5° \times 2.5°$. Twelve pressure levels were used for the Indian Ocean (1000, 925, 850, 700, 600, 500, 400, 300, 250, 200, 150 and 100 hPa). Precipitation data from 839 meteorological stations were provided by the China Meteorological Administration. The surface air temperature and atmospheric pressure were extracted from the Scientific Data Center for the Cold and Arid Regions of China surface meteorological datasets with a temporal and spatial resolution of $0.1° \times 0.1°$. The monthly mean data which were calculated by using these datasets form the basis of the analytical approach.

Taking into consideration the remote connection between the equatorial Indian Ocean's sea surface temperature and the East Asian climate [19], the location of the Indian Ocean was taken as (25° S–30° N, 20° E–125° E) and the location of China as (15° N–55° N, 70° E–145° E). The time period measured was 1981–2010. Due to the horizontal wind at 850 hPa, the vertical motion of air (omega) at 500 hPa and the sea surface temperature play an important role in some regions in the Indian Ocean, while outgoing longwave radiation and the vertical motion of air (omega) at 500 hPa dominate for other regions in the occurrence of extreme rainfall [20]. The sea surface temperature anomaly (dipole event) over the Indian Ocean also has a good correlation with the geopotential height of 500 hPa, and is closely related to the precipitation anomaly in China during summer [21,22]. Therefore, the vertical motions of 500 hPa and 850 hPa over the Indian Ocean are selected. We studied the correlation between vertical movement over the Indian Ocean in January as well as June and the climate in East Asia, since the Indian Ocean summer monsoon erupts in June while January is the beginning of winter [23].

To determine the reliability of the data, we compared the vertical motion of air in the NCEP, ERA-Interim (produced by the European Center for Medium-Range Weather Forecasts) and JRA-55 (from the Japan Meteorological Agency) datasets over the Indian Ocean and the Tibetan Plateau and found that they had almost identical systems and centers [24]. It is therefore reasonable to analyze the vertical motion of air using the NCEP data.

2.2. Methodology

To diagnose and analyze the vertical motion of air over the Indian Ocean, we used empirical orthogonal function (EOF) analysis [25,26] to decompose the vertical velocity in winter (December–February), summer (June–August), January and June. The vertical velocity fields were decomposed into products of space function and time function by EOF decomposition. EOF analysis can be used to decompose the original data field, anomaly field and standardization field of vertical velocity. The results of decomposing different data fields are different in climatic significance. Because we performed orthogonal function decomposition on the original vertical velocity field, the first eigenvector (the spatial distribution) represents the average state (main pattern) of the vertical velocity field in the study area (explanation variance is large), and the corresponding time coefficient represents the time variation characteristics of the main pattern. The calculation of the EOF is as follows:

$$X_{M \times N} = V_{M \times P} \times T_{P \times N} \tag{1}$$

where $X_{M \times N}$ is a data matrix of the original vertical velocity composed of N observations of M spatial points. V are eigenvectors and T are eigenvalues. We used the first eigenvector in the analysis. To determine whether the first eigenvector has a physical meaning, we used the rule suggested by [26] to test the results:

$$e_j = \lambda_j \left(\frac{2}{N}\right)^{\frac{1}{2}} \tag{2}$$

where e_j is the error range of the eigenvalue λ_j and $N = 30$ is the sample size. When the adjacent eigenvalues satisfied $\lambda_j - \lambda_j + 1 \geq e_j$, we considered that the EOFs corresponding to these two eigenvalues were significant.

The time series of the corresponding main pattern in winter and summer was used to extract the periodic variation signals of the spatial distribution pattern using Morlet wavelet analysis [27–29].

The continuous wavelet transform $W_n^X(s)$ on a scale s of a discrete time series x_n ($n = 1, \ldots, N$) with uniform time steps δt was defined as the convolution of x_n with the scaled and translated version of the wavelet function ψ_0:

$$W_n^X(s) = \sqrt{\frac{\partial t}{s}} \sum_{n'=0}^{N-1} x_{n'} \psi_0^* \left[\frac{(n'-n)\partial t}{s}\right] \tag{3}$$

where * indicates the complex conjugate, N is the total number of data points in the time series and $(\partial t/s)^{1/2}$ is the factor used to normalize the wavelet function, such that every wavelet function has a unit energy at each wavelet scale s.

By transforming the wavelet scale s and localizing along the time index n, we obtained a diagram showing the fluctuation characteristics of the time series at a certain scale and its variation with time—that is, the wavelet power spectrum [27,28,30]. The Morlet wavelet is not only non-orthogonal, but is an exponential complex-valued wavelet regulated by a Gaussian distribution defined as:

$$\psi_0(t) = \pi^{-1/4} e^{i\omega_0 t} e^{-t^2/2} \tag{4}$$

where t is the dimensionless time and ω_0 is the dimensionless frequency. When $\omega_0 = 6$, the wavelet scale s is basically equal to the Fourier period ($\lambda = 1.03$ s) [30], so the scale term and the periodic term can be substituted for each other. Then the wavelet power spectrum $\left| W_n^X(s) \right|^2$ is calculated.

To eliminate edge effects (i.e., the cone of influence), we used red noise processes as the background spectrum to test the statistical significance of the wavelet power spectrum [27–29]. Values outside the cone of influence were estimated at the 95% confidence level on each scale. Correlation analyses were conducted between the time series of the primary pattern and the meteorological indices (surface air temperature, atmospheric pressure and precipitation) in January and June; t-tests were used to verify the statistical results.

3. Vertical Motion of Air over the Indian Ocean

3.1. Distribution of the Vertical Velocity of Air

Figure 1a shows that upward motion of atmosphere (negative) is dominant over the Indian Ocean throughout the year. The central axis of upward motion (the connection line of the upward motion center) not only moves along the meridian from north to south, but also rotates with the seasons: the central axis is at about $10°$ S in spring, $7.5°$ S in autumn and $10°$ S in winter. The axis rotates clockwise from winter to summer and counterclockwise from summer to winter.

There are two ascending centers in the Indian Ocean in spring and summer. In spring, the two centers are located at about ($10°$ S, $72.5°$ E) and ($10°$ S, $100°$ E), whereas in summer they are located at (0, $60°$ E) and ($6°$ S, $90°$ E).

There is only one rising center in the Indian Ocean in autumn and winter. From autumn to winter, the rising center of the South Indian Ocean moves not only longitudinally, but also latitudinally. The center moves from ($7.5°$ S, $80.5°$ E) in autumn to ($10°$ S, $72°$ E) in winter. There is a center of subsidence in the northern Arabian Sea in spring, autumn and winter, but not in summer.

In the Bay of Bengal region of the North Indian Ocean, subsidence is dominant in winter and spring, whereas ascent is dominant in summer and autumn. This is because the Bay of Bengal is surrounded on three sides by land: the highest plateau in the world, the Tibetan Plateau, lies to the north; the Indian subcontinent lies to the west; and the Central South Peninsula lies to the east.

3.2. Distribution Characteristics of Atmospheric Vertical Motion over the Indian Ocean

Since an EOF analysis decomposes the original vertical velocity of the air, the spatial distribution of the principal mode represents the average distribution feature of the vertical velocity of air, and its time series represents the time-varying characteristics of the average distribution of the vertical velocity of air. Here, the explanatory variances of the principal modes for vertical motion at 850 and 500 hPa over the Indian Ocean in summer were 93% and 90%, respectively, and therefore their spatial distribution can be used to fully represent the average distribution of vertical motion at these altitudes (Figure 1b).

From Figure 1c, we know that these time coefficients are all greater than zero. This may be because we decomposed the original vertical velocity (X matrix in Equation (1)) by EOF analysis; the time series of the principal mode came to the first quadrant after EOF decomposition (coordinate rotation) [31–35]. Analysis of the principal mode showed that in summer, the vertical motion was relatively weak at 850 hPa and that there was only one center of ascending motion in the east ($6°$ S, $90°$ E). The center of upward motion at 500 hPa was also located at ($6°$ S, $90°$ E) (Figure 1b), which shows that the center between upper and lower layers is symmetric.

The explanatory variances in the principal modes for vertical motion at 850 and 500 hPa over the Indian Ocean in winter are 88 and 85%, respectively. There are two centers of ascending motion at 850 hPa ($8°$ S, $61°$ E) and ($10°$ S, $72°$ E), but only one center of ascending motion at 500 hPa ($12°$ S, $74°$ E) in winter, which shows that they are asymmetric.

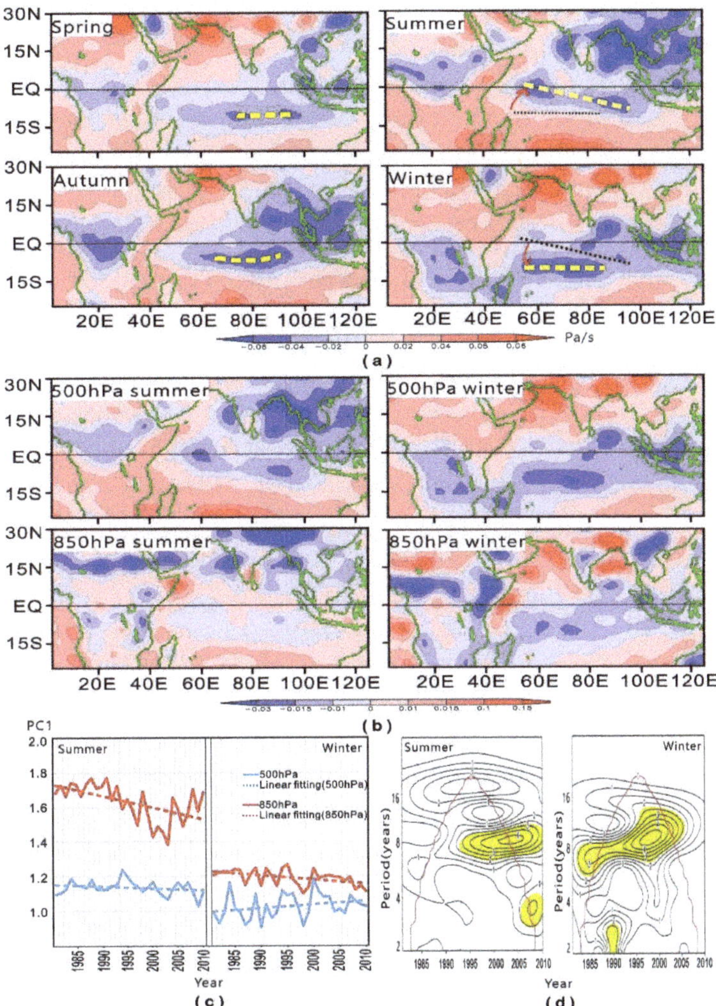

Figure 1. Distribution of the vertical velocity of air over the Indian Ocean. (**a**) Vertical velocity at 500 hPa in spring (March–May), summer (June–August), autumn (September–November) and winter (December–February). The yellow line delineates the central axis of upward motion. (**b**) Spatial distribution of the primary EOF-analyzed pattern for the vertical velocity at 500 and 850 hPa in summer and winter; all values passed North's significance test. (**c**) Temporal variation in the primary pattern of the vertical velocities at 500 and 850 hPa in summer and winter. (**d**) Wavelet power spectrum of the temporal coefficients of the primary pattern of the vertical velocity at 500 hPa in both summer and winter. The red line delineates the cone of influence and the yellow areas show confidence levels > 95%.

The Arabian Sea and the Bay of Bengal are dominated by the upward motion of air in summer, but by the downward motion of air in winter, as they are strongly affected by the thermal differences between the continents and the oceans.

Analysis of the time coefficients of the principal modes (Figure 1c) shows that the vertical motion at 850 and 500 hPa had a downward trend in summer from 1981 to 2010, which indicated that the vertical motion over the Indian Ocean weakened over time, especially at 850 hPa. It has been suggested previously that global warming may weaken atmospheric motion [36]. The situation is different in winter. The vertical upward motion of air over the South Indian Ocean and the vertical subsidence of air over the North Indian Ocean both decreased at 850hPa over time. By contrast, the vertical motion at 500 hPa was enhanced—that is, the upward vertical motion of air in the South Indian Ocean and the downward vertical motion of air in the North Indian Ocean increased.

3.3. Period of Vertical Motion of Air

Because the explanatory variances of the principal modes at 500 hPa in summer and winter are 90 and 85%, respectively, they can be used to fully represent the distribution of the mean vertical motion of air over the Indian Ocean. We used the time coefficients of the principal mode to carry out wavelet analysis to understand the periodic variation in vertical motion over the Indian Ocean.

The distribution of the principal mode of the vertical velocity of air over the Indian Ocean in summer had a period of about 7–9 years from 1990 to 2010. The variances all passed the 95% reliability test. The periodic oscillation in winter was 9–12 years from the early 1990s to around 2003 and 2–3 years from 1987 to 1991. Both variances passed the 95% significance test (Figure 1d).

4. Relationship between the Vertical Motion of Air and the Climate in East Asia

The onset of the Indian monsoon in the Indian Ocean occurs in June. To understand the correlation between the climate in China and the vertical motion of air over the Indian Ocean, we analyzed the relationship between the temporal coefficients of the principal modes of the EOF analysis (explanatory variance 73%) for the vertical motion at 500 hPa over the Indian Ocean and the surface air temperature as well as the atmospheric air pressure in June. In addition, the monthly mean precipitation from the respective 839 meteorological stations in China is used to correlate with the time coefficient of the primary mode of the vertical motion of air at 500 hPa over the Indian Ocean; 839 correlation coefficients are obtained. Then, Figure 2a,b are formed by the correlation coefficients from 839 stations. For comparison, we also studied the correlation in January (explanatory variance 81%) and used the T-test to verify the correlation coefficient between them.

4.1. Vertical Motion of Air over the Indian Ocean and the East Asian Climate in Summer
4.1.1. Vertical Motion and Precipitation in June

The correlation between the vertical motion of air over the Indian Ocean and precipitation in China exceeded the 95% confidence level in a number of areas in June. The negative correlation in the western Tibetan Plateau, the positive correlation in the northeastern Tibetan Plateau, the negative correlation in the north of the lower reaches of Yangtze River and the positive correlation in the south of the lower reaches of Yangtze River (Figure 2a) all exceeded the 95% confidence level. These correlations indicate that with the enhancement of the vertical movement of air in the Indian Ocean in June, the precipitation decreases in the western Tibetan Plateau, increases in the northeastern Tibetan Plateau, decreases in the north and increases in the south of the lower reaches of the Yangtze River, and vice versa (Figure 2a).

4.1.2. Surface Air Temperature, Pressure and the Vertical Motion of Air in June

The vertical motion of air at 500 hPa over the Indian Ocean in June was only sporadically and positively correlated with the surface temperature and pressure over the Tibetan Plateau and northeastern China (Figure 2c,e).

4.2. Vertical Motion of Air over the Indian Ocean and the East Asian Climate in Winter
4.2.1. Vertical Motion of Air and Precipitation in January

Figure 2b shows that the correlation between precipitation in China and the vertical motion of air over the Indian Ocean passed the 95% significance test in five places in China in January: there was a negative correlation with precipitation in southern Xinjiang and four positive correlations in northern Xinjiang, northeastern China, northern China and Sichuan (the upper Yangtze river), indicating that the precipitation in southern Xinjiang decreased, whereas the precipitation in the other four regions increased as the vertical motion of air over the Indian Ocean increased, and vice versa.

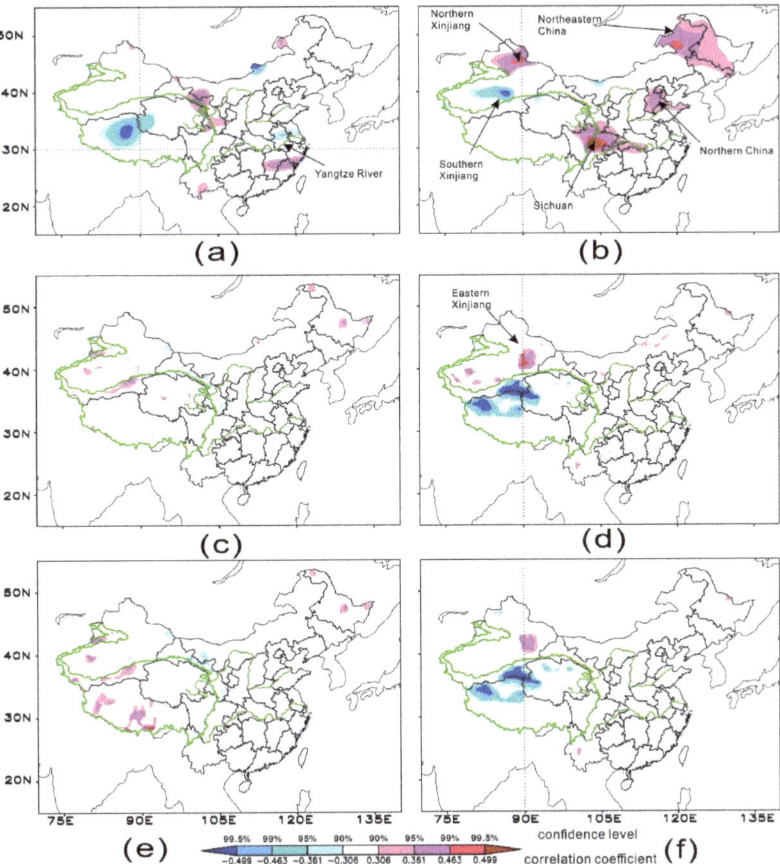

Figure 2. Correlation analysis between the surface meteorological variables and the vertical motion of air over the Indian Ocean. (**a**) Correlation between the precipitation and the vertical velocity in June. (**b**) Correlation between the precipitation and the vertical velocity in January. (**c**) Correlation between the surface pressure and the vertical velocity in June. (**d**) Correlation between the vertical velocity and the surface pressure in January. (**e**) Correlation between the vertical velocity and the surface temperature in June. (**f**) Correlation between the surface temperature and the vertical velocity in January.

4.2.2. Surface Air Temperature, Pressure and Vertical Motion of Air in January

The correlation between the surface air temperature in China and the vertical motion of air over the Indian Ocean in January is positive in eastern Xinjiang and negative in the northwestern Tibetan Plateau. The correlation with surface pressure is the same as that with temperature. This distribution of correlation indicates that as the vertical motion of air over the Indian Ocean increased (weakened), the surface temperature (pressure) in southern Xinjiang increased (weakened), whereas it decreased (increased) in the northwestern part Tibetan Plateau (Figure 2d,f).

5. Discussion

5.1. Relevance to Precipitation

We have shown that the precipitation in the western Tibetan Plateau is negatively correlated with the vertical motion over the Indian Ocean in June (Figure 2a). The negative correlation indicates that the precipitation in the western Tibetan Plateau decreased (enhanced) as the vertical motion over the Indian Ocean strengthened (weakened).

The average meridian circulation along 90° E in June showed that there is vertical ascending motion over both the Indian Ocean and the Tibetan Plateau (Figure 3a). These vertical motions from the Indian Ocean have sunken branches on the north of plateau, and an ascending motion below the descending branch. When the vertical motion from the Indian Ocean strengthened, the sinking motion over the north of the Tibetan Plateau is also strengthened, which suppresses the ascent of the lower layer and weakens precipitation in this area, resulting in a negative correlation.

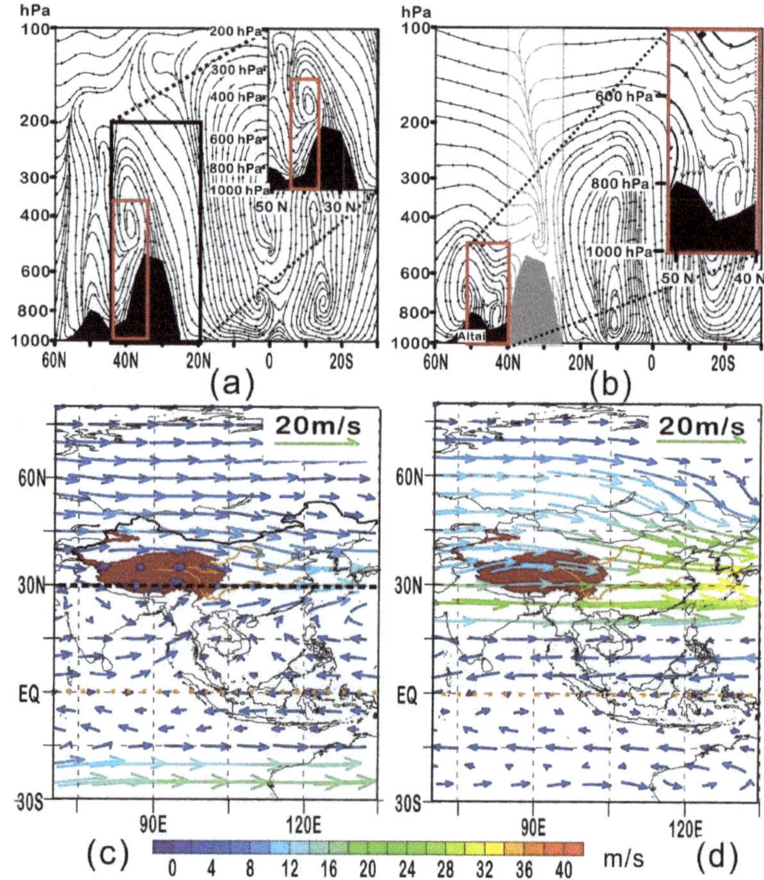

Figure 3. Wind fields. (**a**) Mean meridional circulation at 90° E in June. (**b**) Mean meridional circulation at 90° E in January. (**c**) Wind field at 500 hPa in June. (**d**) Wind field at 500 hPa in January. The red rectangles (**a,b**) represent the intersection of the updraft and the downdraft. The grey area in part (**b**) represents the Tibetan Plateau area.

Figure 2a also shows that there are two correlations between the vertical motion of air over the Indian Ocean and precipitation in the lower reaches of the Yangtze River in June. The correlation is bounded by 30° N; north of 30° N is negative correlation and south of 30°N is a positive correlation in June. Figure 3c shows that the airflow from the Indian Ocean is also bound by 30° N, with a southwesterly airflow (summer monsoon direction) south of 30° N and a northwesterly airflow north of 30° N (in the lower reaches of the Yangtze River). The airflow is strengthened with the increase in the Indian monsoon, transporting abundant amounts of water vapor to the south of 30° N, increasing precipitation. The negative correlation to the north of 30° N (the Yangtze River) and the positive correlation between the vertical motion of air over the Indian Ocean and the precipitation in the northeastern side of the Tibetan Plateau and require further study.

In addition, we have also shown that the vertical motion over the Indian Ocean in January is negatively correlated with precipitation in southern Xinjiang and positively correlated with precipitation in northern Xinjiang, northeastern China, northern China and the Sichuan in China (the upper and middle reaches of the Yangtze River) (Figure 2b).

The negative correlation between the vertical motion of the Indian Ocean and precipitation in southern Xinjiang in January suggests that the precipitation in southern Xinjiang decreased when the vertical motion of the Indian Ocean strengthens. Figure 3b shows that cold air from the north (the high latitude) sinks over southern Xinjiang and leads to an ascending motion of the lower layer (Figure 3b). When the amount of cold air from the north (the high latitude) increases, the ascending motion in the lower layer and precipitation both increase. Cold air from the high latitude meets the air from the Indian Ocean at about 30° N, producing a strong downdraft (Figure 3b). The strengthening cold air inhibits the movement of air from the Indian Ocean, and the vertical motion over the Indian Ocean decreases—that is, precipitation in southern Xinjiang increases with the decrease in vertical motion over the Indian Ocean, leading to a negative correlation (Table 1).

Table 1. Correlation between climate in China and the vertical motion over the Indian Ocean in January.

Cold Air from the High Latitude	Vertical Motion over the Indian Ocean	Precipitation in Southern/ Northern Xinjiang	Upward Motion of Air in Eastern Xinjiang	Temperature/ Pressure in the Eastern Xinjiang	Sinks in the North of the Tibetan Plateau	Temperature/ Pressure in the Tibetan Plateau
Strengthen (+)	Weaken (−)	Increase/decrease (+/−)	Increase (+)	Decrease/decrease (−/−)	Increase (+)	Increase/increase (+/+)
Weaken (−)	Strengthen (+)	Increase/decrease (−/+)	Decrease (−)	Increase/increase (+/+)	Decrease (+)	Decrease/decrease (−/−)

The areas with a positive correlation (northern Xinjiang, northeastern China, northern China and the upper as well as the middle reaches of the Yangtze River) are all basins or plains, and they are all in the path of the cold wave that is invading China [37]. Cold air from the high latitude crosses the mountains (plateaus) before reaching the basin or plain, and then produces a downward airflow in these four areas which decreases the precipitation. The amount of precipitation decreases when the downward airflow strengthens. When cold air from the high latitude strengthens, the vertical upward motion over the Indian Ocean weakens (Table 1)—that is, when the vertical upward motion over the Indian Ocean weakens, the vertical subsidence motion over these four regions increases and the amount of precipitation decreases, leading to a positive correlation. Here, we have only given some explanation for the phenomenon we found, and there may be other reasons to be further researched. We can only show cross-sections along 90° E (Figure 3b), due to the limitation of our drawing level. The above results will be more clearly shown if a flow profile can be drawn from the origin area of high-latitude cold air to these areas with a positive correlation.

5.2. Correlation between Vertical Motion over the Indian Ocean and Surface Temperature and Pressure in China in Winter

Figure 2d,f has shown that there is a negative correlation between the vertical motion of air over the Indian Ocean and the surface temperature and pressure on the Tibetan Plateau in January, whereas there is a positive correlation between the vertical motion of air over the Indian Ocean and the surface temperature and pressure in eastern Xinjiang in January. In January, deep, cold air from the high latitude moves south, and part of this package of air is overturned in the Altai Mountains and then accumulates in the northern Tibetan Plateau (eastern Xinjiang), resulting in an upward motion of air (Figure 3b). This cold air lowers the local surface temperature and pressure due to the ascent of the air. The forward movement of the cold air is then blocked by the Tibetan Plateau, and therefore sinks in the north of the plateau, increasing the surface pressure. When the sinking motion of air increases, the cloud amount decreases, the direct solar radiation increases, and the net radiation on the ground surface increases. Therefore, the radiation energy used to heat the atmosphere increases by sensible heat [38], which increases the surface temperature, and vice versa.

The cold air continues to move southward and meets the air from the Indian Ocean at about 30° N, producing a strong downdraft (Figure 3b). When the cold air from the north (high latitude) strengthens, the intersection of the two strands of air moves southward, inhibiting the vertical motion from the Indian Ocean, which then weakens. Corresponding to this weakened vertical motion over the Indian Ocean (the strengthening of cold air from the origin area of high latitude), the surface temperature and pressure decrease in eastern Xinjiang, leading to a positive correlation. As the sinking of air over northern Tibet Plateau increases, the surface pressure and temperature increase, leading to a negative correlation (Table 1).

5.3. Turning of the Axis of Vertical Motion Center

The axis of vertical motion in the atmosphere over the Indian Ocean turns clockwise from winter to summer, but counterclockwise back to its original position from summer to winter (Figure 1a). We conducted an EOF analysis on the anomaly in the sea–air temperature difference (sea surface temperature minus the air temperature of two meters above the sea) over the Indian Ocean from January to December. The spatial distribution from the first pattern of the anomaly rotates clockwise from January to July (the explanatory variances are 15.9% in January and 16.1% in July) and counterclockwise from summer to winter (Figure 4a,c).

We know from the wind field over the Indian Ocean at 1000 hPa that the northeasterly wind (winter monsoon) from the high latitudes of the northern hemisphere in January crosses the equator and then turns to the northwest and meets the southeasterly wind from the southern hemisphere at about 10° S (Figure 4b). In July, the southeasterly winds from the high latitudes of the southern hemisphere flow to the equator and then become southwesterly near the equator. During this process, the wind speed in the western Indian Ocean is higher than that in the eastern Indian Ocean (Figure 4d). The wind-driven currents on the surface of the Indian Ocean are affected by these changes in wind direction and wind speed. The heating field of the ocean to the atmosphere also changes with the process, and the axis of vertical motion over the Indian Ocean changes to clockwise or counterclockwise.

This study only shows the correlation between the vertical movement of air over the Indian Ocean and China's climate. The mechanism behind this phenomenon will need further study. In addition, we only calculated the monthly and seasonal mean precipitation data for 30 years from 1981 to 2010; their homogeneity was not analyzed. There are some phenomena we cannot explain yet. These deficiencies will need to be addressed in future research.

By investigating the correlation between the vertical motion of air over the Indian Ocean and surface air temperature, atmospheric pressure and precipitation in China, we found that the vertical motion of air associated with the monsoonal circulation over the Indian Ocean plays a certain role in the formation of and change in the climate in China. The results of this research have certain significance and practical value for understanding and forecasting climate in East Asia.

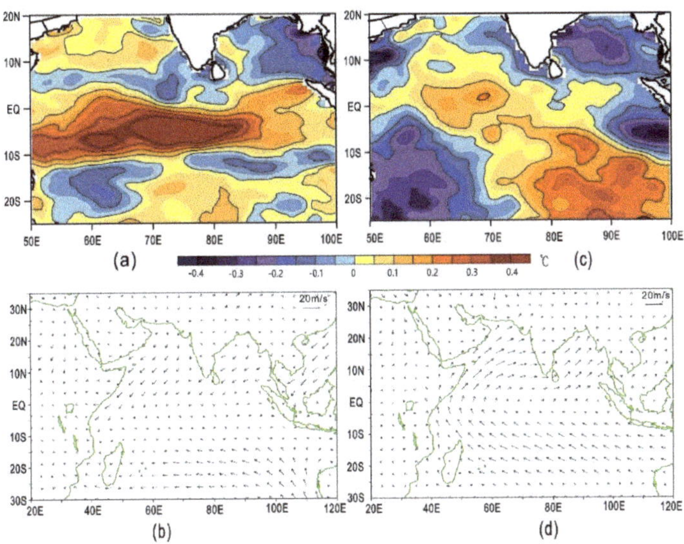

Figure 4. Difference in temperature between the sea and air and the wind field over the Indian Ocean. (**a**) The first pattern of the anomaly in the sea - air temperature difference (sea surface temperature minus the air temperature of two meters above the sea) in January. (**b**) Wind fields at 1000 hPa in January. (**c**) The first pattern of the anomaly in the sea-air temperature difference in July. (**d**) Wind fields at 1000 hPa in July.

6. Conclusions

The vertical motion of the atmosphere over the Indian Ocean is closely related to the climate in some particular regions of China. Climate diagnosis and statistical analysis were used to study the vertical motion of air over the Indian Ocean and its relationship with the climate in China. The following conclusions can be drawn from these results:

(1) The vertical motion of air is negatively correlated with precipitation in the Tibetan Plateau during summer and positively correlated with precipitation in northern Xingjiang, northeast China, northern China and the Sichuan province (the upper and middle reaches of the Yangtze River) during winter. This can be explained by the interaction between the vertical motion of air over the Indian Ocean and cold air from the high latitude of the Northern Hemisphere.

(2) The vertical motion over the Indian Ocean was weakened from 1981 to 2010, except at a height of 500 hPa in winter. The vertical motion of air over the Indian Ocean had a period of 7–9 years in summer and 2–3 in addition to 9–12 years in winter.

(3) The ascending motion of air over the Indian Ocean is dominant throughout the year. The center of ascending air moves and rotates as the seasons change, and the central axis rotates clockwise from winter to summer and counterclockwise from summer to winter. This is because the heating of the atmosphere over the Indian Ocean changes from winter to summer with the East Asian monsoon, and vice versa.

Author Contributions: R.T. conceived the idea and wrote the manuscript. Y.M. and W.M. revised the manuscript. All authors have read and agreed to the published version of the manuscript.

Funding: This research was funded by the Second Tibetan Plateau Scientific Expedition and Research Program (STEP), grant no. 2019QZKK0103, and the National Natural Science Foundation of China (grant no. 41775142).

Institutional Review Board Statement: Not applicable.

Informed Consent Statement: Not applicable.

Data Availability Statement: Data and methods used in the research have been presented in sufficient detail in the paper.

Acknowledgments: This work was supported by the Second Tibetan Plateau Scientific Expedition and Research Program (STEP), grant no. 2019QZKK0103 and the National Natural Science Foundation of China (grant no. 41775142). We acknowledge the use of meteorological data collected from the National Centers for Environmental Prediction/National Center for Atmospheric Research, the China Meteorological Administration and the Scientific Data Center for the Cold and Arid Regions of China. Duo Zha, Xinfang Zhang and Yiwei Ye completed the drawing of this paper. All data in the research can be obtained by contacting the corresponding author, Rongxiang Tian (trx@zju.edu.cn).

Conflicts of Interest: The authors declare no conflict of interest.

References

1. Schott, F.A.; McCreary, J.P., Jr. The monsoon circulation of the Indian Ocean. *Prog. Oceanogr.* **2001**, *51*, 1–123. [CrossRef]
2. Li, C.; Yanai, M. The Onset and Interannual variability of the Asian Summer Monsoon in relation to land-sea thermal contrast. *J. Clim.* **1996**, *9*, 358–375. [CrossRef]
3. Lau, K.M.; Wu, H.T.; Bony, S. The role of large-scale atmospheric circulation in the relationship between tropical convection and sea surface temperature. *J. Clim.* **1997**, *10*, 381–392. [CrossRef]
4. Wang, B.; Wu, R.; Li, T. Atmosphere–warm ocean interaction and its impacts on Asian–Australian monsoon variation. *J. Clim.* **2003**, *16*, 1195–1211. [CrossRef]
5. Gu, Z.C. Analysis and calculation of large-scale vertical movement. *Acta Meteorol. Sin.* **1954**, *25*, 3–20. (In Chinese)
6. Sreelekha, P.N.; Babu, C.A. Organized convection over southwest peninsular India during the pre-monsoon season. *Theor. Appl. Climatol.* **2019**, *135*, 1279–1293. [CrossRef]
7. Hadley, G. Concerning the Cause of the General Trade-Winds. *Philos. Trans.* **1735**, *39*, 58–62.
8. Mathew, S.S.; Kumar, K.K. Characterization of the long-term changes in moisture, clouds and precipitation in the ascending and descending branches of the Hadley Circulation. *J. Hydrol.* **2019**, *570*, 366–377. [CrossRef]
9. Bjerknes, J. Atmospheric teleconnections from the equatorial. *Mon. Weather. Rev.* **1969**, *97*, 163–172. [CrossRef]
10. Wyrtki, K.; Eldyn, G. Equatorial upwelling events in the central pacific. *J. Phys. Oceanogr.* **1982**, *12*, 984–988. [CrossRef]
11. Peixoto, J.P.; Oort, A.H. *Physics of Climate*; American Institute of Physics: New York, NY, USA, 1992; p. 520.
12. Saha, K. Zonal anomaly of sea surface temperature in equatorial Indian Ocean and its possible effect upon monsoon circulation. *Tellus* **1970**, *22*, 403–409. [CrossRef]
13. Bjerknes, P.V.F. Synoptical representation of atmospheric motions. *Q. J. R. Meteorol. Soc.* **1910**, *36*, 267–286. [CrossRef]
14. Kidson, J.W. African rainfall and its relation to upper air circulation. *Q. J. R. Meteorol. Soc.* **1977**, *103*, 441–456. [CrossRef]
15. Long, M.; Entekhabi, D.; Nicholson, S.E. Interannual variability in rainfall, water vapor flux, and vertical motion over West Africa. *J. Clim.* **2000**, *13*, 3827–3841. [CrossRef]
16. Liu, Y.Y.; Ding, Y.H. Teleconnection between the indian summer monsoon onset and the meiyu over the yangtze river valley. *Sci. China Ser. D Earth Sci.* **2008**, *51*, 1021–1035. [CrossRef]
17. Wu, R.G. Relationship between Indian and East Asian Summer Rainfall Variations. *Adv. Atmos. Sci.* **2017**, *34*, 4–5. [CrossRef]
18. Kalnay, E.; Kanamitsu, M.; Kistler, R.; Collins, W.; Deaven, D.; Gandin, L.; Joseph, D. The NCEP/NCAR 40-year reanalysis project. *Bull. Am. Meteorol. Soc.* **1996**, *77*, 437–472. [CrossRef]
19. Wang, B.; Wu, R.; Fu, X. Pacific–East Asian teleconnection: How does ENSO affect East Asian climate? *J. Clim.* **2000**, *13*, 1517–1536. [CrossRef]
20. Swain, M.; Sinha, P.; Mohanty, U.C.; Pattnaik, S. Dominant large-scale parameters responsible for diverse extreme rainfall events over vulnerable Odisha state in India. *Clim. Dyn.* **2020**, *54*, 2221–2236. [CrossRef]
21. Ji, Z.G.; Chao, J.P. Teleconnections of the sea surface temperature in the Indian ocean with sea surface temperature in the eastern equatorial pacific, and with the 500 hPa geopotential height field in the Northern Hemisphere. *Adv. Atmos. Sci.* **1987**, *4*, 343–348. [CrossRef]
22. Jia, X.L.; Li, C.Y. Dipole Oscillation in the Southern Indian Ocean and Its Impacts on Climate. *Chin. J. Geophys.* **2005**, *48*, 1323–1335. [CrossRef]
23. Parthasarathy, B.; Munot, A.A.; Kothawale, D.R. All-India Monthly and Seasonal Rainfall Series 1871–1993. *Theor. Appl. Climatol.* **1994**, *49*, 217–224. [CrossRef]
24. Xu, J.Y.; Tian, R.X.; Feng, S. Comparison of Atmospheric Vertical Motion over China in ERA-Interim, JRA-55, and NCEP/NCAR Reanalysis Datasets. *Asia-Pac. J. Atmos. Sci.* **2021**. [CrossRef]
25. Lorenz, E.N. Statistical forecasting program: Empirical orthogonal functions and statistical weather prediction. *Sci. Rep.* **1956**, *409*, 997–999.
26. North, G.R.; Bell, T.L.; Cahalan, R.F. Sampling errors in the estimation of empirical orthogonal functions. *Mon. Weather Rev.* **1982**, *110*, 699–706. [CrossRef]
27. Torrence, C.; Compo, G.P.A. Practical guide to wavelet analysis. *Bull. Am. Meteorol. Soc.* **1998**, *79*, 61–78. [CrossRef]

28. Grinsted, A.; Moore, J.C.; Jevrejeva, S. Application of the cross wavelet transform and wavelet coherence to geophysical time series. *Nonlinear Process. Geophys.* **2004**, *11*, 561–566. [CrossRef]
29. Furon, A.C.; Wagner-Riddle, C.; Smith, C.R.; Warland, J.S. Wavelet analysis of wintertime and spring thaw CO_2 and N_2O fluxes from agricultural fields. *Agric. For. Meteorol.* **2008**, *148*, 1305–1317. [CrossRef]
30. Torrence, C.; Webster, P.J. Interdecadal Changes in the ENSO-Monsoon system. *J. Clim.* **1999**, *12*, 2679–2690. [CrossRef]
31. Freire, S.L.M. Application of singular value decomposition to vertical seismic profiling. *Geophysics* **1988**, *53*, 778–785. [CrossRef]
32. Henry, E.; Hofrichter, J. Singular value decomposition-application to analysis of experimental-data. *Methods Enzymol.* **1992**, *210*, 129–192.
33. Stork, C. Singular value decomposition of the velocity-reflector depth tradeoff, Part 2: High-resolution analysis of a generic model. *Geophysics* **1992**, *57*, 927–932. [CrossRef]
34. Wallace, J.M.; Smith, C.; Bretherton, C.S. Singular value decomposition of wintertime sea surface temperature and 500-mb height anomalies. *J. Clim.* **1992**, *5*, 561–576. [CrossRef]
35. Kreimer, N.; Sacchi, M.D. A tensor higher-order singular value decomposition for prestack seismic data noise reduction and interpolation. *Geophysics* **2012**, *77*, 113–122. [CrossRef]
36. Kjellsson, J. Weakening of the global atmospheric circulation with global warming. *Clim. Dyn.* **2015**, *45*, 975–988. [CrossRef]
37. Zhu, Q.G.; Lin, J.R.; Shou, S.W.; Tang, D.S. *Principles and Methods of Synoptic Meteorology*; China Meteorological Press: Beijing, China, 2007; pp. 168–213.
38. Cai, W.Y.; Xu, X.D.; Sun, J.H. An investigation into the surface energy balance on the southeast edge of the Tibetan Plateau and the cloud's impact. *Acta Meteorol. Sin.* **2012**, *70*, 837–846. (In Chinese)

water

Article

Impact of Fully Coupled Hydrology-Atmosphere Processes on Atmosphere Conditions: Investigating the Performance of the WRF-Hydro Model in the Three River Source Region on the Tibetan Plateau, China

Guangwei Li [1,2], Xianhong Meng [2,*], Eleanor Blyth [3], Hao Chen [1], Lele Shu [1], Zhaoguo Li [1], Lin Zhao [1] and Yingsai Ma [1,2]

[1] Key Laboratory of Land Surface Process and Climate Change in Cold and Arid Regions, Northwest Institute of Eco-Environment and Resources, Chinese Academy of Science, Lanzhou 730000, China; gwli@lzb.ac.cn (G.L.); chenhao@lzb.ac.cn (H.C.); shulele@lzb.ac.cn (L.S.); zgli@lzb.ac.cn (Z.L.); zhaolin_110@lzb.ac.cn (L.Z.); mayingsai@lzb.ac.cn (Y.M.)
[2] University of Chinese Academy of Science, Beijing 100049, China
[3] UK Centre for Ecology & Hydrology, Wallingford OX10 8BB, UK; emb@ceh.ac.uk
* Correspondence: mxh@lzb.ac.cn; Tel.: +86-0931-4967239

Citation: Li, G.; Meng, X.; Blyth, E.; Chen, H.; Shu, L.; Li, Z.; Zhao, L.; Ma, Y. Impact of Fully Coupled Hydrology-Atmosphere Processes on Atmosphere Conditions: Investigating the Performance of the WRF-Hydro Model in the Three River Source Region on the Tibetan Plateau, China. *Water* **2021**, *13*, 3409. https://doi.org/10.3390/w13233409

Academic Editor: Aizhong Ye

Received: 17 October 2021
Accepted: 24 November 2021
Published: 2 December 2021

Publisher's Note: MDPI stays neutral with regard to jurisdictional claims in published maps and institutional affiliations.

Abstract: The newly developed WRF-Hydro model is a fully coupled atmospheric and hydrological processes model suitable for studying the intertwined atmospheric hydrological processes. This study utilizes the WRF-Hydro system on the Three-River source region. The Nash-Sutcliffe efficiency for the runoff simulation is 0.55 compared against the observed daily discharge amount of three stations. The coupled WRF-Hydro simulations are better than WRF in terms of six ground meteorological elements and turbulent heat flux, compared to the data from 14 meteorological stations located in the plateau residential area and two flux stations located around the lake. Although WRF-Hydro overestimates soil moisture, higher anomaly correlation coefficient scores (0.955 versus 0.941) were achieved. The time series of the basin average demonstrates that the hydrological module of WRF-hydro functions during the unfrozen period. The rainfall intensity and frequency simulated by WRF-Hydro are closer to global precipitation mission (GPM) data, attributed to higher convective available potential energy (CAPE) simulated by WRF-Hydro. The results emphasized the necessity of a fully coupled atmospheric-hydrological model when investigating land-atmosphere interactions on a complex topography and hydrology region.

Keywords: WRF-Hydro model; runoff; precipitation; three river source region

1. Introduction

The hydrological processes at the surface-atmosphere interface are a key process in the terrestrial water cycle and impact weather systems [1–3]. Soil moisture change, transpiration, and runoff are the three main components which affect water resources management, agriculture, ecology, and flood prediction [4–7]. Such hydrological processes are becoming increasingly detailed and precise in regional climate models [8–11], especially as recent studies have pointed out that even with inadequate hyper-resolution meteorological and surface data, it is worthwhile to integrate lateral flow dynamics in hyper-resolution land surface models that have a resolution on the order of 100 m at continental scales [12] to better reflect water and energy heterogeneity [13]. Moreover, several numerical models incorporating lateral flow have been developed [14–16].

The Tibetan Plateau (TP) is the highest landform globally, with an average elevation exceeding 4000 m above sea level [17], and is located in the area where the East Asia monsoon and the South Asian monsoon interact. The TP acts as a strong 'dynamic pump' [18], attracting moist air from low latitudes during the warm season. The intense surface radiation and topographic lifting of TP provides a favorable condition for convection [19,20].

Summer precipitation is about 60% of annual precipitation [21]. In addition, the thermal forcing convective cloud system in the TP usually moves downstream into eastern China, contributing to heavy rain and severe convective storms in China [22–26]. Precipitation in winter is stored in solid form as snow and ice and released as meltwater when the temperature rises during spring and summer [27]. The huge number of glaciers, snow-packs, lakes, rivers, and a large amount of water storage in TP serves as 'the world's water tower' [20].

The source region of the Three River basins (SRTR), located in the TP, is the source region of the three largest rivers in China, i.e., the Yangtze, Lancang, and Yellow River. The region is suitable for performing the coupled atmospheric-hydrological modeling due to its vital geographic location and essential hydrological functions [28]. Based on meteorological observations and ice core records, the intensified global water cycle [29], especially precipitation and evaporation, might be enhanced in TP because TP is more sensitive to climate change due to high altitudes [30]. Rapid warming in TP [31] could lead to notable changes in the phase and intensity of surface precipitation, which impacts evapotranspiration and runoff generation [32].

Numerical models are powerful tools for studying the hydrology of the SRST, and each of the models has its advantages [33]. There are many models that focus either on the hydrology or on land-atmosphere interactions only. For example, the soil and water assessment tool (SWAT) is a conceptual, distributed parameter model [34], suitable for long-term runoff and pollution transfer simulation in large river basins. Variable infiltration capacity (VIC) is a physically-based macroscale hydrological model, developed to solve water and energy balances [35]. The community land model (CLM) [36] robustly simulates the exchange of water, energy and carbon and nitrogen between the land and the atmosphere. The community Noah land surface model with multi-parameterization options (Noah-MP) [37] is an enhanced version of Noah similar to CLM, but more oriented to applications in regional climate models. These models have been instrumental in the study of the hydrology and atmosphere of SRTR [38–42]. However, due to the design orientation, SWAT and VIC cannot be coupled directly to the atmospheric model. At the same time, the physical processes of lateral terrestrial water flow are absent in the CLM, Noah, and Noah-MP models. As a result, the impacts of the coupled hydrology and atmosphere processes are still less investigated in the SRTR.

The hydrologically enhanced version of the weather research and forecasting modeling system (WRF-Hydro) [43] is a high-resolution model that explicitly describes the surface overland flow. And it is a fully coupled atmospheric-hydrological model, since the hydrological processes are added in the WRF model (using Noah-MP as the land surface module). WRF-Hydro contains physical processes including the re-infiltration, surface overland flow and lateral flow of water within the soil layers [11]. The changes in soil moisture due to lateral flow are fed back to the atmosphere. Therefore, the precipitation simulation is different from that of WRF [44–46]. Besides, the accuracy of turbulent heat fluxes increased in Germany using WRF-Hydro [1]. The WRF-Hydro revealed good performance in simulating the diurnal cycle of land surface states and fluxes during the North American monsoon [47]. Coupled WRF-Hydro slightly outperforms the WRF stand-alone concerning sensible and ground heat flux, near-surface mixing ratio and temperature, boundary layer profiles of temperature [1]. The hydrologically enhanced process of WRF-Hydro showed an increased precipitation recycling rate in the inland area of China [48] but decreased slightly in East Africa [49]. In addition, streamflow corresponds with observations at monthly timescales on the south side of the Himalayas [50].

In this paper, the fully coupled WRF-Hydro is chosen to carry out modeling experiments of SRTR. The aim of this work is (1) to validate the applicability of WRF-Hydro in the runoff simulation in the SRTR, (2) in order to investigate the effects of lateral flow on soil moisture and near-surface meteorological variables in a coupled atmospheric-hydrological model, (3) to investigate the effects of coupled atmospheric-hydrological processes versus the uncoupled ones on the boundary layer and regional precipitation.

2. Study Area, Model Description and the Numerical Experiment Design

2.1. Study Area

The SRTR is located northeast of the TP, with an altitude average of 4510 m and ranging from 2600 m to 6500 m a.s.l. (Figure 1). It is an adjacent area consists of the source region of the Yellow River basin (YRB, 1.22×10^5 km^2), the Yangtze River basin (YARB, 1.37×10^5 km^2), and the Lancang River basin (LRB, 5.3×10^4 km^2). The SRST is mainly covered by grassland [51], and the soil type is loam. The YRB contributes about 35% of the river flow [52]. Glacier cover fractions of YARB and YRB are 0.95% and 0.11% [33], respectively. The proportion of melt contributing to runoff ranges from 3.9 to 6% in four subbasins of the YARB [42]. During 1956–2012, the annual runoff in LRB and YARB increased, while YRB slightly decreased [53]. Under the impact of climate change, frozen ground degradation increased the groundwater discharge rate in winter [54]. Direct snowmelt runoff coefficients are mainly controlled by the air temperature freezing index [55].

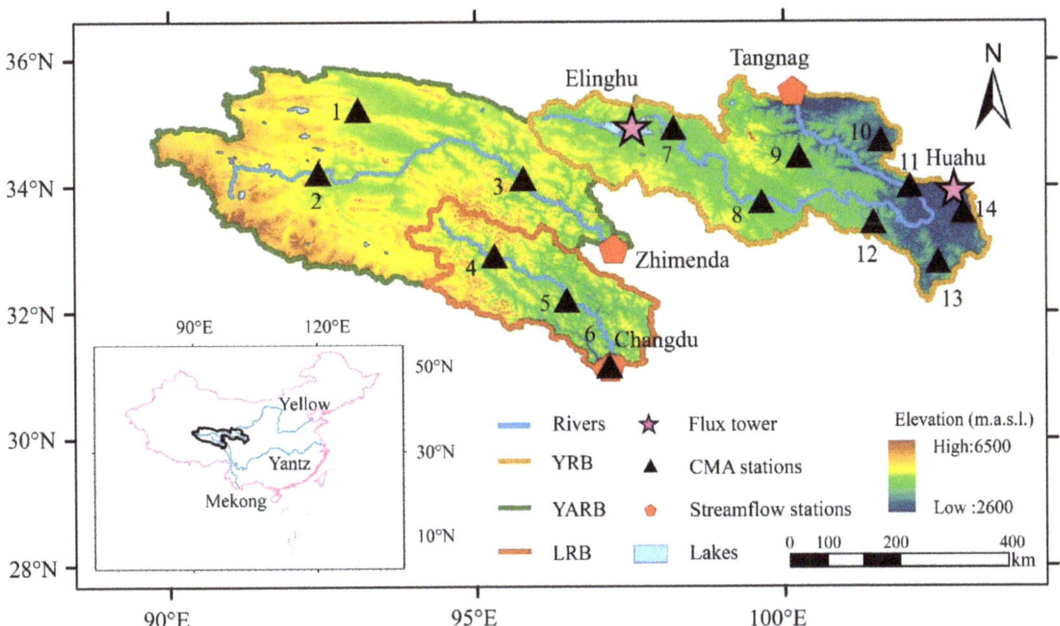

Figure 1. Geographic locations of the observational stations and source region of Three River basins (SRTR) include the Yellow River basin (YRB), the Yangtze River basin (YARB) and the Lancang River basin (LRB). The coordinates of streamflow stations, flux towers and meteorological stations are shown in Table 1.

2.2. Model Description

2.2.1. WRF-ARW and NOAH-MP

The WRF-ARW model [56] is a time-split non-hydrostatic atmospheric model. It is widely used in the study of land-atmosphere interactions [6], dynamic downscaling [57,58], weather and climate research [59,60]. The ability of WRF to resolve strongly nonlinear small-scale phenomena such as stratiform precipitation and convective precipitation makes it suitable for the coupling study of atmospheric and hydrological processes.

Noah-MP [37] is an enhanced version of the Noah land surface model, and introduces a framework for multiple schemes. Due to the enhancement of biophysical and soil freeze-thaw processes, the Noah-MP has been widely used in the TRSR [41,61,62]. Noah-MP can be selected as the land surface process module of the WRF model.

2.2.2. WRF-Hydro

WRF-Hydro is a modeling framework to facilitate the coupling of WRF and hydrological models. The interactions between hydrological and land surface processes are calculated as follows [43]. Land surface states and fluxes computed by Noah-MP are disaggregated to the high-resolution terrain routing grid. Then the physical hydrologic process is calculated on the routing grid. After that, land surface states and fluxes aggregated from the routing grid are updated to the Noah-MP model grid. Therefore, it can be used as a land surface model offline or fully coupled to the WRF model. The uncoupled WRF-Hydro model is good at spin-up, model calibration, and data assimilation. In contrast, the coupled WRF-Hydro model is used for hydrology-atmosphere coupling research.

There are five major parts in the hydrologic processes of WRF-Hydro [43], namely subsurface flow, overland flow, channel, lake, and a conceptual base flow module. The subsurface flow process calculates subsurface lateral flow [63] and exfiltration from a supersaturated soil column. During overland flow, the infiltration excess and exfiltration calculated in the previous step flow into the river by solving the diffusive wave formula [64]. The channel process simulates the flow in the river network by either fine grid routing using a diffusive wave equation or on a vectorized network of channel reaches by solving the Muskingum or Muskingum-Cunge equation [65]. Besides, WRF-Hydro provides a simple mass balance, a level-pool lake/reservoir routing module, and a conceptual exponential bucket base flow module [43].

2.3. Numerical Experiment Design

To investigate the performance of WRF-Hydro, uncoupled and coupled simulations were performed over SRST. The first is a WRF-Hydro uncoupled experiment (WRF-H-UP, hereafter), used to find out the calibrated hydrological parameters and test the simulation performance of runoff generation. Secondly, a group of coupled WRF-Hydro (WRF-H, hereafter) and WRF-ARW experiments (WRF-S, hereafter) were conducted to test the effects of hydrological processes on the atmosphere simulation.

2.3.1. Model Calibration

A WRF-Hydro simulation is set up in offline mode (WRF-H-UP) for parameter calibration and runoff simulation. Since it is still a challenge to reproduce daily runoff using fully coupled models [48], only runoff from the uncoupled simulation is analyzed in the study. The domain of WRF-H-UP corresponds to the area of the inner domain (d02) used by the subsequent coupling run, shown in Figure 2. The simulations of WRF-ARW drive the WRF-H-UP to verify the forecasting ability of WRF-Hydro on the runoff discharge. The runoff simulations are produced after the parameters have been calibrated. Because WRF-Hydro includes many parameterized nonlinear physical processes, the default parameters given by WRF-Hydro are only valid over a small region; calibration of related model parameters is often required to use in the new domain [66].

The parameters are calibrated manually in this study. Most of the parameters are insensitive during the calibration, except those that control base flow (maximum depth, Zmax). By increasing the Zmax from 10 to 50, WRF-Hydro can calibrate errors caused by overestimated precipitation by storing water in conceptual groundwater buckets. Other parameters involved in the calibration process include the control of water movement in the soil, such as reference infiltration factor (REFKDT), soil evaporation exponent (FXEXP_DATA), deep drainage coefficient (SLOPE), saturated soil hydraulic conductivity (SATDK), saturated hydraulic conductivity coefficient of lateral flow (LKSATFAC), porosity (MAXSMC), and filed capacity (REFSMC). Parameters are relevant to overland flow, that is, the roughness coefficient (SFC_ROUGH), and Manning's roughness coefficient (MannN). Parameters required for lake routing include coefficient (WeirC), weir length (WeriL), orifice coefficient (OrificeC), orifice area (OrificeA). Groundwater parameters include the coefficient of baseflow (Coeff) and exponent (Expon) of the bucket model. However, adjusting these parameters had little effect in the runoff simulations, so all default values were used in coupled simulations except for Zmax, which was calibrated from 10 to 50.

2.3.2. Coupled Simulations

The WRF-ARW (WRF-S, hereafter) model is configured standalone on a two-way nested domain of 20.4 km cell resolution, covering east Asia and SRTR, respectively (Figure 2). The simulation has 33 vertical layers up to 50 hPa and is driven by the European Centre for Medium-Range Weather Forecasts Reanalysis (ERA) Interim, which has a resolution of 0.75 degrees. The land surface static physiographic input is generated by WRF Preprocessing Tools, in which the MODIS IGBP 21-category data is used to interpolate land use categories. Soil data is from a comprehensive 30 arc-second resolution grided soil characteristics data of China [67]. Grell-Devenyi [68] is selected as Cumulus parameterization options in domain1, and the convection-permitting module is used for domain2. The free-drainage approach [8,69] is selected as the lower boundary condition. A previous study shows that parameter calibration cannot resolve the deficiency of groundwater table-based parameterizations in simulating runoff [62]. Other selected parameterization schemes in this study are shown in Table S1 (in Supplementary Materials). The hindcast simulation period is from 1 June 2018 to 30 November 2018, with the first 45 days for warm-up and the rest for comparison. The reason for setting such a long warm-up time is that 1.5 spin-up days are needed for precipitation, but a much more extended period is necessary for discharge due to the influence of soil moisture [70].

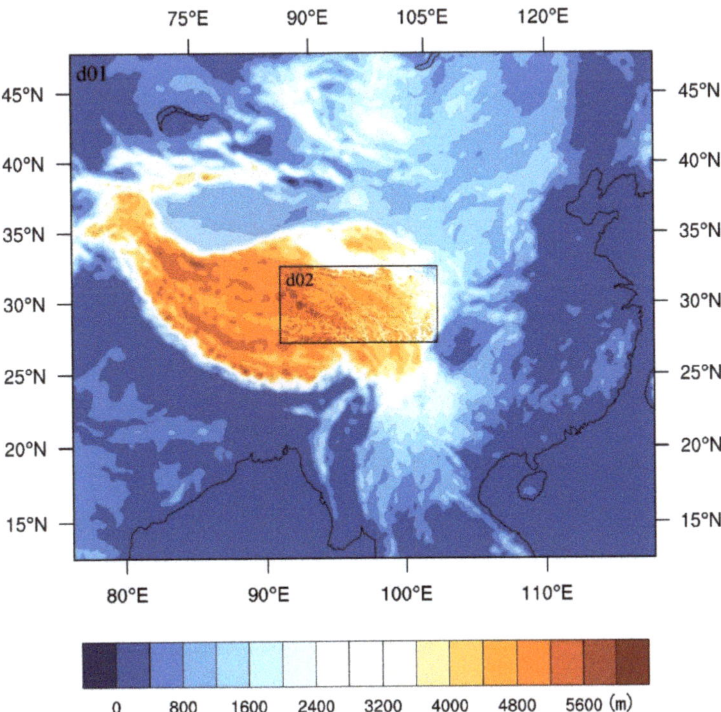

Figure 2. The terrain elevation map of the WRF nested domains. WRF-Hydro has the same settings as WRF, except the inner domain (d02) is coupled with a routing subgrid in WRF-Hydro.

To investigate the influence of the WRF-Hydro extension, the coupled WRF-Hydro model (WRF-H, hereafter) is set using the same parameters, driven by the same forcing data in the same hindcast period with the WRF-Standalone model (WRF-S, hereafter), except the inner domain of WRF-H is coupled with a sub-grid at a 400-m resolution to compute overland and river flow. The refined routing grids in the WRF-H extension are created by WRF-Hydro GIS Pre-Processing Tools, Version5.1, with the aggregation set to 10. The input elevation data is from the Shuttle Radar Topography Mission (SRTM), which has a spatial resolution of 90 m [71]. The simulations start from 1 June using the same initial and boundary conditions, and the data before 15 July are used for spin up.

3. Validation Data and Analysis Metrics

3.1. Validation Data

For model validation, in situ data were collected from three hydrological discharge stations, two eddy covariance stations, and fourteen meteorological stations (Figure 1). The primary information of the observation data is shown in Table 1, and the geographical coordinates of the stations are shown in Table 2. The runoff data was observed at 12:00 a.m. each day. Data periods are from July to November 2018. The Northwest Institute of Eco-Environment and Resources, Chinese Academy of Science in SRTR, set up two eddy-covariance systems to record the sensible and latent heat fluxes every 30 min. Meteorological station data were supplied by the China Meteorological Administration (CMA), and the observations were made at 0:00, 6:00, 12:00, and 18:00 UTC, and data were processed as daily averages.

Table 1. Information of validation data, including China Meteorological Administration observation (CMA), Hydrologic data, Eddy covariance data, Global the Precipitation Mission (GPM) level_3 IMERG final product, the Soil Moisture Active Passive (SMAP) mission level-4 product, Global Land Evaporation Amsterdam Model (GLEAM), and China Meteorological Forcing Dataset (CMFD).

Data Set	Type	Variable	Spatial Resolution	Time Interval
CAM (China Meteorological Administration observation data)	in situ	Precipitation, surface skin temperature and pressure, 2 m air temperature and humidity, wind speed	point	six-hourly
Hydrologic data	in situ	Discharge	point	daily
Eddy covariance	in situ	Heat flux	point	30 min
GPM level_3 IMERG final	remote sensing	Precipitation	0.1°	30 min
SMAP level-4	remote sensing	Soil moisture	9 km	three-hourly
GLEAM	remote sensing	Evapotranspiration	0.25°	daily
CMFD	Fusion of reanalysis and in-situ data	Air temperature at 2 m	0.1°	three-hourly

In addition, the grided dataset in Table 1 is used for validation. The GPM level_3 IMERG final product [72] is the rainfall estimates combining data from all passive-microwave instruments in the GPM constellation [73]. The Soil Moisture Active Passive (SMAP) Mission is a satellite-based soil moisture product [74]. SMAP mission level-4 product is a modeled product. Based on brightness temperature observations (SMAP L1C_TB), it assimilates SMAP L1C_TB into the NASA catchment land surface model using a spatially distributed ensemble Kalman filter [75,76], producing estimations of surface (0–5 cm) and root zone (0–100 cm) soil moisture estimation. GLEAM (Global Land Evaporation Amsterdam Model) is a set of algorithms [77] providing evapotranspiration and estimating its components separately. The dataset of version 3.3b [78] is mainly based on satellite data. The China Meteorological Forcing Dataset (CMFD) [79] is a gridded near-surface meteorological dataset, which is made by fusion ground-based observations with several gridded datasets from remote sensing and reanalysis. The CMFD dataset was only used for verification in this study.

Table 2. The coordinates of hydrographic stations (red pentagons in Figure 1), turbulent heat flux stations (pink pentagrams in Figure 1), and China Meteorological Administration ground observation (CMA) stations (black triangles in Figure 1).

Station ID	Stations	Latitude (° N)	Longitude (° E)	Elevation (m)
Runoff gauge1	Changdu	31.13	97.18	3306
Runoff gauge2	Tangnag	35.5	100.15	4989
Runoff gauge3	Zhimenda	33.01	97.25	3531
Eddy covariance1	Elinghu	34.91	97.55	4322
Eddy covariance2	Huahu	33.92	102.82	3432
CMA 1	Wudaoliang	35.22	93.08	4612
CMA 2	Tuotuohe	34.22	92.43	4533
CMA 3	Qumalai	34.13	95.78	4175
CMA 4	Zaduo	32.9	95.3	4066
CMA 5	Nangqian	32.2	96.48	3644
CMA 6	Chagndu	31.15	97.17	3306
CMA 7	Maduo	34.92	98.22	4272
CMA 8	Dari	33.75	99.65	3968
CMA 9	Guoluo	34.47	100.25	3719
CMA 10	Henan	34.73	101.6	3530
CMA 11	Maqu	34	102.08	3471
CMA 12	Jiuzhi	33.43	101.48	3629
CMA 13	Hongyuan	32.8	102.55	3492
CMA 14	Ruoergai	33.58	102.97	3440

GPM is reliable for rainfall and cold season solid precipitation estimates [80], making it suitable for evaluating precipitation in SRTR [81–84]. Previous assessments evaluating the SMAP against two soil moisture observation networks on TP showed that the SMAP retrievals could well capture the amplitude and temporal variation of the soil moisture [85]. When atmospheric water balances, evapotranspiration was taken as the evapotranspiration (ET) baseline, GLEAM matches low ET estimates, and errors are amplified when ET is higher [86]. Previous studies showed that the surface air temperature downscaled from CMFD is more accurate than ERA-interim data over the TP [87]. This dataset is used for spatial validation after compensating for the deficiencies of the sparse observation of CMA weather stations in western China and systematic bias of reanalysis/remote sensing datasets [48].

3.2. Skill Metrics

The Pearson correlation coefficient (CC, range −1 to +1) and root mean square error (RMSE, range 0 to ∞) evaluate the model simulation effectiveness against the in situ observations or remote sensing data. In addition to RMSE, the Nash-Sutcliff Efficiency (NSE, range −∞ to 1) is used in runoff assessment. The formula of the criterion at a given site or grid point is:

$$CC = \frac{cov(V_{obs}, V_{mod})}{\sigma_{obs}\sigma_{mod}} \tag{1}$$

$$RMSE = \sqrt{\frac{1}{n}\sum_{i=1}^{n}\left(V_{mod}^{i} - V_{obs}^{i}\right)^2} \tag{2}$$

$$NSE = 1 - \frac{\sum_{i=1}^{n}\left(V_{mod}^{i} - V_{obs}^{i}\right)^2}{\sum_{i=1}^{n}\left(V_{mod}^{i} - \mu_{obs}\right)^2} \tag{3}$$

where V_{obs} and V_{mod} denote the observation and simulation values. $cov(V_{obs}, V_{mod})$ is the covariance of V_{obs} and V_{mod}. σ_{obs} and σ_{mod} are the standard deviation of the observation and simulation. μ_{obs} is the average value of the observation. V_{obs}^{i} and V_{mod}^{i} are the observation and simulation of the *i*th time step. In addition, Taylor diagrams providing summary statistics of CC, RMSE and standard deviation (SD) were used for model performance analysis [88].

The *t*-statistic tests the statistical significance of the linear correlation coefficient:

$$t = \frac{r\sqrt{n-2}}{\sqrt{1-r^2}} \tag{4}$$

where r is the linear correlation coefficient, n is the sample size. The significance level in this study is set at 95%.

Uncentered anomaly correlation coefficient (a number range 0 to 1) [89,90] is used in the verification of spatial fields.

$$ACC = \frac{\sum_{m=1}^{M} y'_m o'_m}{\sqrt{\sum_{m=1}^{M} \left(y'_m\right)^2 \sum_{m=1}^{M} \left(o'_m\right)^2}} \tag{5}$$

where y' and o' are the departure of simulation and verifying data from a reference state at each grid point (m).

The final metric is a quantitative method to study the precipitation frequency-intensity structure proposed by [91]. In this method, two quantitative parameters are obtained by fitting a double exponential function with the statistical results of the frequency of different rainfall intensities. The equation is:

$$\text{Fr}(I) + 1 = \exp\left[\exp\left(\alpha - \frac{1}{\beta}I\right)\right] \tag{6}$$

where $\text{Fr}(I)$ is the frequency when the precipitation intensity is in I (mm) categories. α and β are the parameters we want. For the left and right sides of Equation (6), take two natural logarithms at the same time to get:

$$\ln(\ln(\text{Fr}(I) + 1)) = \alpha - \frac{1}{\beta}I \tag{7}$$

That is to say, the double logarithm of the occurrence frequency of a certain precipitation intensity is a linear function of precipitation intensity. Previous studies have shown that the precipitation distribution of meteorological station observations, satellite remote sensing, and numerical simulation obey this law well [92,93]. In this function, α and β are related to weak and intense precipitation occurrence frequency, respectively. Numerical simulations are associated with an increased frequency of light precipitation but decreased frequency in heavy precipitation, resulting in a larger α and smaller β in simulation than observation. [92].

4. Results

4.1. Streamflow Simulation

The performance of the runoff simulation was evaluated using the parameters obtained from the calibration procedure in Section 3.2, i.e., Zmax = 50. The NSE of the source region of the Yangtze, Lancang and Yellow rivers are 0.12, 0.19 and 0.36, respectively, based on daily observations from hydrological stations (Figure 3). The effect of setting Zmax to 50 is to take the overestimation of precipitation in the WRF-S and store it in the conceptual groundwater bucket to keep the simulated streamflow close to the observations. Such results were consistent with the study in the eastern Alps [45], where after parameter-calibrated that the bias in precipitation does not deteriorate the reproduction of runoff discharges.

As the three river sources are interconnected, considering them as a whole reduces errors in runoff simulation caused by the errors in the precipitation fall zone. This precipitation fall zone error is generated by the GCM and introduced into the flow production simulation system by the precipitation as driving information. The simulated and observed total flows of the three rivers are obtained by adding up the flows of Zhimenda, Changdu and Tangnag. Then the Nash-Sutcliff Efficiency (NSE) and root mean square error (RMSE) between simulation and observation of runoff in SRTR is 0.55 and 324.2 m s^{-1}, respectively (Figure 3d). This considerable improvement in NSE suggests that the overestimation of precipitation by the GCM and the error in the precipitation fall zone are sources of error in flow production.

Nevertheless, The WRF-Hydro produces high temporal resolution estimates of yield flow with acceptable results. And it illustrates that the bias from meteorology is essentially addressed through calibration. The accuracy of model simulations on the daily scale will be further improved as the driving information error decreases due to more observations.

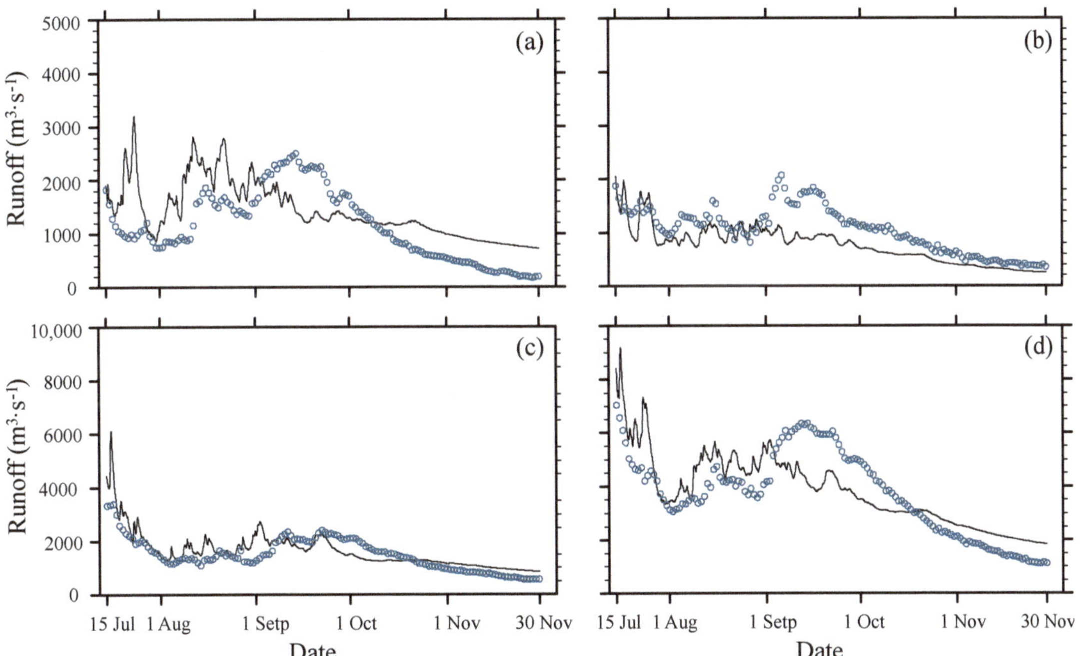

Figure 3. Time series of observed and simulated (uncoupled WRF-Hydro) discharge of (**a**) Zhimenda station; (**b**) Changdu station; (**c**) Tangnag station; (**d**) amount of three stations.

4.2. Comparison with In-Situ Observations

4.2.1. Comparison with CMA Data

As the first step toward evaluating the coupled WRF-Hydro modeling system's performance, we present comparisons between observed and simulated variables at fourteen individual CMA stations in Figure 1. The compared variables are listed in Table 1. Simulations in the WRF-S/WRF-H model grid nearest to the CMA stations are converted from hourly data to daily for comparison with observations.

Taylor diagrams for different variables are shown in Figure 4. As thermodynamic variables, both the air temperature and surface pressure are simulated with a slight standard deviation. These two variables and surface skin temperature strongly correlate with observations due to daily and seasonal variation. The simulated wind speed has shown the worst correlation among all variables. The positions of the stations on the Taylor map are essentially the same on the humidity map as they are on the precipitation map. The precipitation RMSE between WRF-H/WRF-S and observations is mainly derived from the uncertainties in the model, forcing the heavily overestimated precipitation in ERA-Interim over the TP [94]. In general, the simulation performance varies more between observations than between models.

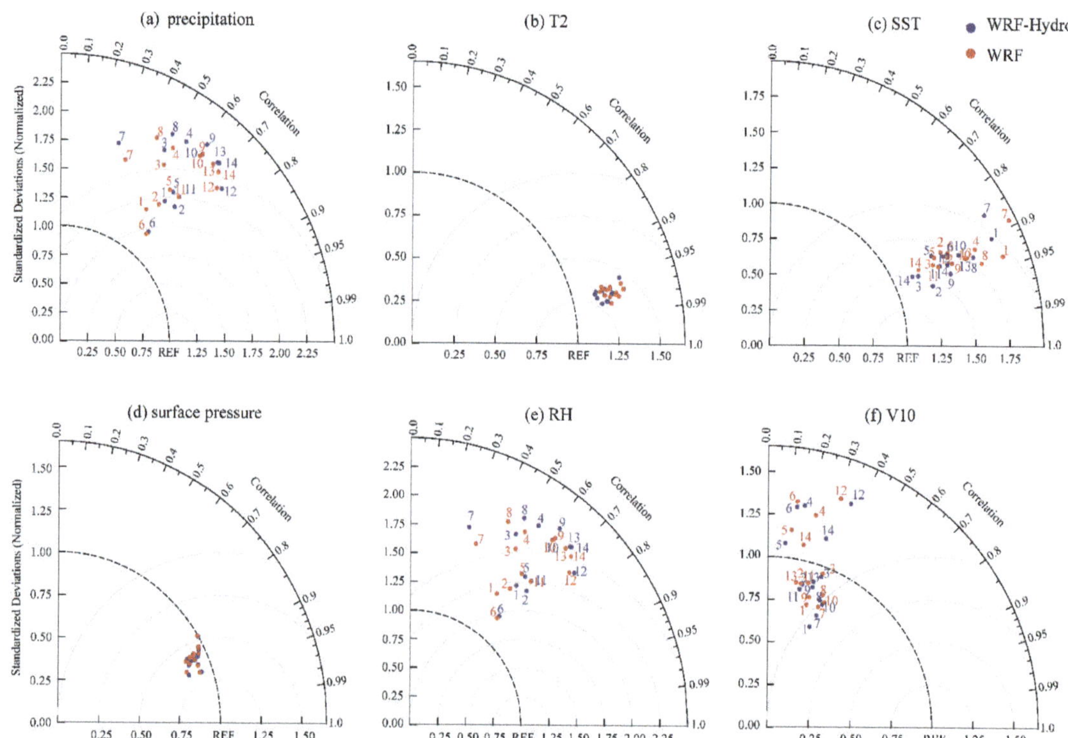

Figure 4. Taylor diagram for simulations compared with 14 CMA observational daily data for; (**a**) precipitation; (**b**) air temperature at 2 m (T2); (**c**) surface skin temperature (SST); (**d**) surface pressure; (**e**) relative humidity at 2 m (RH); (**f**) wind speed at 10 m (V10). The number near the dot in the figure is the ID of the stations listed in Table 2.

Table 3 shows the average value of each variable for 14 stations. By comparison, all six variables from WRF-H have a smaller RMSE than WRF-S, and five variables have a better correlation with observations in WRF-H. WRF-H made an improvement of 13.48% in RMSE and 8.31% in CC than WRF-S.

Table 3. Averaged Root-Mean-square Error (RMSE) and Correlation Coefficient (CC) between the measurements collected from the 14 China Meteorological Administration (CMA) stations and the simulated near-surface and land surface meteorological variables for the period from 15 July 2009 to 30 November 2018. The last column is the number of sites that passed the significance test across the 14 observatories.

Variables	RMSE		CC		Number of Sites	
	WRF-H	**WRF-S**	**WRF-H**	**WRF-S**	**WRF-H**	**WRF-S**
Precipitation (mm)	5.9	6.1	0.33	0.23	13	11
Air temperature at 2 m (k)	3.1	3.2	0.97	0.97	14	14
Surface skin temperature (k)	6.8	7.0	0.91	0.91	14	14
Surface pressure (hPa)	15.7	15.8	0.91	0.91	14	14
Relative humidity (%)	15.7	15.7	0.6	0.59	14	14
Wind speed (ms^{-1})	1.0	1.1	0.29	0.28	12	12

4.2.2. Comparison with Turbulent Heat Flux Observation

We also evaluated the WRF-H/WRF-S simulations against turbulent flux measurements shown in Figure 1. Table 4 shows the mean RMS error and CC in all comparison periods. All correlation coefficients in the table pass the significance test with a confidence level of 95%. The result at Elinghu shows that WRF-H simulates a larger CC than the WRF-S in both latent and sensible heat flux. For RMSE, the WRF-H and WRF-S are superior to each other in the simulation of sensible heat flux and latent heat flux. The simulation of WRF-H increases RMSE by 15.91 Wm^{-2} for the latent heat flux and reduces it by 23.77 Wm^{-2} for sensible heat flux. In contrast to the Elinghu station, the WRF-H simulates a smaller RMSE than WRF-S in both latent and sensible heat flux but a larger CC for latent heat simulations and smaller CC for sensible heat simulations.

Table 4. Root-Mean-Square Error (RMSE) and correlation coefficient (CC) computed between WRF-Hydro/WRF and observation from 15 July to 30 November 2018, where the correlation coefficient passed the significance test.

Stations	Variables	RMSE		CC	
		WRF-H	WRF-S	WRF-H	WRF-S
Elinghu	LH	55.7	39.8	0.75	0.74
	HFX	45.9	69.7	0.76	0.74
Huahu	LH	77.4	80.4	0.73	0.72
	HFX	67.9	74.4	0.78	0.79

4.3. Pattern Difference and Time Variation

4.3.1. Spatial Validation

Figure 5 shows the CC and RMSE distribution of soil moisture, evaporation, and air temperature at 2 m against the remote sensing or reanalysis data from 15 July to 30 November. The simulations and validation data are processed as daily averages to avoid the daily cycle dominating the evaluation measure. And bilinear interpolation in space was performed for the simulations to match with the validation data.

When comparing the soil moisture, WRF-H exhibits a lower correlation and a larger RMSE than WRF-S in most areas, especially in the west part of SRTR (Figure 5a–d). Due to the wetter soil moisture simulation, more evapotranspiration was simulated by WRF-H. Result in a slightly low CC and high RMSE simulated by WRF-H than WRF-S (Figure 5e–h). But the WRF-H exhibits advantages in simulation temperature at 2 m, with lower RMSE at the same CC. From the spatial distribution of soil moisture and evaporation, the simulations are similar in the eastern part of the Three Rivers. In contrast, the simulation of WRF-H is worse than WRF-S in the central and western regions.

After spatial validation, WRF-Hydro improved the simulation of temperature and deteriorated the humidity and evapotranspiration. The WRF-Hydro does not show the advantages in space scale as it does at the point scale. It is probably because the CMA stations were built in the more inhabitable areas of the plateau, such as in the valleys. These places are usually crossed by rivers and have a high soil moisture content. And these wetter areas cannot be resolved by GLEAM data at 0.25° resolution but can be identified by the 4 km WRF-Hydro model.

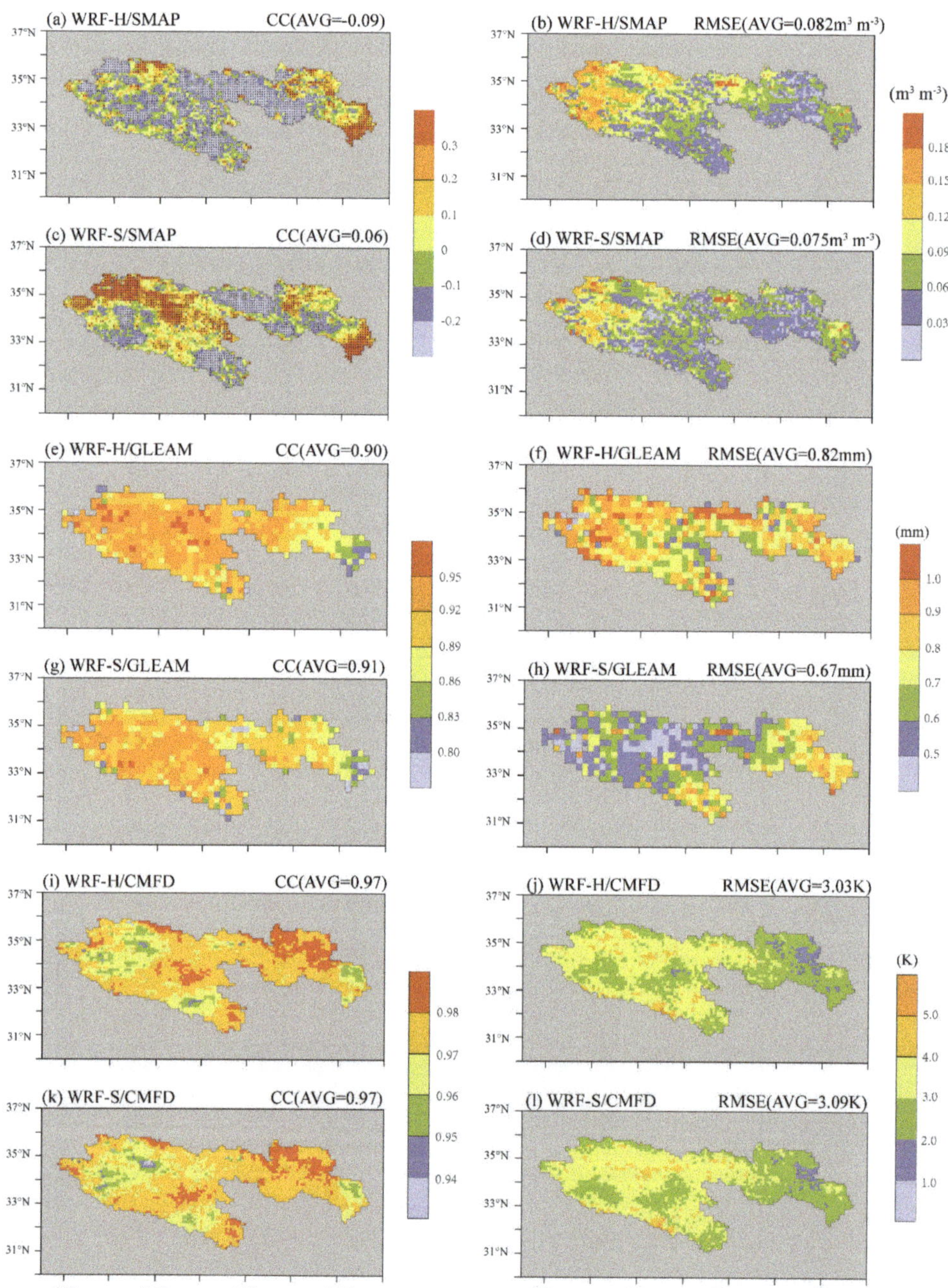

Figure 5. Evaluation of soil moisture, evapotranspiration, and air temperature at 2 m over the SRTR. (**a**) Spatial distribution of correlation coefficient (CC) between satellite retrievals or reanalysis data and soil moisture simulations from WRF-H during the study period; (**b**) Same as (**a**), but for spatial distribution of RMSE; (**c,d**) Same as (**a,b**), but simulations are from WRF-S; (**e–h**) Same as (**a–d**), but for evapotranspiration; (**i–l**) Same as (**a–d**), but for air temperature at 2 m. The dots in (**a**) and (**c**) denote the correlation coefficient and are significantly above the 95% confidence level, and the correlation coefficient in each grid point of (**e,g,i,k**) passed the significance test.

The anomaly correlation is convenient to detect similarities in the patterns of departures. And it avoids the influence of bias in the grid-based validations as they are essentially a modeled product. Figure 6a illustrates that WRF-H exhibits a higher anomaly correlation coefficient (0.955 versus 0.941). They reflect the spatial distribution of soil moisture improvements due to lateral soil water flow simulated by WRF-Hydro. Models that do not incorporate this process may lack the ability to reproduce anomaly patterns of soil moisture. Figure 6b,c show that the anomaly correlation coefficient scores achieved by WRF-H and WRF-S were close, and scores will be lower in winter. Soil water lateral flow has little effect on evapotranspiration and 2 m air temperature patterns.

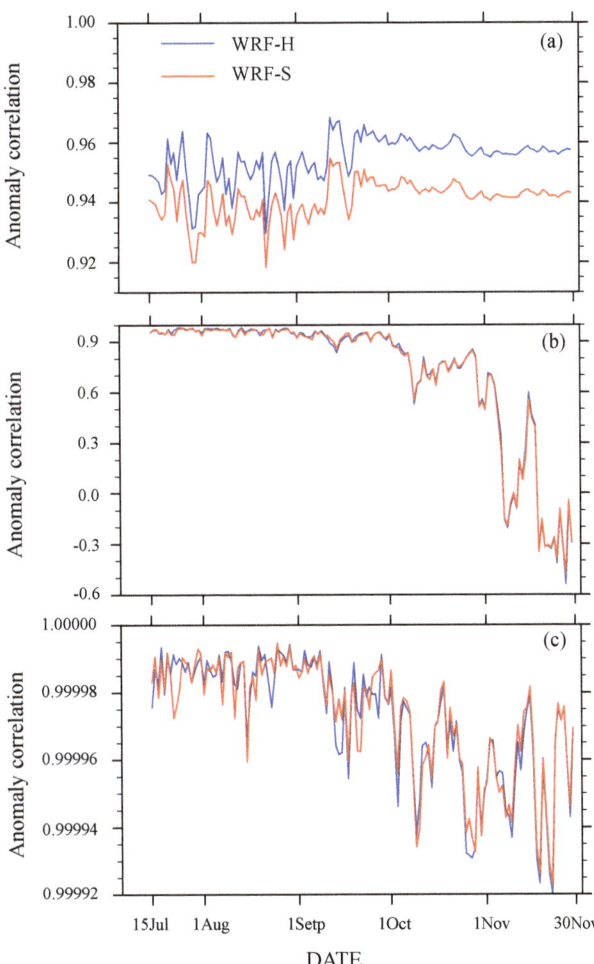

Figure 6. Anomaly correlation coefficients of (**a**) soil moisture against SMAP, (**b**) evapotranspiration against GLEAM, (**c**) air temperature at 2 m against CMFD.

4.3.2. Spatial Distribution

Figure 7 compares the total precipitation, average soil moisture, latent heat and sensible heat fluxes simulated by WRF-H and WRF-S. The spatial distribution of the two sets of the models is similar (Figure 7a,b). The difference (Figure 7c) is prominent in local areas but small when averaged across the region, consistent with other study [43]. The characteristics of the soil moisture, in contrast, are opposite to the spatial distribution of precipitation. WRF-H spatially overestimates soil moisture throughout the SRTR and, more significantly, in the northern part of the SRTR (Figure 7f). The result is that WRF-S simulates soil moisture as wet in the southeast and dry in the northwest (Figure 7e), while WRF-H will hamper such a spatial distribution feature (Figure 7d).

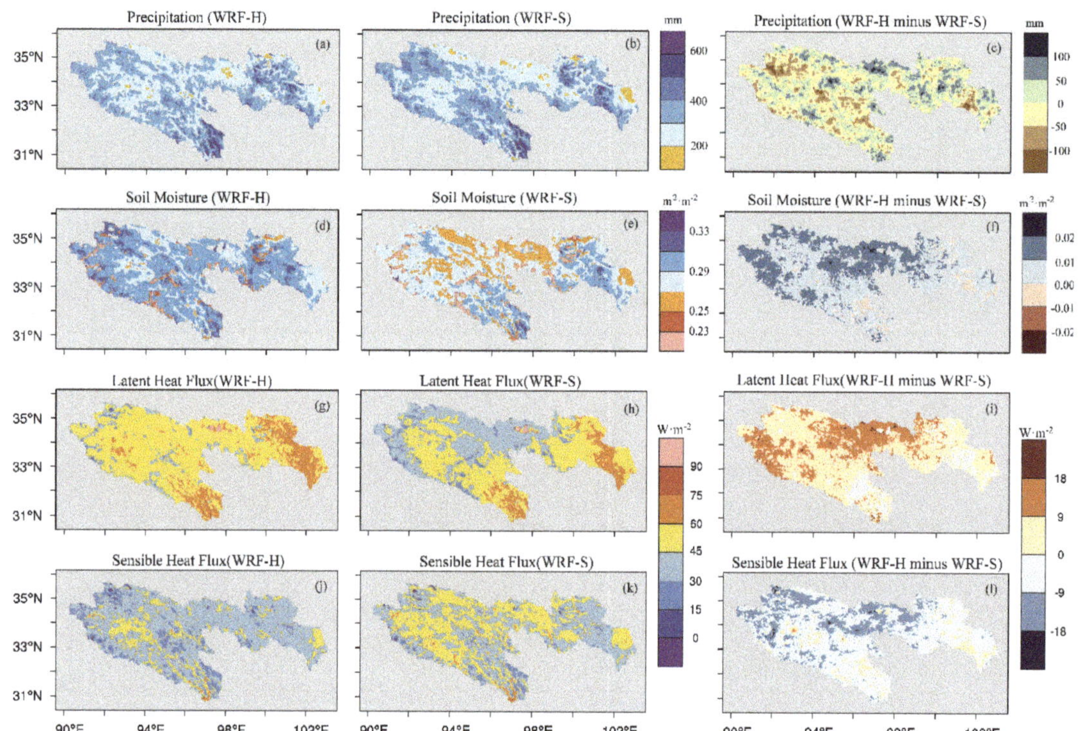

Figure 7. Spatial distribution of precipitation amounts from (**a**) WRF-H, (**b**) WRF-S, (**c**) WRF-h minus WRF-S over the study period. (**d–f**), (**g–i**) and (**j–l**) are the same as (**a–c**), but for averaged soil moisture (0~10 cm), averaged latent heat flux and sensible heat flux, respectively.

The spatial distribution of latent heat has the same characteristics as soil moisture (Figure 7g–i), while the sensible heat flux is the opposite (Figure 7j–l). Changes in the distribution of sensible and latent heat fluxes affect the boundary layer development and influence the precipitation structure. Overall, the pattern of spatial changes in soil moisture, sensible and latent heat flux is quite similar. The temporal variation will be analyzed next.

4.3.3. Time Variation

Figure 8 demonstrates simulated daily variable averages on the TRSR from 15 July to 30 November. Although the precipitations from both models match over the entire analysis period, the soil moisture simulated by WRF-H is more humid than WRF-S during the analysis period. Latent and sensible heat flux declines with seasonal changes, and even negative values emerge in sensible heat flux. The heat fluxes in the Tibetan Plateau are more susceptible to the freeze-thaw process than the high-latitude frozen soil regions [95]. The latent heat flux simulated by WRF-H is greater than WRF-S before the soil freezes, but the sensible heat flux is more petite than WRF-S. After the soil is frozen, WRF-H coincides with the curve of the WRF-S simulation. Due to the energy balance, the surface skin temperature simulation also showed differences before freezing.

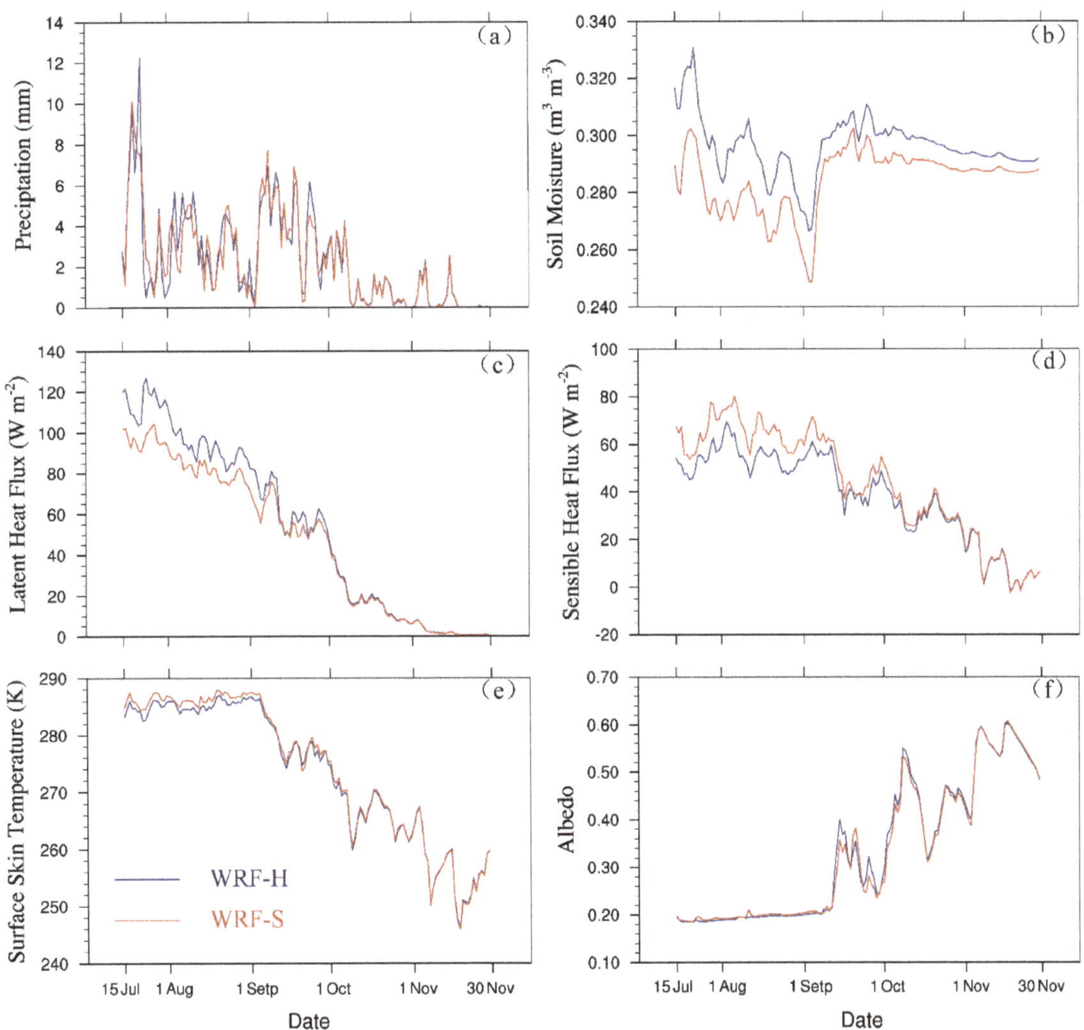

Figure 8. Temporal variations of the SRTR basin average in (**a**) precipitation, (**b**) soil moisture in the first soil layer (0~10 cm), (**c**) latent heat flux, (**d**) sensible heat flux, (**e**) surface skin temperature, (**f**) albedo.

The albedo map shows that the two models' albedo is relatively stable in summer and dramatic in autumn and winter. The steady increase of albedo in summer is caused by the decrease in soil moisture and phenology. The change in albedo due to inconsistent precipitation simulation time is small—the accumulation and sublimation of snowfall cause the fluctuated albedo. In September, due to differences in surface temperature simulations, the albedo of the WRF-H is higher than the WRF-S, then the albedo of the two models coincide as the temperature difference decreases.

WRF-H is more consistent with the WRF-S in winter because of the weakening of overland flow and subsurface lateral flow. The decrease of soil hydraulic conductivity of frozen soil reduced the subsurface flow. Furthermore, most winter precipitation falls to the ground in snowfall and dissipates through sublimation, rarely infiltrating into the soil layer. The snowmelt will contribute to discharges mainly during the rainy and peak flow periods [70].

4.4. Precipitation Structure and Boundary Layer Variables

4.4.1. Precipitation Frequency-Intensity Structure

Figure 9 shows the fitting results of GPM precipitation, WRF-H, and WRF-S. The circle points represent the natural double pairs with different rainfall intensities, and the straight lines represent the fitted curves. The simulations of WRF-H and WRF-S are similar, but WRF-H is slightly closer to the GPM. According to Equation (7), the fitted α is 2.39, 2.74 and 2.8, and the fitted β is 26.63, 9.9 and 9.12 for GPM, WRF-H and WRF-S, respectively. The difference between simulation and GPM, WRF-H is reduced by 15% compared to WRF-S in a and β is reduced by 5%. The fitting results show that soil water lateral flow in WRF-H makes the frequency-intensity structure closer to the observation. Since convective-permitting is used in both models, the boundary layer that triggers convection needs to be further evaluated.

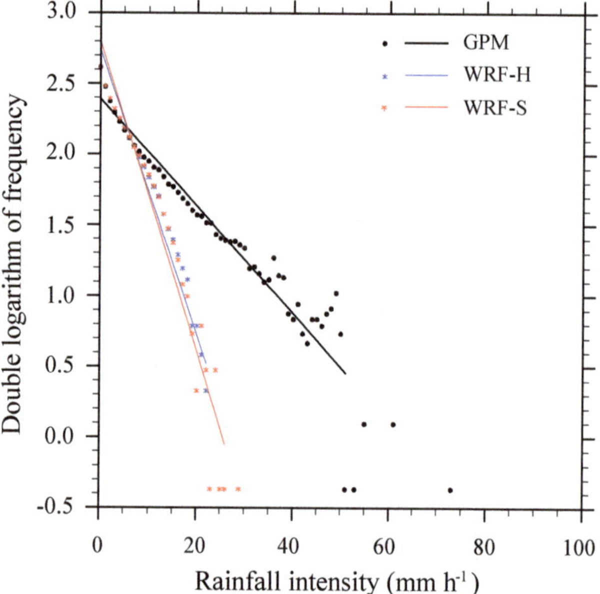

Figure 9. The precipitation frequency over the SRTR is binned by the hourly intensity and its double exponential fit line.

4.4.2. Boundary Layer Variables

In Figure 10a, WRF-H shows a higher convective available potential energy (CAPE)in most of the unfreezing period, and also that coupling hydrology processes make convectional energy more readily available in SRTR. The amplitude of convective inhibition (CIN) did not change in the unfreezing or freezing period between the two simulations. WRF-H exhibits a slightly lower lifting condensation level (LCL) in most of the unfrozen period, indicating a low cloud base. The level of free convection (LFC) fluctuates violently in the unfrozen season but becomes stable when it uplifts to the top of model layers in the frozen season. The improvement of precipitation structure is mainly attributed to the increase of CAPE caused by wetter soil.

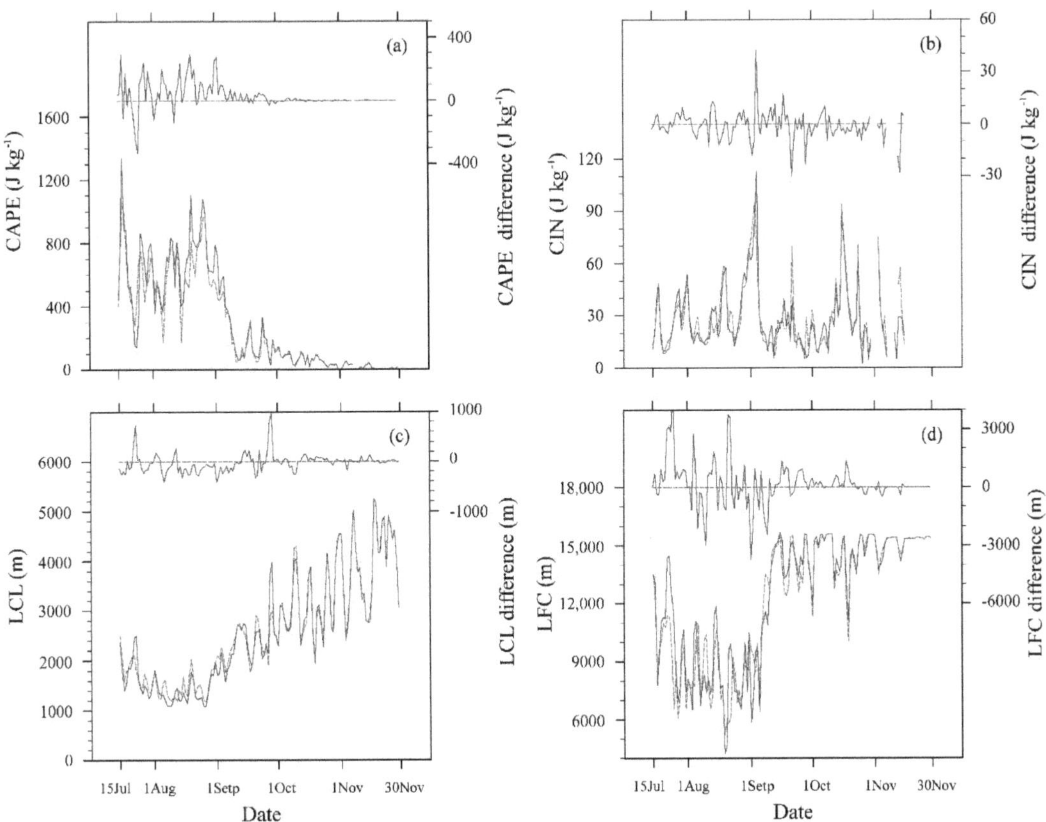

Figure 10. Temporal variations of the TRSR basin averaged values for the boundary layer elements (**a**) CAPE, (**b**) CIN, (**c**) LCL, (**d**) LFC. The curve on the lower half corresponds to the Y−axis on the left, with the red line representing WRF−H and the blue line representing WRF−S. The curve on the upper half corresponds to the y−axis on the right side, and the red line represents the bias of WRF−H relative to WRF−S.

5. Discussion

We evaluated the runoff simulation skills of uncoupled WRF-Hydro and the effect of lateral flow of soil water on the atmosphere by an enhanced fully coupled WRF-Hydro model in SRTR. This is one of the earliest studies using WRF-Hydro in the SRTR, which is located on the Tibetan Plateau and is an important watershed for ecological conservation and water management in China. The results reveal the following main conclusions:

1. The reproduction of the daily discharge amount of the three stations has a Nash-Sutcliffe model efficiency generally above 0.55, demonstrating the potential of WRF-Hydro for hydrological forecasting in SRTR.
2. In the coupled experiment, WRF-Hydro made an improvement of 13.48% in RMSE and 8.31% in CC above WRF-ARW when compared against CMA data, and an improvement of 6.6% in RMSE and 1% in CC when compared against turbulent heat fluxes observation. WRF-Hydro tends to overestimate the soil moisture of the western part of the SRTR, but less so in the eastern part and areas with wetter soil moisture. This difference occurs mainly during the period when the soil is not frozen. Although the root mean square error of WRF-Hydro is larger than that of WRF-ARW compared against SMAP and GLEAM, WRF-Hydro scores higher in the anomaly

correlation coefficient. These findings show the significance of lateral flow in soil moisture simulation.

3. The coupled WRF-Hydro results in an increase in latent heat flux, a decrease in sensible heat flux, and a decrease in soil surface temperature due to the moist soil. The change in turbulent heat flux gives the WRF-Hydro simulation an enormous CAPE and easier convection, reducing precipitation intensity-frequency errors.

The NSE of runoff in this study is not high, perhaps because the agility of WRF-Hydro might be unnecessarily constrained by its complex process [96]. This is probably the reason that NSE is low in some watersheds in similar studies [97–100]. The calibrated model parameters (Zmax = 50) in this study are to offset the impact of continuously excessive precipitation on runoff production. The extra precipitation stored in the conceptual groundwater bucket during the simulation period will deteriorate runoff simulation in the next period, for the discharge of baseflow will increase the runoff in winter. If the simulation time is increased and the accumulated precipitation overestimation is large enough, then the WRF-Hydro cannot obtain an accurate runoff through calibration and the NSE will decrease as a result. It suggests that the output of the GCM is not recommended as a driver for multi-year runoff hindcasting in SRTR. Accurate precipitation-driven data, such as CMFD, is necessary for such simulation. However, using Zmax = 50 does not worsen the coupled simulation, as the parameter does not affect soil moisture and cannot further influence weather processes.

The lateral flow from WRF-Hydro leads to wetter soils in the SRTR, similar to the semi-arid environment in the USA [101]. For areas where WRF-Hydro tends to overestimate the soil moisture, it is recommended to calibrate the parameters affecting soil moisture simulation, such as the reference infiltration factor (REFKDT). This study did not calibrate soil moisture because the calibrated parameters could artificially lead to better simulation results for one variable than the other.

The limitations of the study are mainly threefold. First, the overestimation of GCM precipitation in SRTR [102] resulted in its inability to be used as a force to simulate multi-year runoff. Secondly, although the soil texture dataset [67] has the highest resolution in China right now, the small number of sampling points on SRTR leads to a larger uncertainty in soil texture than in other regions in China. Thirdly, the CMA station was established in a habitable place on the plateau. These areas tend to be low-lying, with moist soils and rivers passing through them. This means that the stations are located where the lateral flow of soil water flows in, not out. More observations are needed where lateral soil water flows out.

The low but acceptable NSE of runoff simulation in the study shows the potential of WRF-hydro in the hydrological simulation of the SRTR. Meanwhile, the high anomaly correlation coefficient scores achieved by WRF-Hydro in soil moisture simulations, as well as the closer to observed precipitation intensity-frequency structure, suggest that the WRF-Hydro module is worthy of being incorporated in convective-scale simulations. Reducing the precipitation overestimation of the GCM in SRTR is urgently needed for the future application of the coupled WRF-Hydro model in SRST.

6. Conclusions

In the present study, we set up two sets of experiments to investigate the prospects of WRF-Hydro in the hydrological forecasting and its impacts on land-atmosphere interaction of SRTR. Results reveal the WRF-Hydro has shown potential in runoff prediction in SRTR. Coupled WRF-Hydro with soil water lateral flow increases wet soil bias in the western part of the SRTR, but improves the soil moisture anomaly pattern. The coupled model also enhances CAPE, and produces a more reliable precipitation intensity-frequency structure. Overall, our results illustrate the effect of WRF-Hydro in the coupled hydrology-atmosphere simulation system. GCM with less overestimation of precipitation in SRTR to drive coupled WRF-Hydro is desirable for future work.

Supplementary Materials: The following are available online at https://www.mdpi.com/article/10.3 390/w13233409/s1, Table S1. Primary WRF, Noah-MP, and WRF-Hydro parameters were used in the simulation [103–110].

Author Contributions: Conceptualization, X.M.; Data curation, G.L., H.C., Z.L., Y.M. and L.Z.; Formal analysis, G.L.; Methodology, G.L., X.M. and E.B.; Project administration, X.M.; Resources, X.M.; Software, G.L.; Validation, X.M., L.S., Z.L. and L.Z.; Visualization, G.L.; Writing—original draft, G.L.; Writing—review & editing, X.M. All authors have read and agreed to the published version of the manuscript.

Funding: This research was funded by the Chinese National Science Foundation Programs: 41930759, the auspices of the Strategic Priority Research Program of the Chinese Academy of Sciences (XDA2006010202), Chinese National Science Foundation Programs: 41822501,the Science and Technology Research Plan of Gansu Province (20JR10RA070), the Chinese Academy of Youth Innovation and Promotion, CAS (Y201874), iLEAPs (integrated Land Ecosystem-Atmosphere Processes Study-iLEAPS), and the National Key Research and Development Program of China (2016YFB0501303).

Institutional Review Board Statement: Not applicable.

Data Availability Statement: The numerical model simulations upon which this study is based are too extensive to archive or transfer. Instead, we provided all the information needed to replicate the simulations; we used model version WRF-Hydro v5.0.3. The model code, compilation script, initial and boundary condition files, WRF-Hydro routing grid files, and the settings (namelist) are available from the authors.

Acknowledgments: We acknowledge the modeling support from the WRF-Hydro Community. We thank the team of China Meteorological Administration, the team of Global Precipitation Measurement (GPM) Mission, the team of Global Land Evaporation Amsterdam Model (GLEAM), the team of The Soil Moisture Active Passive Mission (SMAP), and the National Tibetan Plateau Data Center for providing data for model validation. As well as thanks to the European Centre for Medium-Range Weather Forecasts for providing reanalysis data.

Conflicts of Interest: The authors declare no conflict of interest.

References

1. Fersch, B.; Senatore, A.; Adler, B.; Arnault, J.; Mauder, M.; Schneider, K.; Voelksch, I.; Kunstmann, H. High-resolution fully coupled atmospheric-hydrological modeling: A cross-compartment regional water and energy cycle evaluation. *Hydrol. Earth Syst. Sci.* **2020**, *24*, 2457–2481. [CrossRef]
2. Oleson, K.W.; Niu, G.Y.; Yang, Z.L.; Lawrence, D.M.; Thornton, P.E.; Lawrence, P.J.; Stoeckli, R.; Dickinson, R.E.; Bonan, G.B.; Levis, S.; et al. Improvements to the Community Land Model and their impact on the hydrological cycle. *J. Geophys. Res. Biogeosci.* **2008**, *113*, G01021. [CrossRef]
3. Arnault, J.; Wagner, S.; Rummler, T.; Fersch, B.; Bliefernicht, J.; Andresen, S.; Kunstmann, H. Role of runoff-infiltration partitioning and resolved overland flow on land-atmosphere feedbacks: A case study with the WRF-hydro coupled modeling system for west Africa. *J. Hydrometeorol.* **2016**, *17*, 1489–1516. [CrossRef]
4. Seneviratne, S.I.; Corti, T.; Davin, E.L.; Hirschi, M.; Jaeger, E.B.; Lehner, I.; Orlowsky, B.; Teuling, A.J. Investigating soil moisture-climate interactions in a changing climate: A review. *Earth Sci. Rev.* **2010**, *99*, 125–161. [CrossRef]
5. Naabil, E.; Lamptey, B.L.; Arnault, J.; Olufayo, A.; Kunstmann, H. Water resources management using the WRF-Hydro modelling system: Case-study of the Tono dam in West Africa. *J. Hydrol. Reg. Stud.* **2017**, *12*, 196–209. [CrossRef]
6. Shen, M.; Piao, S.; Jeong, S.-J.; Zhou, L.; Zeng, Z.; Ciais, P.; Chen, D.; Huang, M.; Jin, C.-S.; Li, L.Z.X.; et al. Evaporative cooling over the Tibetan Plateau induced by vegetation growth. *Proc. Natl. Acad. Sci. USA* **2015**, *112*, 9299–9304. [CrossRef] [PubMed]
7. Avolio, E.; Cavalcanti, O.; Furnari, L.; Senatore, A.; Mendicino, G. Brief communication: Preliminary hydro-meteorological analysis of the flash flood of 20 August 2018 in Raganello Gorge, southern Italy. *Nat. Hazards Earth Syst. Sci.* **2019**, *19*, 1619–1627. [CrossRef]
8. Schaake, J.C.; Koren, V.I.; Duan, Q.Y.; Mitchell, K.; Chen, F. Simple water balance model for estimating runoff at different spatial and temporal scales. *J. Geophys. Res. Atmos.* **1996**, *101*, 7461–7475. [CrossRef]
9. Niu, G.-Y.; Yang, Z.-L. Effects of frozen soil on snowmelt runoff and soil water storage at a continental scale. *J. Hydrometeorol.* **2006**, *7*, 937–952. [CrossRef]
10. Swenson, S.C.; Lawrence, D.M.; Lee, H. Improved simulation of the terrestrial hydrological cycle in permafrost regions by the Community Land Model. *J. Adv. Model. Earth Syst.* **2012**, *4*, M08002. [CrossRef]
11. Clark, M.P.; Fan, Y.; Lawrence, D.M.; Adam, J.C.; Bolster, D.; Gochis, D.J.; Hooper, R.P.; Kumar, M.; Leung, L.R.; Mackay, D.S.; et al. Improving the representation of hydrologic processes in Earth System Models. *Water Resour. Res.* **2015**, *51*, 5929–5956. [CrossRef]

12. Wood, E.F.; Roundy, J.K.; Troy, T.J.; Van Beek, L.P.H.; Bierkens, M.F.P.; Blyth, E.; De Roo, A.; Doll, P.; Ek, M.; Famiglietti, J. Hyperresolution global land surface modeling: Meeting a grand challenge for monitoring Earth's terrestrial water. *Water Resour. Res.* **2011**, *47*, W05301. [CrossRef]

13. Ji, P.; Yuan, X.; Liang, X.-Z. Do lateral flows matter for the hyperresolution land surface modeling? *J. Geophys. Res. Atmos.* **2017**, *122*, 12077–12092. [CrossRef]

14. Wagner, S.; Fersch, B.; Yuan, F.; Yu, Z.; Kunstmann, H. Fully coupled atmospheric-hydrological modeling at regional and long-term scales: Development, application, and analysis of WRF-HMS. *Water Resour. Res.* **2016**, *52*, 3187–3211. [CrossRef]

15. Arnault, J.; Wei, J.; Rummler, T.; Fersch, B.; Zhang, Z.; Jung, G.; Wagner, S.; Kunstmann, H. A Joint soil-vegetation-atmospheric water tagging procedure with WRF-hydro: Implementation and application to the case of precipitation partitioning in the upper danube river basin. *Water Resour. Res.* **2019**, *55*, 6217–6243. [CrossRef]

16. Yuan, X.; Ji, P.; Wang, L.; Liang, X.-Z.; Yang, K.; Ye, A.; Su, Z.; Wen, J. High-resolution land surface modeling of hydrological changes over the sanjiangyuan region in the eastern Tibetan Plateau: 1. Model development and evaluation. *J. Adv. Model. Earth Syst.* **2018**, *10*, 2806–2828. [CrossRef]

17. Yang, K.; Wu, H.; Qin, J.; Lin, C.; Tang, W.; Chen, Y. Recent climate changes over the Tibetan Plateau and their impacts on energy and water cycle: A review. *Glob. Planet. Chang.* **2014**, *112*, 79–91. [CrossRef]

18. Wu, G.X.; Zhang, Y.S. Tibetan Plateau forcing and the timing of the monsoon onset over South Asia and the South China Sea. *Mon. Weather Rev.* **1998**, *126*, 913–927. [CrossRef]

19. Wang, Y.; Yang, K.; Zhou, X.; Chen, D.; Lu, H.; Ouyang, L.; Chen, Y.; Lazhu; Wang, B. Synergy of orographic drag parameterization and high resolution greatly reduces biases of WRF-simulated precipitation in central Himalaya. *Clim. Dyn.* **2020**, *54*, 1729–1740. [CrossRef]

20. Xu, X.; Lu, C.; Shi, X.; Gao, S. World water tower: An atmospheric perspective. *Geophys. Res. Lett.* **2008**, *35*, L20815. [CrossRef]

21. Liu, X.D.; Yin, Z.Y. Spatial and temporal variation of summer precipitation over the eastern Tibetan Plateau and the North Atlantic oscillation. *J. Clim.* **2001**, *14*, 2896–2909. [CrossRef]

22. Tao, S.Y.; Ding, Y.H. Observational evidence of the influence of the Qinghai-Xizang (Tibet) Plateau on the occurrence of heavy rain and severe convective storms in China. *Bull. Am. Meteorol. Soc.* **1981**, *62*, 23–30. [CrossRef]

23. Xu, X.; Zhou, M.; Chen, J.; Bian, L.; Zhang, G.; Liu, H.; Li, S.; Zhang, H.; Zhao, Y.; Suolongduoji; et al. A comprehensive physical pattern of land-air dynamic and thermal structure on the Qinghai-Xizang Plateau. *Sci. China Ser. D* **2002**, *45*, 577–594. [CrossRef]

24. Wang, J.Z.; Yang, Y.Q.; Xu, X.D.; Zhang, G.Z. A monitoring study of the 1998 rainstorm along the Yangtze River of China by using TIPEX data. *Adv. Atmos. Sci.* **2003**, *20*, 425–436. [CrossRef]

25. Yasunari, T.; Miwa, T. Convective cloud systems over the Tibetan Plateau and their impact on meso-scale disturbances in the Meiyu/Baiu frontal zone—A case study in 1998. *J. Meteorol. Soc. Jpn.* **2006**, *84*, 783–803. [CrossRef]

26. Zhao, Y.; Xu, X.; Chen, B.; Wang, Y. The upstream "strong signals" of the water vapor transport over the Tibetan Plateau during a heavy rainfall event in the Yangtze River Basin. *Adv. Atmos. Sci.* **2016**, *33*, 1343–1350. [CrossRef]

27. Lu, C.; Yu, G.; Xie, G. Tibetan plateau serves as a water tower. In Proceedings of the 2005 IEEE International Geoscience and Remote Sensing Symposium, IGARSS'05, Seoul, Korea, 25–29 July 2005; pp. 3120–3123. [CrossRef]

28. Wang, J.; Zheng, Y. Preliminary applicable analysis of SWAT model in the three rivers source area. *Meteorol. Environ. Sci.* **2014**, *37*, 102–107. (In Chinese) [CrossRef]

29. Huntington, T.G. Evidence for intensification of the global water cycle: Review and synthesis. *J. Hydrol.* **2006**, *319*, 83–95. [CrossRef]

30. Yao, T.; Liu, X.; Wang, N. The magnitude of climate change in the Tibet Plateau. *Chin. Sci. Bull.* **2000**, *45*, 98–106. [CrossRef]

31. Prein, A.F.; Heymsfield, A.J. Increased melting level height impacts surface precipitation phase and intensity. *Nat. Clim. Chang.* **2020**, *10*, 771–776. [CrossRef]

32. FU, Y.; Liu, Q.; ZI, Y.; Feng, S.; LI, Y.; Liu, G. Summer precipitation and latent heating over the Tibetan Plateau based on TRMM measurements. *Plateau Mt. Meteorol. Res.* **2008**, *28*, 8–18. (In Chinese) [CrossRef]

33. Bai, P.; Liu, X.; Yang, T.; Liang, K.; Liu, C. Evaluation of streamflow simulation results of land surface models in GLDAS on the Tibetan plateau. *J. Geophys. Res. Atmos.* **2016**, *121*, 12180–12197. [CrossRef]

34. Arnold, J.G.; Srinivasan, R.; Muttiah, R.S.; Williams, J.R. Large area hydrologic modeling and assessment—Part 1: Model development. *J. Am. Water Resour. Assoc.* **1998**, *34*, 73–89. [CrossRef]

35. Liang, X.; Lettenmaier, D.P.; Wood, E.F.; Burges, S.J. A simple hydrologically based model of land-surface water and energy fluxes for general-circulation models. *J. Geophys. Res. Atmos.* **1994**, *99*, 14415–14428. [CrossRef]

36. Lawrence, D.M.; Oleson, K.W.; Flanner, M.G.; Thornton, P.E.; Swenson, S.C.; Lawrence, P.J.; Zeng, X.; Yang, Z.-L.; Levis, S.; Sakaguchi, K.; et al. Parameterization improvements and functional and structural advances in Version 4 of the community land model. *J. Adv. Model. Earth Syst.* **2011**, *3*, M03001. [CrossRef]

37. Niu, G.; Yang, Z.; Mitchell, K.E.; Chen, F.; Ek, M.B.; Barlage, M.; Kumar, A.; Manning, K.; Niyogi, D.; Rosero, E.; et al. The community Noah land surface model with multiparameterization options (Noah-MP): 1. Model description and evaluation with local-scale measurements. *J. Geophys. Res. Atmos.* **2011**, *116*, D12109. [CrossRef]

38. Zhang, X.; Srinivasan, R.; Debele, B.; Hao, F. Runoff simulation of the headwaters of the Yellow River using the SWAT model with three snowmelt algorithms. *J. Am. Water Resour. Assoc.* **2008**, *44*, 48–61. [CrossRef]

39. Cuo, L.; Zhang, Y.X.; Gao, Y.H.; Hao, Z.C.; Cairang, L.S. The impacts of climate change and land cover/use transition on the hydrology in the upper Yellow River Basin, China. *J. Hydrol.* **2013**, *502*, 37–52. [CrossRef]
40. Sheng, M.Y.; Lei, H.M.; Jiao, Y.; Yang, D.W. Evaluation of the runoff and river routing schemes in the community land model of the Yellow River Basin. *J. Adv. Modeling Earth Syst.* **2017**, *9*, 2993–3018. [CrossRef]
41. Zheng, D.; van der Velde, R.; Su, Z.; Wen, J.; Wang, X.; Yang, K. Impact of soil freeze-thaw mechanism on the runoff dynamics of two Tibetan rivers. *J. Hydrol.* **2018**, *563*, 382–394. [CrossRef]
42. Fang, Y.; Zhang, X.; Niu, G.; Zeng, W.; Zhu, J.; Zhang, T. Study of the spatiotemporal characteristics of meltwater contribution to the total runoff in the upper changjiang river basin. *Water* **2017**, *9*, 165. [CrossRef]
43. Gochis, D.J.; Barlage, M.; Cabell, R.; Casali, M.; Dugger, A.; FitzGerald, K.; McAllister, M.; McCreight, J.; RafieeiNasab, A.; Read, L.; et al. The NCAR WRF-Hydro®Modeling System Technical Description, NCAR Technical Note. 2020. 107p. Available online: https://ral.ucar.edu/sites/default/files/public/WRFHydroV511TechnicalDescription.pdf (accessed on 24 January 2020). [CrossRef]
44. Senatore, A.; Mendicino, G.; Gochis, D.J.; Yu, W.; Yates, D.N.; Kunstmann, H. Fully coupled atmosphere-hydrology simulations for the central Mediterranean: Impact of enhanced hydrological parameterization for short and long time scales. *J. Adv. Model. Earth Syst.* **2015**, *7*, 1693–1715. [CrossRef]
45. Wang, W.; Liu, J.; Li, C.; Liu, Y.; Yu, F.; Yu, E. An evaluation study of the fully coupled WRF/WRF-hydro modeling system for simulation of storm events with different rainfall evenness in space and time. *Water* **2020**, *12*, 1209. [CrossRef]
46. Rummler, T.; Arnault, J.; Gochis, D.; Kunstmann, H. Role of lateral terrestrial water flow on the regional water cycle in a complex terrain region: Investigation with a fully coupled model system. *J. Geophys. Res. Atmos.* **2019**, *124*, 507–529. [CrossRef]
47. Xiang, T.; Vivoni, E.R.; Gochis, D.J.; Mascaro, G. On the diurnal cycle of surface energy fluxes in the North American monsoon region using the WRF-Hydro modeling system. *J. Geophys. Res. Atmos.* **2017**, *122*, 9024–9049. [CrossRef]
48. Zhang, Z.; Arnault, J.; Wagner, S.; Laux, P.; Kunstmann, H. Impact of lateral terrestrial water flow on land-atmosphere interactions in the Heihe River Basin in China: Fully Coupled Modeling and Precipitation Recycling Analysis. *J. Geophys. Res. Atmos.* **2019**, *124*, 8401–8423. [CrossRef]
49. Kerandi, N.; Arnault, J.; Laux, P.; Wagner, S.; Kitheka, J.; Kunstmann, H. Joint atmospheric-terrestrial water balances for East Africa: A WRF-Hydro case study for the upper Tana River basin. *Theor. Appl. Climatol.* **2018**, *131*, 1337–1355. [CrossRef]
50. Li, L.; Gochis, D.J.; Sobolowski, S.; Mesquita, M.D.S. Evaluating the present annual water budget of a Himalayan headwater river basin using a high-resolution atmosphere-hydrology model. *J. Geophys. Res. Atmos.* **2017**, *122*, 4786–4807. [CrossRef]
51. Zhao, Z.; Liu, J.; Shao, Q. Characteristic analysis of land cover change in nature reserve of three river's source regions. *Sci. Geogr. Sin.* **2010**, *30*, 415–420. (In Chinese) [CrossRef]
52. Zheng, H.; Zhang, L.; Liu, C.; Shao, Q.; Fukushima, Y. Changes in stream flow regime in headwater catchments of the Yellow River basin since the 1950s. *Hydrol. Process.* **2007**, *21*, 886–893. [CrossRef]
53. Jiang, C.; Ll, F.; Gao, Y.; Wang, D.; Zhang, L.; Guo, Y. Streamflow variation in the three-river headwaters region during 1956–2012. *Res. Environ. Sci.* **2017**, *30*, 30–39. (In Chinese) [CrossRef]
54. Liu, L.; Zhang, W.J.; Lu, Q.F.; Jiang, H.R.; Tang, Y.; Xiao, H.M.; Wang, G.X. Hydrological impacts of near-surface soil warming on the Tibetan Plateau. *Permafr. Periglac. Process.* **2020**, *31*, 324–336. [CrossRef]
55. Wu, P.; Liang, S.; Wang, X.-S.; McKenzie, J.M.; Feng, Y. Climate change impacts on cold season runoff in the headwaters of the Yellow River considering frozen ground degradation. *Water* **2020**, *12*, 602. [CrossRef]
56. Skamarock, W.C.; Klemp, J.B. A time-split nonhydrostatic atmospheric model for weather research and forecasting applications. *J. Comput. Phys.* **2008**, *227*, 3465–3485. [CrossRef]
57. Maussion, F.; Scherer, D.; Moelg, T.; Collier, E.; Curio, J.; Finkelnburg, R. Precipitation seasonality and variability over the Tibetan Plateau as resolved by the high asia reanalysis. *J. Clim.* **2014**, *27*, 1910–1927. [CrossRef]
58. Gao, Y.; Xiao, L.; Chen, D.; Chen, F.; Xu, J.; Xu, Y. Quantification of the relative role of land-surface processes and large-scale forcing in dynamic downscaling over the Tibetan Plateau. *Clim. Dyn.* **2017**, *48*, 1705–1721. [CrossRef]
59. Huang, H.; Wang, C.; Chen, G.T.; Carbone, R.E. The role of diurnal solenoidal circulation on propagating rainfall episodes near the eastern Tibetan Plateau. *Mon. Weather Rev.* **2010**, *138*, 2975–2989. [CrossRef]
60. Wang, Z.; Duan, A.; Wu, G. Time-lagged impact of spring sensible heat over the Tibetan Plateau on the summer rainfall anomaly in East China: Case studies using the WRF model. *Clim. Dyn.* **2014**, *42*, 2885–2898. [CrossRef]
61. Gao, Y.; Li, K.; Chen, F.; Jiang, Y.; Lu, C. Assessing and improving Noah-MP land model simulations for the central Tibetan Plateau. *J. Geophys. Res. Atmos.* **2015**, *120*, 9258–9278. [CrossRef]
62. Zheng, D.; Van der Velde, R.; Su, Z.; Wen, J.; Wang, X. Assessment of Noah land surface model with various runoff parameterizations over a Tibetan river. *J. Geophys. Res. Atmos.* **2017**, *122*, 1488–1504. [CrossRef]
63. Koch, M. Dupuit-Forchheimer formulation of a transport finite element model: Application to remediation of a shallow, unconfined aquifer. *Water Stud.* **1998**, *26*, 509–518. [CrossRef]
64. Julien, P.Y.; Saghafian, B.; Ogden, F.L. Raster-based hydrologic modeling of spatially-varied surface runoff. *Water Resour. Bull. AWRA* **1995**, *31*, 523–536. [CrossRef]
65. Garbrecht, J.; Martz, L.W. Automated channel ordering and node indexing for raster channel networks. *Comput. Geosci.* **1997**, *23*, 961–966. [CrossRef]

66. Wang, J.; Wang, C.; Rao, V.; Orr, A.; Yan, E.; Kotamarthi, R. A parallel workflow implementation for PEST version 13.6 in high-performance computing for WRF-Hydro version 5.0: A case study over the midwestern United States. *Geosci. Model Dev.* **2019**, *12*, 3523–3539. [CrossRef]

67. Shangguan, W.; Dai, Y.; Liu, B.; Zhu, A.; Duan, Q.; Wu, L.; Ji, D.; Ye, A.; Yuan, H.; Zhang, Q.; et al. A China data set of soil properties for land surface modeling. *J. Adv. Model. Earth Syst.* **2013**, *5*, 212–224. [CrossRef]

68. Grell, G.A.; Devenyi, D. A generalized approach to parameterizing convection combining ensemble and data assimilation techniques. *Geophys. Res. Lett.* **2002**, *29*, 38-1–38-4. [CrossRef]

69. Gao, Y.; Cheng, G.; Cui, W.; Chen, F.; David, G.; Yu, W. Coupling of enhanced land surface hydrology with atmospheric mesoscale model and its implement in Heihe River Basin. *Adv. Earth Sci.* **2006**, *21*, 1283–1293. (In Chinese) [CrossRef]

70. Li, L.; Pontoppidan, M.; Sobolowski, S.; Senatore, A. The impact of initial conditions on convection-permitting simulations of a flood event over complex mountainous terrain. *Hydrol. Earth Syst. Sci.* **2020**, *24*, 771–791. [CrossRef]

71. Farr, T.G.; Rosen, P.A.; Caro, E.; Crippen, R.; Duren, R.; Hensley, S.; Kobrick, M.; Paller, M.; Rodriguez, E.; Roth, L.; et al. The shuttle radar topography mission. *Rev. Geophys.* **2007**, *45*, RG2004. [CrossRef]

72. Huffman, G.J.; Stocker, E.F.; Bolvin, D.T.; Nelkin, E.J.; Tan, J. *GPM IMERG Final Precipitation L3 1 Day 0.1 Degree x 0.1 Degree V06*; Andrey Savtchenko, A., Ed.; Goddard Earth Sciences Data and Information Services Center (GES DISC): Greenbelt, MD, USA, 2019. [CrossRef]

73. Tapiador, F.J.; Turk, F.J.; Petersen, W.; Hou, A.Y.; Garcia-Ortega, E.; Machado, L.A.T.; Angelis, C.F.; Salio, P.; Kidd, C.; Huffman, G.J.; et al. Global precipitation measurement: Methods, datasets and applications. *Atmos. Res.* **2012**, *104*, 70–97. [CrossRef]

74. Entekhabi, D.; Njoku, E.G.; O'Neill, P.E.; Kellogg, K.H.; Crow, W.T.; Edelstein, W.N.; Entin, J.K.; Goodman, S.D.; Jackson, T.J.; Johnson, J.; et al. The soil moisture active passive (SMAP) mission. *Proc. IEEE* **2010**, *98*, 704–716. [CrossRef]

75. Reichle, R.H.; De Lannoy, G.J.M.; Liu, Q.; Ardizzone, J.V.; Colliander, A.; Conaty, A.; Crow, W.; Jackson, T.J.; Jones, L.A.; Kimball, J.S.; et al. Assessment of the SMAP Level-4 surface and root-zone soil moisture product using in situ measurements. *J. Hydrometeorol.* **2017**, *18*, 2621–2645. [CrossRef]

76. Reichle, R.H.; Liu, Q.; Koster, R.D.; Crow, W.; De Lannoy, G.J.M.; Kimball, J.S.; Ardizzone, J.V.; Bosch, D.; Colliander, A.; Cosh, M.; et al. Version 4 of the SMAP Level-4 soil moisture algorithm and data product. *J. Adv. Model. Earth Syst.* **2019**, *11*, 3106–3130. [CrossRef]

77. Miralles, D.G.; De Jeu, R.A.M.; Gash, J.H.; Holmes, T.R.H.; Dolman, A.J. Magnitude and variability of land evaporation and its components at the global scale. *Hydrol. Earth Syst. Sci.* **2011**, *15*, 967–981. [CrossRef]

78. Martens, B.; Miralles, D.G.; Lievens, H.; van der Schalie, R.; de Jeu, R.A.M.; Fernandez-Prieto, D.; Beck, H.E.; Dorigo, W.A.; Verhoest, N.E.C. GLEAM v3: Satellite-based land evaporation and root-zone soil moisture. *Geosci. Model Dev.* **2017**, *10*, 1903–1925. [CrossRef]

79. He, J.; Yang, K.; Tang, W.; Lu, H.; Qin, J.; Chen, Y.; Li, X. The first high-resolution meteorological forcing dataset for land process studies over China. *Sci. Data* **2020**, *7*, 25. [CrossRef]

80. Hou, A.Y.; Kakar, R.K.; Neeck, S.; Azarbarzin, A.A.; Kummerow, C.D.; Kojima, M.; Oki, R.; Nakamura, K.; Iguchi, T. The global precipitation measurement mission. *Bull. Am. Meteorol. Soc.* **2014**, *95*, 701–722. [CrossRef]

81. Ma, Y.; Tang, G.; Long, D.; Yong, B.; Zhong, L.; Wan, W.; Hong, Y. Similarity and error intercomparison of the GPM and its predecessor-TRMM multisatellite precipitation analysis using the best available hourly gauge network over the Tibetan Plateau. *Remote Sens.* **2016**, *8*, 569. [CrossRef]

82. Xu, R.; Tian, F.; Yang, L.; Hu, H.; Lu, H.; Hou, A. Ground validation of GPM IMERG and TRMM 3B42V7 rainfall products over southern Tibetan Plateau based on a high-density rain gauge network. *J. Geophys. Res.-Atmos.* **2017**, *122*, 910–924. [CrossRef]

83. Lu, D.; Yong, B. Evaluation and hydrological utility of the latest GPM IMERG V5 and GSMaP V7 precipitation products over the Tibetan Plateau. *Remote Sens.* **2018**, *12*, 2022. [CrossRef]

84. Ma, L.; Zhao, L.; Tian, L.-M.; Yuan, L.-M.; Xiao, Y.; Zhang, L.-L.; Zou, D.-F.; Qiao, Y.-P. Evaluation of the integrated multi-satellite retrievals for global precipitation measurement over the Tibetan Plateau. *J. Mt. Sci.* **2019**, *16*, 1500–1514. [CrossRef]

85. Chen, Y.; Yang, K.; Qin, J.; Cui, Q.; Lu, H.; La, Z.; Han, M.; Tang, W. Evaluation of SMAP, SMOS, and AMSR2 soil moisture retrievals against observations from two networks on the Tibetan Plateau. *J. Geophys. Res. Atmos.* **2017**, *122*, 5780–5792. [CrossRef]

86. Li, X.; Long, D.; Han, Z.; Scanlon, B.R.; Sun, Z.; Han, P.; Hou, A. Evapotranspiration estimation for Tibetan Plateau headwaters using conjoint terrestrial and atmospheric water balances and multisource remote sensing. *Water Resour. Res.* **2019**, *55*, 8608–8630. [CrossRef]

87. Ding, L.; Zhou, J.; Zhang, X.; Liu, S.; Cao, R. Downscaling of surface air temperature over the Tibetan Plateau based on DEM. *Int. J. Appl. Earth Obs. Geoinf.* **2018**, *73*, 136–147. [CrossRef]

88. Taylor, K.E. Summarizing multiple aspects of model performance in a single diagram. *J. Geophys. Res. Atmos.* **2001**, *106*, 7183–7192. [CrossRef]

89. DelSole, T.; Shukla, J. Specification of wintertime north american surface temperature. *J. Clim.* **2006**, *19*, 2691–2716. [CrossRef]

90. Miyakoda, K.; Hembree, G.D.; Strickler, R.F.; Shulman, I. Cumulative results of extended forecast experiments.1. model performance for winter cases. *Mon. Weather Rev.* **1972**, *100*, 836–855. [CrossRef]

91. Li, J.; Yu, R.C. A Method to Linearly evaluate rainfall frequency-intensity distribution. *J. Appl. Meteorol. Climatol.* **2014**, *53*, 928–934. [CrossRef]

92. Li, J.; Yu, R.; Yuan, W.; Chen, H.; Sun, W.; Zhang, Y. Precipitation over East Asia simulated by NCAR CAM5 at different horizontal resolutions. *J. Adv. Model. Earth Syst.* **2015**, *7*, 774–790. [CrossRef]
93. Li, J. Hourly station-based precipitation characteristics over the Tibetan Plateau. *Int. J. Climatol.* **2018**, *38*, 1560–1570. [CrossRef]
94. Meng, X.; Lyu, S.; Zhang, T.; Zhao, L.; Li, Z.; Han, B.; Li, S.; Ma, D.; Chen, H.; Ao, Y.; et al. Simulated cold bias being improved by using MODIS time-varying albedo in the Tibetan Plateau in WRF model. *Environ. Res. Lett.* **2018**, *13*, 044028. [CrossRef]
95. Abbaszadeh, P.; Gavahi, K.; Moradkhani, H. Multivariate remotely sensed and in-situ data assimilation for enhancing community WRF-Hydro model forecasting. *Adv. Water Resour.* **2020**, *145*, 103721. [CrossRef]
96. Mendoza, P.A.; Clark, M.P.; Barlage, M.; Rajagopalan, B.; Samaniego, L.; Abramowitz, G.; Gupta, H. Are we unnecessarily constraining the agility of complex process-based models? *Water Resour. Res.* **2015**, *51*, 716–728. [CrossRef]
97. Sarkar, S.; Himesh, S. Evaluation of the skill of a fully-coupled atmospheric-hydrological model in simulating extreme hydrometeorological event: A case study over Cauvery river catchment. *Pure Appl. Geophys.* **2021**, *178*, 1063–1086. [CrossRef]
98. Lin, P.; Yang, Z.-L.; Gochis, D.J.; Yu, W.; Maidment, D.R.; Somos-Valenzuela, M.A.; David, C.H. Implementation of a vector-based river network routing scheme in the community WRF-Hydro modeling framework for flood discharge simulation. *Environ. Model. Softw.* **2018**, *107*, 1–11. [CrossRef]
99. Yang, Y.; Yuan, H.; Yu, W. Uncertainties of 3D soil hydraulic parameters in streamflow simulations using a distributed hydrological model system. *J. Hydrol.* **2018**, *567*, 12–24. [CrossRef]
100. Arnault, J.; Rummler, T.; Baur, F.; Lerch, S.; Wagner, S.; Fersch, B.; Zhang, Z.; Kerandi, N.; Keil, C.; Kunstmann, H. Precipitation sensitivity to the uncertainty of terrestrial water flow in WRF-Hydro: An ensemble analysis for central Europe. *J. Hydrometeorol.* **2018**, *19*, 1007–1025. [CrossRef]
101. Lahmers, T.M.; Castro, C.L.; Hazenberg, P. Effects of lateral flow on the convective environment in a coupled hydrometeorological modeling system in a semiarid environment. *J. Hydrometeorol.* **2020**, *21*, 615–642. [CrossRef]
102. Gao, Y.; Xu, J.; Chen, D. Evaluation of WRF mesoscale climate simulations over the Tibetan Plateau during 1979–2011. *J. Clim.* **2015**, *28*, 2823–2841. [CrossRef]
103. Hong, S.Y.; Dudhia, J.; Chen, S.H. A revised approach to ice microphysical processes for the bulk parameterization of clouds and precipitation. *Mon. Weather Rev.* **2004**, *132*, 103–120. [CrossRef]
104. Collins, W.; Rasch, P.J.; Boville, B.A.; McCaa, J.; Williamson, D.L.; Kiehl, J.T.; Briegleb, B.P.; Bitz, C.; Lin, S.-J.; Zhang, M.; et al. Description of the NCAR community atmosphere model (CAM 3.0) (No. NCAR/TN-464+STR). *Univ. Corp. Atmos. Res.* **2004**, *226*, 1326–1334. [CrossRef]
105. Hong, S.-Y.; Noh, Y.; Dudhia, J. A new vertical diffusion package with an explicit treatment of entrainment processes. *Mon. Weather Rev.* **2006**, *134*, 2318–2341. [CrossRef]
106. Ball, J.T.; Woodrow, I.E.; Berry, J.A. A model predicting stomatal conductance and its contribution to the control of photosynthesis under different environmental conditions. *Prog. Photosynth. Res.* **1987**, *5*, 221–224. [CrossRef]
107. Brutsaert, W. Energy fluxes at the earth's surface. In *Evaporation into the Atmosphere*; Springer: Dordrecht, The Netherlands, 1982; pp. 128–153. [CrossRef]
108. Chen, F.; Dudhia, J. Coupling an advanced land surface-hydrology model with the Penn State-NCAR MM5 modeling system. Part I: Model implementation and sensitivity. *Mon. Weather Rev.* **2001**, *129*, 569–585. [CrossRef]
109. Verseghy, D.L. Class—A canadian land surface scheme for gcms 1. soil model. *Int. J. Climatol.* **1991**, *11*, 111–133. [CrossRef]
110. Jordan, R. *A One-Dimensional Temperature Model for a Snow Cover: Technical Documentation for SNTHERM. 89*; Special Report 91-16; US Army Corps of Engineers: Washington, DC, USA, 1991; Volume 89.

water

Article

The Impact of Climate Warming on Lake Surface Heat Exchange and Ice Phenology of Different Types of Lakes on the Tibetan Plateau

Jiahe Lang [1,2], Yaoming Ma [2,3,4,*], Zhaoguo Li [1] and Dongsheng Su [1,2]

1   Key Laboratory of Land Surface Process and Climate Change in Cold and Arid Regions, Northwest Institute of Eco-Environment and Resources, Chinese Academy of Sciences, Lanzhou 730000, China; langjiahe@itpcas.ac.cn (J.L.); zgli@lzb.ac.cn (Z.L.); sds@lzb.ac.cn (D.S.)
2   University of Chinese Academy of Sciences, Beijing 100049, China
3   Key Laboratory of Tibetan Environment Changes and Land Surface Processes, Institute of Tibetan Plateau Research, Chinese Academy of Sciences, Beijing 100101, China
4   CAS Center for Excellence in Tibetan Plateau Earth Sciences, Chinese Academy of Sciences, Beijing 100101, China
*   Correspondence: ymma@itpcas.ac.cn; Tel.: +86-10-8409-7108

Abstract: Increasing air temperature is a significant feature of climate warming, and is cause for some concern, particularly on the Tibetan Plateau (TP). A lack of observations means that the impact of rising air temperatures on TP lakes has received little attention. Lake surfaces play a unique role in determining local and regional climate. This study analyzed the effect of increasing air temperature on lake surface temperature (LST), latent heat flux (LE), sensible heat flux (H), and ice phenology at Lake Nam Co and Lake Ngoring, which have mean depths of approximately 40 m and 25 m, respectively, and are in the central and eastern TP, respectively. The variables were simulated using an adjusted Fresh-water Lake (FLake) model (FLake_α_ice = 0.15). The simulated results were evaluated against in situ observations of LST, LE and H, and against LST data derived from the Moderate Resolution Imaging Spectroradiometer (MODIS) for 2015 to 2016. The simulations show that when the air temperature increases, LST increases, and the rate of increase is greater in winter than in summer; annual LE increases; H and ice thickness decrease; ice freeze-up date is delayed; and the break-up date advances. The changes in the variables in response to the temperature increases are similar at the two lakes from August to December, but are significantly different from December to July.

Keywords: Tibetan Plateau; climate warming; lake surface temperature; heat exchange; lake ice phenology

Citation: Lang, J.; Ma, Y.; Li, Z.; Su, D. The Impact of Climate Warming on Lake Surface Heat Exchange and Ice Phenology of Different Types of Lakes on the Tibetan Plateau. *Water* **2021**, *13*, 634. https://doi.org/10.3390/w13050634

Academic Editor: Thomas Meixner

Received: 26 December 2020
Accepted: 23 February 2021
Published: 27 February 2021

Publisher's Note: MDPI stays neutral with regard to jurisdictional claims in published maps and institutional affiliations.

1. Introduction

Climate change has received much attention in recent decades. The Intergovernmental Panel on Climate Change (IPCC) Working Group I's contribution to the IPCC Fifth Assessment Report (WGI AR5) [1], specifically the chapter "Observations: atmosphere and surface", shows that the global average surface temperature warmed by 0.85 °C (0.65 to 1.06 °C) between 1880 and 2012. Global surface temperature is likely to change by more than 1.5 °C by the end of the 21st century under almost all representative concentration pathway (RCP) scenarios, and the change is likely to exceed 2 °C under RCP8.5 [1]. Lakes have a near-global distribution. Differences between land and lake surfaces mean that lakes play a unique role in determining local and regional climate, for example through their low albedo, small roughness length, and high heat capacity [2–4]. They are considered to be sentinels of climate change [5]. Recent studies have found significant warming for lakes throughout the world, with a mean increasing trend of 0.34 °C per decade between 1985 and 2009. In many cases, the observed rate of lake warming is more rapid than the air temperature increases seen at high latitudes, which may indicate a close relationship between these [6–9]. Lake surface temperature is an important indicator of the lake state and the evaporation process occurs only on the skin layer (~10 μm) of the water body [9,10]. The lake surface temperature is affected not only by the interactions between the atmosphere and the skin layer, but also between the skin layer and the water column [11]. The former interaction is governed by four primary meteorological variables: solar radiation, atmospheric humidity, air temperature, and wind speed; while the latter is controlled by the energy stored in the water body [12,13]. Trends towards earlier break-up and later freeze-up dates for lake ice have been observed for most lakes in Canada between 1951 and 2012, with the latter showing a lower degree of temporal coherence than the former [14]. Studies have shown that a shortening of the frozen period can lead to an earlier establishment of the summer thermocline in lakes, which accelerates the warming of the upper lake water [9,15]. The freezing date of a lake is significantly affected by the lake's heat storage, air temperature and other climatic variables such as weed speed, while the melting of lake ice is mainly determined by air temperature and solar radiation [16,17]. Several additional lake responses to climate change have also been widely observed globally, such as more stable stratification and shallower thermocline depths [18–20].

The Tibetan Plateau (TP) is a large distinct geographic region, with a mean elevation that exceeds 4000 m. The TP is sometimes called the Asian Water Tower, and is more sensitive to climate change than global land surfaces are (0.46 °C per decade, 1984–2009) [21–23]. The TP has experienced rapid climate change, with surface air warming and moistening, solar dimming, and wind stilling. Precipitation is increasing over the central TP, and is decreasing in the Himalayan region [23,24]. The environmental changes experienced on the TP are mostly associated with rapid surface warming [23]. The annual mean air temperature between 1980 and 2018 obtained from the 95 China Meteorological Administration (CMA) weather stations on the TP is 4.1 °C [25]. For a global average temperature increase between 1.5 °C and 2.0 °C at the end of the century (2100), the increases in the maximum temperatures on the TP are projected to be between 2.34 °C and 3.20 °C under RCP4.5, and between 2.34 °C and 3.14 °C under RCP8.5, respectively [26]. These projections are helpful for the sensitivity experiments in our study.

The TP includes 57.2% of all lakes in China, and these account for ~1.9% of the total global lake surface area [27–29]. The lakes on the TP are completely covered by ice for 5–7 months each year, with a total ice area of approximately 5×10^4 km^2 in 2018 [29,30]. Mixing in ice-covered lakes can be caused by through-flow currents, by oscillations in the ice cover, by heat flow from the sediments, or by solar radiation penetrating the ice (influenced by the ice albedo). In contrast to high-latitude areas, solar radiation is very strong on the TP and there is little snow, and there is almost none on lake surfaces, where snow is blown by the wind, meaning that convection, driven by penetrating solar radiation, is more effective, particularly in the ice melt period [31]. The impact of the rising air temperature on lake ice phenology may be clearer for lakes on the TP. Studies have shown that the area, level and volume of lakes on the TP decreased slightly from 1976 to the mid-1990s, and then increased rapidly [25]. These lakes play an important role in the global water cycle and the Asian monsoon system. They are highly sensitive to global climate change and are therefore an effective indicator of the absence of direct anthropogenic influences [24,32–36]. Lakes can affect local land–atmosphere interactions and regional heat and water budgets, and can impact on local atmosphere boundary layer processes, as shown in numerical climate model simulations [25,37]. The characteristics of the surface energy balance and turbulent exchange differ from lake to lake, depending on the lake area and depth, as well as on meteorological conditions [38,39]. Long-term changes in lake evaporation, and in the latent heat fluxes (LE) and sensible heat fluxes (H), have been shown for Lake Nom Co and Lake Ngoring [40–44].

Almost all the previously mentioned research on lakes on the TP has focused either on lake–atmosphere interactions at a single lake in ice-free conditions by seasonal in situ observation, or on the response of lake area, level and volume to climate change. The responses of lake surface heat exchange and ice phenology to increasing air temperature on the TP have rarely been addressed [25].

In this study, our goal is to investigate the response of lake ice phenology, lake surface temperature (LST), sensible heat flux (H) and latent heat flux (LE) to rising air temperature for two different lakes (Lake Nam Co and Lake Ngoring) on the TP. Our results were simulated using the Fresh-water Lake (FLake) model, developed by Mironov [45], and the model was driven by a dataset of long-term in situ observations. We adjusted the model parameterization scheme for lake ice albedo and improved the accuracy of the winter simulations. The simulated results were evaluated against in situ observed LST, H and LE data, and against LST data derived from Moderate Resolution Imaging Spectroradiometer (MODIS) observations. Then, lake ice phenology, LST, H and LE were investigated under different air temperature scenarios. Finally, we analyzed the maximum possible impact of future air temperature increases on these two lakes on the TP.

2. Study Area, Data and Methods

2.1. Study Area

The Lake Nam Co and Lake Ngoring basins are characterized by a cold and semi-arid continental climate, and the thermal structure of the lakes means that they belong to the dimictic lake type. However, the two lakes differ in surface area, depth, latitude, and altitude. Lake Ngoring is the highest large freshwater lake in China, with a mean depth of 17 m and a surface area of 610 km^2, and is located in the Yellow River source region on the eastern TP (34.46–35.4° N, 97.3–97.55° E; 4274 m a.m.s.l.; Figure 1). The average precipitation between early December and early April is only 28.16 mm (1954–2014) [46]. The minimum and maximum air temperatures occurred in January (−14.2 °C) and August (9.0 °C), respectively, and the annual mean air temperature was approximately −1.9 °C (2011–2016). Lake Ngoring is usually completely ice-covered from early December to early April [15]. The thickest ice appears in late February, and was ~0.7 m in 2013 and 2016 [46]. Lake Nam Co is the highest large lake on Earth, with an area of 2021.3 km^2 as of 2010 [47], and its mean depth is approximately 40 m. It is located on the southern part of the TP (30.5–30.95° N, 90.2–91.05° E; 4710 m a.m.s.l.; Figure 1). The average annual precipitation observed at Nam Co Station amounts to more than 400 mm, and mainly occurs from May to October (Figure 1) [48]. The minimum and maximum air temperatures occurred in January (4 °C) and July (9.3 °C), respectively, and the annual mean air temperature was approximately 0.5 °C (2011–2016). Lake Nam Co is usually completely ice-covered from early January to late March. Since 1978, the persistence of full ice cover for Nam Co Lake has decreased by 19 to 20 days [34].

2.2. Data

2.2.1. In Situ Measurements

There are four sets of observations (two weather stations and two monitoring stations) available for the two lakes. The station data (2011–2016) are used as long-term forcing to drive the FLake models, which were derived from the Lake Nam Co station and Lake Ngoring station on the lakeshore. The monitoring stations' data (2015–2016) are used to evaluate the model results. The observation site at Nam Co provides LST, H and LE data (2015–2016), and the observation site at Ngoring provides H and LE data (2015–2016), as well as lake ice albedo data (2014).

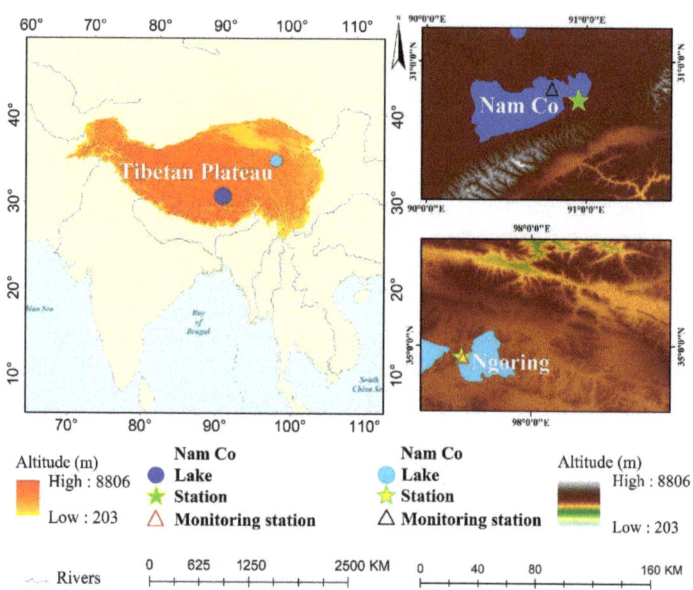

Figure 1. The topography around Lake Nam Co and Lake Ngoring, and the location of Nam Co station (upper right), and of Ngoring station (lower right). The location of the monitoring station at Lake Ngoring for ice albedo, and at Lake Nam Co for lake surface temperature (LST), sensible heat flux (H) and latent heat flux (LE) is also marked.

The weather station at Lake Nam Co was established on the eastern shore of the lake, about 1.5 km from the shoreline, in 2005 (green five-pointed star in Figure 1). The automatic weather station (AWS) tower records wind speed, wind direction (WD), air temperature, humidity, pressure, precipitation, and downward short- and long-wave radiation. The in situ observation site for Lake Nam Co is situated on an island (an area of approximately 0.18 km^2, shown as a black triangle in Figure 1). An eddy covariance (EC) system (4.5 m above the island surface) and AWS tower (1.52m and 9.52m above the land surface) were established on the island, which is about 10 m from the shore, on July 28th, 2015. The EC system has been applied to all kinds of lakes including several lakes of TP to measure turbulent fluxes (momentum, sensible heat and latent heat flux) [40,49]. The EC system consisted of an open-path CO_2/H_2O infrared gas analyzer (LI-7500A, LI-COR Biosciences) and a three-dimensional sonic anemometer (CSAT3, Campbell Scientific, Inc.). Temperature, humidity and three-dimension wind speeds are measured at a frequency of 10 Hz by the gas analyzer and ultrasonic anemometer, respectively. A water temperature profiler, measuring to a depth of 0.5 m, was installed (90.7979° E, 30.8107° N) from July 28th to November 19th in 2015, and from July 7th to November 18th in 2016. The temperature at the 0.5 m depth is used as the LST in our study (Ts, °C).

The lake station at Ngoring was installed in June 2011. Initially, an AWS tower and an EC observation system were situated in the northwestern part of the lake, standing on a submerged rock about 200 m from the shore (35.038° N, 97.658° E) [37,41]. These were damaged by ice in the winter of 2011–2012. Another two systems were located on the southwestern lake shore (yellow five-pointed star in Figure 1, 34.918° N, 97.558° E) to ensure continuity with records from 2012, and these then developed into Ngoring Station. Air temperature and humidity were measured with a temperature and humidity probe (HMP45C, Vaisala) at a 3 m height. The incoming and outgoing shortwave and longwave radiations were measured with a net radiometer (CNR-1, Kipp and Zonen) 1.5 m above the lake surface. The sensor signals were recorded by a data logger (CR5000, Campbell

Scientific, Inc.) at a frequency of 10 Hz. The AWS tower measures air temperature, humidity, precipitation, pressure, wind speed and direction, and downward short- and long-wave radiation. In this study, the AWS data are used as long-term forcing data to drive the lake model for Lake Ngoring. The EC observation system measures H and LE and the data are used to evaluate the model results.

Half-hourly data for H and LE at Nam Co and Ngoring lakes were calculated by processing the high-frequency EC system observations using the "Turbulence Knight 3" software (Windows TK311) (https://zenodo.org/record/20349#, accessed on 27 February 2021). All the relevant corrections (spike removal, time lag compensation, spectral correction, planar fit rotation, and Webb–Pearman–Leuning density correction) were included [40,50]. Sporadic missing data were replaced through interpolation combined with other meteorological variables such as radiation, wind speed, and temperature (specific humidity) differences between the lake surface and air [49]. Sensible heat flux (H) and latent heat flux (LE) are given by Equations (1) and (2), respectively.

$$H = \rho_a c_p \overline{w\prime T\prime} \tag{1}$$

$$LE = L_v \rho_a \overline{w\prime q\prime} \tag{2}$$

Here, ρ_a is the air density, c_p is the specific heat of air at a constant pressure, L_v is the latent heat of vaporization, $w\prime$ is fluctuation of the vertical wind component, and $T\prime$ and $q\prime$ are the temperature and specific humidity fluctuations. To ensure the data quality of the EC observations, we considered many criteria [43,49]. Biermann et al. [13] showed the in situ-observed turbulent heat flux for conditions with wind direction (WD) from the lake surface. Footprint analysis [51] was used to identify observations collected when Lake Ngoring and Lake Nam Co Lake were the dominant source areas. So, we discarded the turbulent heat flux data when wind direction (WD) criteria were not met (Nam Co WD < 135° and WD > 270°, Ngoring 35° < WD < 215°), since these fluxes are contaminated with land. In this study, EC system observation data from 2015–2016 are used to validate our lake model results.

In situ observations of ice albedo were recorded over west Ngoring Lake from 3 to 6 January 2014 (red triangle in Figure 1). The instrument used for this observation is a Kipp & Zonen 4-Component Net Radiometer (CNR4) (1.20 m above the ice surface), in which the pyrgeometer and pyranometers measure longwave (4500–42,000 nm) and shortwave (300–2800 nm) infrared radiation, respectively [15]. The albedo (α) measured by the pyranometers over the lake ice surface is obtained from the ratio: $\alpha = R_S\_dw / R_S\_up$, where R_S_dw and R_S_up are the measured upward and downward shortwave radiances, respectively. In this study, the data were used with the MODIS-observed data to modify the parameterization scheme used to represent lake ice albedo.

2.2.2. MODIS Lake Surface Temperature, Ice Albedo and Snow/Ice Cover Ratio

The MODIS albedo dataset, MCD43A3, from October 2014 to June 2015, was used to adjust the parameterization scheme for lake ice albedo, using the in situ observation data in this study. The MODIS data include albedo measurements at a 500 m spatial resolution and are updated every 8 days. The two lakes are large enough (the area of Lake Nam Co is 2021.3 km^2 while Lake Ngoring is 610 km^2) that multiple MODIS footprints can fit. The pixels that cover the two lakes are carefully selected (two pixels within the lake boundary are removed) to ensure that land contamination is not an issue. The information in this product is detailed in Ref. [52]. The MCD43A product used in this study is the White Sky Albedo (WSA), from the short waveband product. Studies have shown that the snow and ice albedo obtained from MCD43A3 have good accuracy for individual regions, with mean biases ranging from 0.06 to 0.07, when compared with in situ observations [15,53]. MCD43A2 is the quality assurance product for MCD43A3, and it records whether each pixel has snow-cover or not (ice-cover). The MCD43A2 data for winters in 2012–2013,

2013–2014, and 2014–2015 are used to analyze the snow/ice cover ratio for Lake Ngoring and Lake Nam Co during the frozen period.

In this study, the MODIS LST product (MOD11A2) is used to evaluate the results simulated by the lake model for 2015–2016. The MOD11A2 product includes two instantaneous observations each day (approximately 11:00 and 21:00 local time), with a spatial resolution of 1 km. Many studies have shown that the MOD11A2 product has good accuracy for individual regions, with mean biases ranging from 0.2 °C to 1.05 °C [12,54,55]. The water surface temperature from the MODIS observations of the skin layer is generally lower than the in situ observation of the mixed layer [56–58], because of the cool skin effect [12,59].

2.3. Lake Model and Numerical Experiment Design

One-dimensional (1-D) lake models have demonstrated their ability of simulating the lake thermodynamics [60,61]. The driving assumption of the 1-D lake model is horizontal homogeneity [62]. The one-dimensional lake surface energy balance is given by $Q^* = Q_{LE} + Q_H + Q_S + Q_{GL}$, where Q^* is net radiation, Q_{LE} is the latent heat flux (evaporative heat flux), Q_H is the sensible heat flux, and Q_{GL} is the heat flux across the lake bottom. The heat storage in the lake is determined as $Q_S = C_W \Delta T_W / \Delta t$, where C_W is the heat capacity of water. The temperature change with time, $\Delta T_W / \Delta t$, is integrated over the total depth of the lake. Certain 1-D lake models with different degrees of sophistication in physical processes have been widely developed, including the 2-layer FLake model based on similarity theory [62,63]; turbulence lake models [64,65]; and radiation-diffusion lake models [11]. Many studies have conducted a series of intercomparisons of the available 1-D lake models' performances in simulating the thermal features of different lakes during the past several years [66–70]. Based on these studies, much considerable progress has been made to improve the 1-D lake models [61,70–73].

Compared to the relatively comprehensive studies on the evaluation and development of the 1-D lake models over lowlands and wet regions, few annual modeling studies being have previously been done for TP lakes because of the harsh environment conditions in winter. Recently, the lakes on TP have received increasing attention [37,43,70,74,75]. The competences of the FLake, WRF-Lake, and CoLM-Lake models in simulating the thermal features of Lake Nam Co have been evaluated, and FLake performs the best in simulating the temporal evolution and intensity of temperature in the shallow layers [70]. Huang et al. [70] also adjusted three key parameters (temperature of maximum water density, light extinction coefficient, surface roughness lengths) within FLake and improved the model results. Li et al. [75] evaluated the effects of ice and snow albedo, ice and water extinction coefficients on the lake ice phenology, water temperature, and sensible and latent heat fluxes using the LAKE2.0 model. The computation of the FLake is efficient, due to its relatively simple construction, and it performs reasonably across lake categories in predicting both surface temperatures and ice characteristics [61]. In terms of accuracy, previous studies have found small positive biases in the H and LE simulated in the FLake model, as well as good correlations between the LST from the FLake model and in situ observations, and between FLake LST and MODIS observations, for the ice-free period [76,77]. Results from simulations for the freezing period show that the lake temperature in winter has a negative bias [15,42,67,70].

FLake has been applied to typical temperature freezing lakes of North America and Alaska (Bear Lake) for several sensitivity analyses of lake depth, water transparency, explicit snow and snow/ice albedo, snow density and heat conductivity [68,78]. FLake has also been driven by regional climate scenarios, applied to small lakes in Berlin, Germany, for modeling the impact of global warming on water temperature and seasonal mixing regimes [79,80]. However, FLake has never previously been applied to Lake Nam Co and Lake Ngoring for assessing air temperature sensitivity to compare the effects of different warming degrees on the lake.

2.3.1. FLake Model

The lake model used in this study is the 1-D Mironov FLake model, which is a freshwater model, and is suggested to be appropriate for lakes with depths of less than 50 m because of its relatively simple stratification scheme. The model requires a small number of lake parameters to be specified: the lake depth and the optical characteristics of the lake. The FLake model is based on the concept of self-similarity (assumed shape) for the vertical temperature profile in the thermocline, and simulate lakes' temperature profiles and surface heat fluxes [62,81,82]. The essence of the concept of self-similarity of the temperature profile $\theta(z, t)$ in the thermocline is that the dimensionless temperature profile in the thermocline can be parameterized through a "universal function of dimensionless depth" with reasonable accuracy, that is [62]:

$$\frac{\theta_s(t) - \theta(z, t)}{\Delta\theta(t)} = \Phi_\theta(\zeta) \ at \ h(t) \leq z \leq h(t) + \Delta h(t), \tag{3}$$

where t is time, z is depth, $\theta_s(t)$ is the temperature of the upper mixed layer of depth $h(t)$, $\Delta\theta(t) = \theta_s(t) - \theta_b(t)$ is the temperature difference across the thermocline of depth $\Delta h(t)$, $\theta_b(t)$ is the temperature at the bottom of the thermocline, and $\Phi_\theta \equiv [\theta_s(t) - \theta(z, t)]/\Delta\theta(t)$ is a dimensionless "universal function of dimensionless depth", $\zeta \equiv [z - h(t)]/\Delta h(t)$, that satisfies the boundary conditions $\Phi_\theta(0) = 0$ and $\Phi_\theta(1) = 1$. A snow layer, a lake ice layer, an upper mixed layer, a thermocline layer and a lake sediment layer are considered in the FLake model, and the concept of self-similarity is applied to all layers except the upper mixed layer. The water temperature of the upper mixed layer is nearly vertically uniform. The depth of the mixed layer is described with an entrainment equation for convective conditions and a relaxation-type equation for stable conditions.

The water surface albedo, with respect to solar radiation, is set to 0.07 in the default configuration, and the albedo of the ice surface is given by [15]:

$$\alpha_{ice} = a_{max} - (\alpha_{max} - \alpha_{min})exp\left(-95.6\left(T_{f0} - T_i\right)/T_{f0}\right) \tag{4}$$

where α_{max} is white ice albedo, α_{max} = 0.6. The α_{min} is blue ice albedo, α_{min} = 0.1. T_{f0} is the freezing temperature (273.15 K), T_i is the ice surface temperature. As the ice surface temperature (T_i) approaches the freezing temperature (T_{f0}), the albedo approaches 0.10. The solar radiation transfer between the water and the snow or ice is calculated using a one-band exponential approximation of the Beer–Lambert decay law, with an extinction coefficient of 3 m^{-1} for water, and 1.0×10^7 m^{-1} for both ice and snow, in the default configuration. The parameterized scheme for the turbulent fluxes of momentum, and sensible and latent heat at the lake surface is adopted in the FLake model [62].

2.3.2. Modification of the Lake ice Albedo Parameterization Scheme in the FLake Model

Previous studies have found small positive biases in H and LE simulated in the FLake model, as well as good correlations between the LST from the FLake model and in situ observations, and between FLake LST and MODIS observations, for the ice-free period [76,77]. Results from simulations for the freezing period show that the lake temperature in winter has a negative bias [15,42,67,70]. Since our study focuses on changes to lake ice phenology in different air temperatures, the accuracy of the model LST and ice phenology in winter is very important. Since there is almost no snow on lake surfaces on the TP in winter, and albedo is a key parameter for calculating lake ice phenology, we modified the parameterization scheme for lake ice albedo to improve the negative model bias for LST in winter. The MODIS-observed average ice surface albedo is 0.15 for lakes on the TP [15]. Based on results from our previous study, we calculated the lake ice albedo from a new albedo parameterization scheme (Equation (4)), and tested the value for α_{max} by replacing the constant value set for the maximum ice albedo (α_{max} = 0.6, Equation (4)) to 0.4 and 0.15 for the two lakes. We conducted a further two tests of Equation (4) by replacing the maximum ice albedo (α_{max} = 0.6, Equation (4)) with 0.15, and setting α_{ice} to 0.15. The in situ observation data for the ice albedo were used to evaluate the results from each test for daily changes at Lake Ngoring (Figure 2a), and the MODIS observation data were used to evaluate the results of the two tests for the frozen period at Lake Ngoring (Figure 2b) and at Lake Nam Co (Figure 2c). It can be seen from Figure 2 that the albedo parameterization scheme in the FLake model greatly overestimated the ice albedo, and did not describe daily changes well. It is very difficult to accurately describe daily changes in ice albedo. Since the lake ice albedo values from the in situ observations are concentrated around 0.15 for most of the daytime, the ice albedo had a large positive bias when $\alpha_{max} = 0.4$ (open down triangle in Figure 2), or $\alpha_{max} = 0.6$ (solid up triangle in Figure 2). The results of the parameterized scheme show a small negative bias when $\alpha_{max} = 0.15$ (asterisk in Figure 2), compared with the MODIS observation data. The results when $\alpha_{ice} = 0.15$ (red line in Figure 2a, red dot in Figure 2b,c) are the most accurate, and we therefore selected this scheme as the final albedo parameterization scheme and used this in all following simulations, after evaluating the model results for LST, H and LE for this scheme against the in situ and MODIS observations.

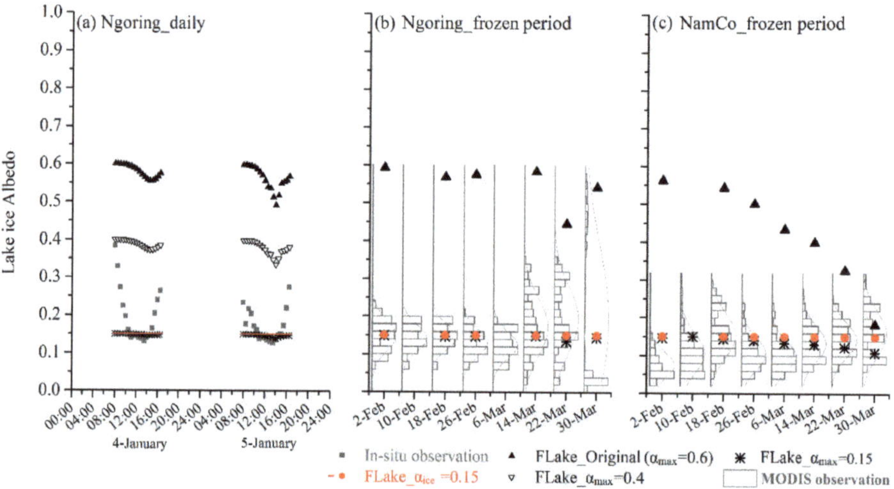

Figure 2. Lake ice albedo from the in situ and MODIS observations, and calculated using the different albedo parameterization schemes, at Lake Ngoring and Lake Nam Co for (**a**) 4 to 5 January 2014, and (**b,c**) 1 February to 31 March 2015. Time is shown in local time.

2.3.3. Numerical Experiment Design

In this study, we applied the FLake model to evaluate the influence of air temperature on simulations for two different lakes: Lake Ngoring and Lake Nam Co. To evaluate the influence of different degrees of air temperature rise (1.5 °C to 3.5 °C) on LST, H, LE, and ice phenology at the two lakes, we first carried out a control experiments (CTRL) by running the FLake model with the default model configuration. We then calculated an additional simulation to test the modified ice albedo parameterization scheme (same in both of the two lakes). Based on the results from the previous tests of the ice albedo parameterization scheme, we conducted a series of sensitivity experiments by increasing the air temperature of the input data by different amounts (1.50 °C, 1.75 °C, 2.00 °C, 2.25 °C, 2.50 °C, 2.75 °C, 3.00 °C, 3.25 °C, and 3.50 °C) to investigate the influence of increasing air temperature on the two lakes. The model lake depth is set to equal the mean depths of the lakes (25 m for Lake Ngoring, 40 m for Lake Nam Co). The initial water temperature for the bottom of Lake Nam Co was set to 276.65 K. The value is close to the temperature that has been observed at a 44 m depth in other years. Other parameters that it is not essential to specify will be set as default with no corresponding observations. The forcing data are derived from the observations made at the stations at Lake Ngoring and Lake Nam Co. The main input variables include the surface air temperature, relative humidity, wind speed, air pressure, downward shortwave and longwave radiation, and precipitation from 1 July 2011 to 1 January 2017. The simulations began in July 2011 and ended in December 2016. The integration time step was 30 min. The simulations began in July 2011, instead of January 2012, to allow for model spin-up.

3. Results

3.1. Evaluation of the Simulated Results

Observations of LST, H and LE from 2015 to 2016 were used to validate the results from both the original FLake model and for the FLake_α_{ice} = 0.15 model, for Nam Co and Ngoring, as shown in Figure 3 (Nam Co) and Figure 4 (Ngoring). The results show that the FLake model's performance is sensitive to the ice albedo parameterization scheme, especially in winter. Setting the ice albedo to 0.15, instead of 0.6, in the FLake model consistently led to the improved simulation of H and LE during winter and the ice melt period for both lakes, and improved the simulated LST throughout the year at Lake Nam Co. Reducing the ice albedo in the FLake model results in fewer frozen days and a smaller maximum ice thickness, leading to a better agreement with the in situ observations reported in other studies [33,37,83]. Further intercomparison indicates that FLake_α_{ice} = 0.15 performs better than the standard FLake model in simulating lake ice phenology for both of the two lakes, and in simulating LST for Lake Nam Co; it also simulates LST, H and LE better than the FLake model during the ice melt period for Lake Ngoring.

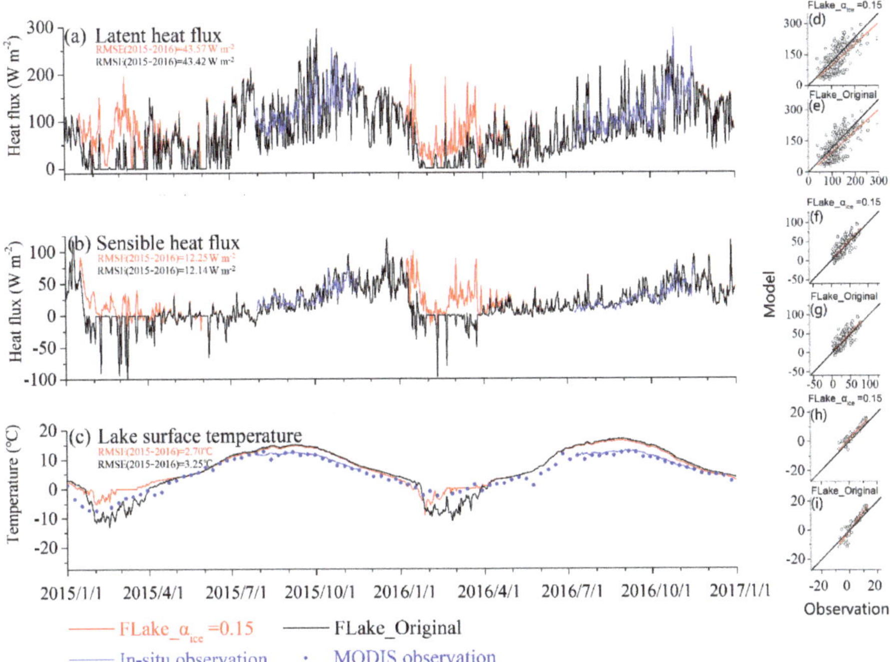

Figure 3. (**a–c**) Comparison and (**d–i**) evaluation of simulated LST, H and LE with in situ and MODIS observation data for Nam Co: (**a,d,e**) LE, (**b,f,g**) H, (**c,h,i**) LST. A fitted curve is shown for each variable, where the x represents the in situ observations and the y represents the simulated results. The comparison period is from January 2015 to December 2016.

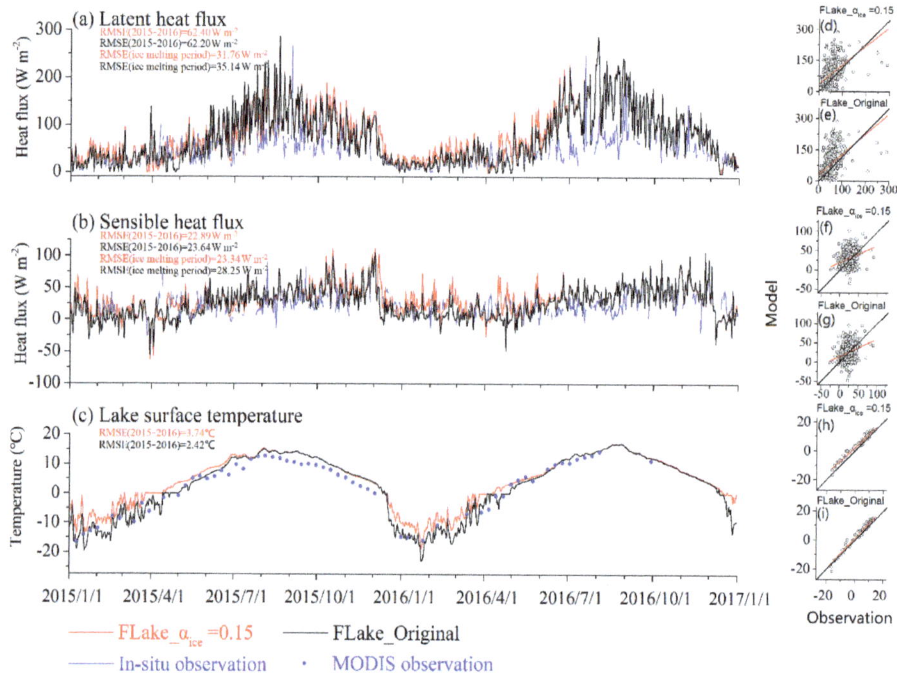

Figure 4. (**a–c**) Comparison and (**d–i**) evaluation of simulated LST, H and LE with in situ and MODIS observation data for Ngoring: (**a,d,e**) LE, (**b,f,g**) H, (**c,h, i**) LST. A fitted curve is shown for each variable, where the x represents the in situ observations and the y represents the simulated results. The comparison period is from January 2015 to December 2016.

3.1.1. Lake Surface Temperature

The simulated LSTs from FLake_$\alpha_{ice} = 0.15$ are higher in winter, and lower in summer, at both lakes, relative to the original FLake model. The FLake_$\alpha_{ice} = 0.15$ model simulates LST better than the FLake model at Lake Nam Co, with a mean bias of 2.22 °C, relative to the bias in the FLake model bias of 2.69 °C. The root-mean-square error (RMSE) values for the LSTs simulated at Lake Nam Co in the FLake_$\alpha_{ice} = 0.15$ and FLake models are 2.70 °C and 3.25 °C, respectively. At Lake Ngoring, the mean LST bias is 3.04 °C for the original FLake model, and 1.94 °C for FLake_$\alpha_{ice} = 0.15$, and the RMSEs are 3.74 °C and 2.42 °C, respectively. One possible reason for the warm biases simulated for Lake Ngoring in winter is that the ice at Lake Ngoring is thicker than at Lake Nam Co, and the frozen period is longer. This effectively inhibits energy exchange between the lake water and the atmosphere. In the model, all the solar radiation entering the lake via the lake ice remains in the ice layer, resulting in the model's overestimation of the lake surface temperature in winter (ice layer surface temperature). Another possible reason is that Lake Nam Co is deeper than Lake Ngoring. Mixing in ice-covered lakes is caused by many factors, and convection driven by penetrating solar radiation is one effective driver (Bengtsson, 1996). The shallower the lake, the greater the impact of the through-ice solar radiative flux on mixing. Lake Ngoring is much shallower than Lake Nam Co, so it is more sensitive to changes in ice albedo.

3.1.2. Latent Heat Flux and Sensible Heat Flux

The model can simulate seasonal variations in H and LE at the lake surface. The simulated H and LE in FLake_$\alpha_{ice} = 0.15$ are higher in winter and spring for both lakes (Lake Nam Co: January to May; Lake Ngoring: December to June) than in the original

FLake model (Figures 3 and 4). The FLake_α_{ice} = 0.15 model simulates LE and H better than the FLake model during the melt period (March to April) at Lake Ngoring, with RMSEs of 31.76 W m^{-2} and 23.34 W m^{-2}, respectively. The RMSEs for the FLake model simulations of LE and H are 35.14 W m^{-2} and 28.25 W m^{-2}, respectively. There are no in situ EC observation data for the corresponding ice melt period at Lake Nam Co to facilitate a comparison. Ref. [40] indicated that sublimation during winter is non-zero over Lake Nam Co, as LE was observed to be very high during the ice-covered period in the EC data. In simulations from the original FLake model, LE is zero throughout frozen period, and simulations from FLake_α_{ice} = 0.15 significantly improve on this bias.

3.1.3. Lake Ice Phenology

The simulated number of frozen lake ice days is lower for both lakes in the FLake_α_{ice} = 0.15 simulations than in the original FLake simulations. At Lake Nam Co, the number of frozen days in simulations from FLake_α_{ice} = 0.15 is 82, and there are 101 frozen days in simulations from FLake. At Lake Ngoring, the number of frozen days is 139 in FLake_α_{ice} = 0.15, and 154 in FLake. The simulated lake ice freeze-up date is several days earlier in FLake than in FLake_ α_{ice} = 0.15, and the onset of ice melt occurs half a month earlier in FLake_ α_{ice} = 0.15 than in the original FLake model for both lakes. In situ observations show that freeze-up generally occurs in mid-January at Lake Nam Co, and that ice break-up occurs in early April [33,83]. The average duration of complete freezing (i.e., complete ice cover) at Nam Co is 58.5 days [84]. The lake ice thickness at Lake Nam Co peaks in February, and the maximum thickness in the in situ observations is about 0.4 m, which occurs relatively near to the Nam Co station [83]. The in situ observations of ice thickness at Lake Ngoring show that it remained above 0.6 m from mid-January to early March, with a maximum (0.73 m) measured in late February [37]. Lake ice thickness at Ngoring was measured from December 2012 to March 2013 near the lakeshore, very close to the AWS. In Figure 5, the simulated lake ice phenology is shown to agree well with the in situ observations reported in other studies.

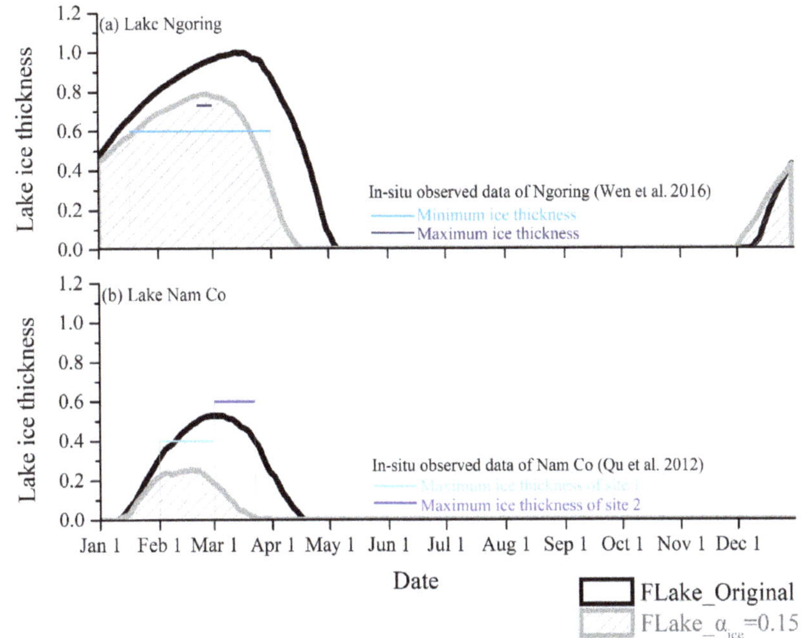

Figure 5. Seasonal variations in lake ice thickness from this study, from Qu et al. 2012 [83] (averaged over 2006–2011), and from Wen et al. 2016 [37] (averaged over December 2012 to March 2013) (**a**) for Lake Ngoring and (**b**) for Lake Nam Co.

In summary, the comparison of the simulations against in situ and MODIS observations shows that the FLake_$\alpha_{\text{ice}} = 0.15$ model can approximately reproduce the observed lake ice phenology, and improves upon the original FLake model for the freezing and melting periods, but there is no significant improvement in the simulation results for the ice-free period. The simulated LST, H and LE values at Ngoring are higher than the observed values, with mean biases of 3.04 °C, 17.61 W m^{-2}, 45.17 W m^{-2}, respectively.

3.2. The Influence of Rising Air Temperature at the Two Different Lakes

All the above-mentioned positive biases in LST and LE for Lake Ngoring are present in both the FLake and the FLake_$\alpha_{\text{ice}} = 0.15$ simulations, and have relatively little influence on seasonal variations for the lake. FLake_$\alpha_{\text{ice}} = 0.15$ simulates the lake ice freezing and melting periods at Lake Ngoring, and the LST and frozen period at Lake Nam Co, better than the original FLake model. We therefore assume that the FLake_$\alpha_{\text{ice}} = 0.15$-simulated LST, H, LE and ice phenology data can be used to analyze changes in seasonal variations at the two lakes.

Although global climate models (such as CMIP5 and CMIP6) can provide climatic variables which can be used as a future climate scenario, it is well known that the simulation errors of global climate system models for the TP are large [23,25]. The forcing data of the control experiment in our study are in situ-observed climatic variables data; these data can provide a more realistic driving field of the model. Because the environmental changes experienced on the TP are mostly associated with rapid surface warming [23], we use offline simulation and only change the air temperature, mainly to compare the effects of different warming degrees on the lake. Here, we use simulations from FLake_$\alpha_{\text{ice}} = 0.15$ to analyze the effects of rising air temperatures on LST, H, LE and ice phenology.

3.2.1. Seasonal Variations in the Effect of Rising Air Temperature on the Lake Surface Temperature and on Heat Fluxes

Figure 6 shows seasonal variations in LST, H and LE after air temperatures have risen by 1.50 °C, 2.25 °C, 3.00 °C, and 3.50 °C at Lake Nam Co and Lake Ngoring, from 2012 to 2016. In ten groups of sensitivity tests, we selected four modes (1.50 °C, 2.25 °C, 3.00 °C, 3.50 °C), for which there were clear changes, to analyze the response of the seasonal variability at the two lakes to the increase in air temperature.

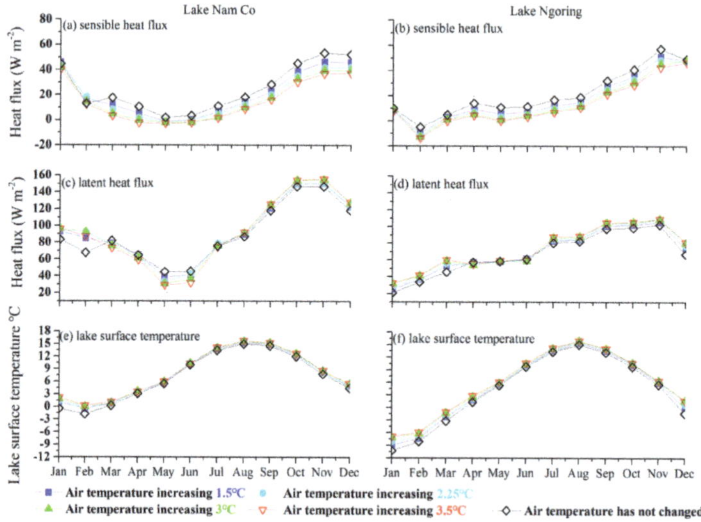

Figure 6. Seasonal variations in LST, H and LE after the air temperature has risen by 1.5, 2.25, 3 and 3.5 °C (averaged from January 2012 to December 2016) (**a,c,e**) at Lake Nam Co and (**b,d,f**) at Lake Ngoring.

H (Figure 6a,b) is the most sensitive parameter to changes in air temperature, with a decreasing trend for every month as air temperature rises, except for February at Lake Nam Co and December in Lake Ngoring. The difference in H simulated under different air temperatures reaches a maximum in November (before freeze-up), and suddenly decreases when the two lakes begin to freeze. The minimum decreases in H at Lake Nam Co reach $-16.45\,\mathrm{Wm^{-2}}$ and $-14.94\,\mathrm{Wm^{-2}}$ in November and December, respectively. After Lake Nam Co has frozen, the minimum decrease is in January, and is $-4.46\,\mathrm{Wm^{-2}}$. The minimum decreases at Lake Ngoring reach $-11.97\,\mathrm{Wm^{-2}}$ and $-14.34\,\mathrm{Wm^{-2}}$ in October and November, respectively. After Lake Ngoring has frozen, the minimum decrease is in December, and is $-2.49\,\mathrm{Wm^{-2}}$. LST rises rapidly in winter with the increasing air temperature, and the change in H at Lake Nam Co in February (Lake Ngoring in December) is opposite to that for other months.

With increasing air temperature, LE (Figure 6c,d) generally decreases from the end of the frozen period to a period after ice break-up (March to June at Lake Nam Co, April to June at Lake Ngoring), and increases in the other months, especially during the frozen period. The maximum decreases at Lake Nam Co reach $16.00\,\mathrm{Wm^{-2}}$ and $13.45\,\mathrm{Wm^{-2}}$ in May and June, respectively, and the maximum increase reaches $25.89\,\mathrm{Wm^{-2}}$ in February. The maximum decrease at Lake Ngoring reaches $4.18\,\mathrm{Wm^{-2}}$ in April, while the maximum increase reaches $14.88\,\mathrm{Wm^{-2}}$ in December.

LST (Figure 6e,f) increases every month as the air temperature rises, and two relatively large peaks appear in winter and summer, with the largest increase in winter. The maximum increases at Lake Ngoring reach $3.02\,°\mathrm{C}$ and $3.40\,°\mathrm{C}$ in December and January, respectively, while the maximum increase at Lake Nam Co reaches $2.64\,°\mathrm{C}$ in January.

Figures 7 and 8 show the impact of different temperature increases on the seasonal changes in LE, H, LST, and lake mixed layer temperature (MLT). Here, we use the differences between the different sensitivity tests (line in Figures 7 and 8) to analyze the influence of different increase intervals on seasonal variations in the four variables. The differences before and after the frozen period fluctuate significantly. The variability in the sensitivities for the four variables in the ice-free period is roughly the same at Lake Ngoring and at Nam Co, but the magnitude of the sensitivities is small. Since the frozen period is longer at Lake Ngoring, we can see differences between the variations in the sensitivities at the two lakes during the frozen period. In the middle of the frozen period, there is a relatively regular one-month (February) change in LST, LE and H with increasing air temperature at Lake Ngoring, and we do not see this at Nam Co. The thicker ice at Lake Ngoring means that the MLT is unaffected by the increase in air temperature for the three months after freeze-up. However, the MLT at Lake Nam Co changes significantly once the air temperature rises beyond $2.75\,°\mathrm{C}$. At both lakes, there is a clear difference between the simulations with unchanged air temperature, and the simulations when the air temperature has been increased by $1.5\,°\mathrm{C}$, shown with a black line in Figures 7 and 8. The next obvious difference is between the simulations calculated with temperature increases of $2.75\,°\mathrm{C}$ and $3.00\,°\mathrm{C}$ at Lake Nam Co, shown with blue line in Figures 7 and 8 ($2.25\,°\mathrm{C}$ and $2.50\,°\mathrm{C}$ at Lake Ngoring, shown with a red line in Figures 7 and 8).

Differences between the sensitivity tests can be used to analyze the rate of change for the lake surface and MLT for different intervals of increasing temperature. The larger the positive (negative) value, the faster the LST rises (drops) over the temperature interval. The variations in the differences for LST and MLT are the same in the ice-free period, with relatively regular changes except for the months before and after the ice has frozen. After freeze-up, the thick ice layer isolates the lake body from the air, and the responses of the mixed layer to different air temperature increases become increasingly similar, until they become the same, i.e., the sensitivity to the magnitude of the change in air temperature disappears. This happens at Lake Ngoring, but differences appear again between the mixed layer responses to different air temperature increases at Lake Nam Co once the air temperature rises beyond $2.75\,°\mathrm{C}$. When the air temperature rises beyond $2.75\,°\mathrm{C}$ at Lake

Nam Co, the lake ice is thin enough, and solar radiation will be absorbed by the lake water through the ice layer, and then affect the mixed layer temperature.

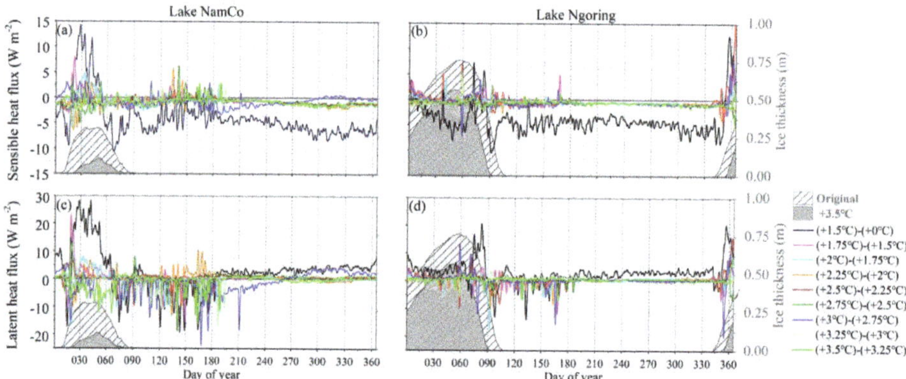

Figure 7. Seasonal variations in lake ice thickness (shading), and differences between the sensitivity of H and LE to different air temperature increases (lines), averaged from January 2012 to December 2016 (**a,c**) at Lake Nam Co and (**b,d**) at Lake Ngoring. The plotted differences are the differences between the sensitivities found for different air temperature changes, as detailed in the legend, i.e., (+1.5 °C)–(+0 °C) refers to the difference between the sensitivity found for a temperature increase of 1.5 °C, and for a temperature increase of 0. The y coordinates for each line show how much the variable has changed within this air temperature increase interval. The gray-shaded area and the left-slash-filled area show the simulated values for no change in air temperature, and for an air temperature increase of 3.5 °C, respectively.

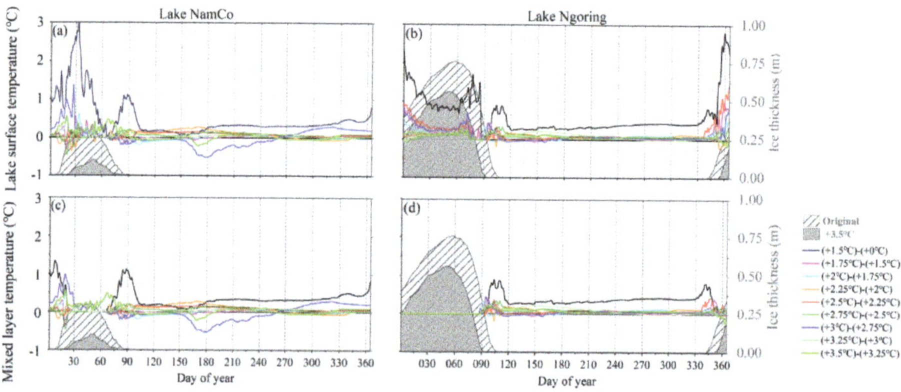

Figure 8. Seasonal variations in the lake ice thickness (shading), and the difference between LST and MLT sensitivities to different temperature increases (lines), averaged from January 2012 to December 2016 (**a,c**) at Lake Nam Co and (**b,d**) at Lake Ngoring. The details of the legend are the same as in Figure 7.

Changes in H are closely related to changes in LST. The LST for Lake Nam Co from the end of July to mid-January follows a generally increasing trend, when the air temperature increases from 2.00 °C to 2.75 °C, and differences in LST are more obvious with temperature increases of between 0 °C and 1.5 °C. When the air temperature rises from 1.50 °C to 2.00 °C, and above 2.75 °C, the LST decreases with the increasing air temperature for three months after the ice break-up, and the most significant decrease occurs in the last month. When the lake begins to freeze-up in January, the differences appear to be irregular, and fluctuate until the end of February. The LST at Lake Ngoring from the end of June to the end of November follows a steadily increasing trend for air temperature increases from 0 °C to 3.5 °C, and the differences in LST are more obvious for a temperature increase of 0 °C to

1.5 °C. When the air temperature rises from 1.5 °C to 2.25 °C, and from 2.75 to 3 °C, LST decreases with increasing air temperature during two months after ice break-up. When the lake begins to freeze-up in November, the differences appear to fluctuate until the end of March. By February, the rising LST trend has stabilized, and then there is a rapid change in March, when the ice is about to begin break-up.

3.2.2. The Effect of Rising Air Temperature on Lake Surface Temperature and Ice Phenology

Figure 9 shows the impact of the change in air temperature on the simulated LST in summer and winter, and Table 1 shows the impact on the simulated lake ice phenology. As expected, the LST rises with increasing air temperature. When the air temperature rises, the number of frozen days decreases and the maximum ice thickness decreases, meaning the freeze-up date of lake ice will be delayed, and the break-up date will be advanced.

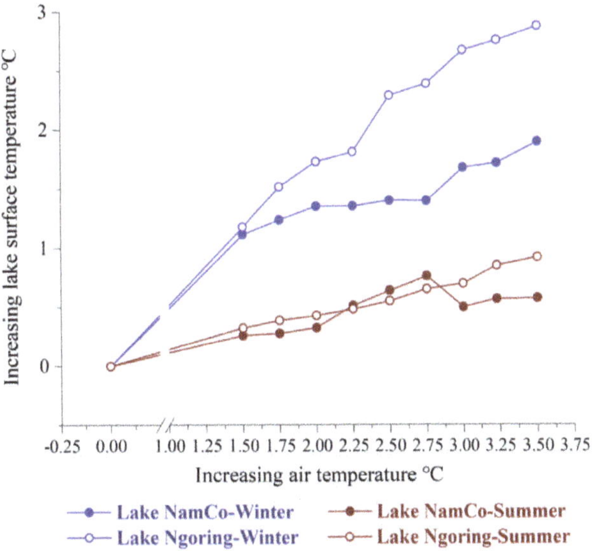

Figure 9. Plot of changes in LST in summer and winter for air temperature increases of between 1.5 °C and 3.5 °C at Lake Nam Co and Lake Ngoring (summer data are averaged from June, July, and August 2012–2016, and winter data are averaged from December, January, and February 2012–2016).

The LST in summer and winter increases with the increases in air temperature. However, the LST rises much faster in winter than in summer, and the increase is greater at Lake Ngoring than at Lake Nam Co in both seasons. When the air temperature rises by 3.50 °C, the LST at Lake Nam Co rises by 1.90 °C in winter, and by 0.57 °C in summer, while the LST at Lake Ngoring rises by 2.88 °C in winter, and by 0.92 °C in summer. The pattern with which LST rises with increasing air temperature is different in winter and summer. Once the winter air temperature rises beyond 2.75 °C at Lake Nam Co, LST rises faster with increasing air temperature. The same thing occurs at Lake Ngoring when the winter air temperature rises beyond 2.25 °C. However, when the air temperature increases from 2.75 °C to 3.00 °C at Lake Nam Co, the LST decreases from 0.76 °C to 0.50 °C, and at the same time, the air temperature begins to affect the temperature of the mixed layer.

Table 1. Changes in lake ice thickness, the number of frozen days, and dates for the onset of freeze-up and melt when the air temperature increases by between 1.5 °C and 3.5 °C at Lake Nam Co and Lake Ngoring (averaged from January 2012 to December 2016).

Lake	Increasing Air Temperature (°C)	Freezing Date	Melting Date	Frozen Days	Maximum Ice Thickness (m)
Lake Nam Co	0	9 January	1 April	83	0.300
	0.150	11 January	23 March	72	0.226
	0.175	17 January	22 March	65	0.216
	0.200	18 January	21 March	63	0.197
	0.225	11 January	21 March	70	0.195
	0.250	11 January	21 March	70	0.192
	0.275	11 January	21 March	70	0.196
	0.300	11 January	20 March	69	0.159
	0.325	16 January	20 March	64	0.141
	0.350	18 January	20 March	62	0.103
Lake Ngoring	0	6 December	17 April	133	0.761
	0.150	11 December	15 April	126	0.678
	0.175	11 December	15 April	126	0.656
	0.200	13 December	15 April	124	0.645
	0.225	11 December	15 April	126	0.638
	0.250	20 December	6 April	108	0.607
	0.275	20 December	5 April	107	0.602
	0.300	21 December	3 April	104	0.580
	0.325	21 December	4 April	105	0.573
	0.350	21 December	3 April	104	0.56409

When the air temperature rises from 0 °C to 3.50 °C, the number of frozen days at Lake Nam Co decreases from 83 days to 62 days, while the number of frozen days at Lake Ngoring decreases from 133 days to 104 days; the maximum ice thickness at Lake Nam Co reduces from 0.3 m to 0.1 m while the maximum ice thickness at Lake Ngoring reduces from 0.76 m to 0.56 m; ice freeze-up is delayed by 9 days, and the onset of ice melt advances by 12 days at Lake Nam Co, while freeze-up is delayed by 15 days and the onset of ice melt advances by 14 days at Lake Ngoring. When the air temperature rises from 2.25 °C to 2.50 °C, the frozen period at Lake Ngoring changes more obviously than it does between cooler temperatures. Specifically, when the air temperature at Lake Ngoring rises from 0 °C to 2.25 °C, and from 2.50 °C to 3.50 °C, the number of frozen days decreases by 7 days and 4 days, respectively, and when the air temperature rises from 2.25 °C to 2.50 °C, the number of frozen days decreases by 18 days. There is no obvious rule to relate the change in the duration of the frozen period at Lake Nam Co with rising air temperature. The processes that determine the duration of the frozen period are relatively complicated, particularly around the time of freeze-up and break-up. Lake Ngoring has a relatively long and stable freezing period, and we speculate that temperature increases must reach a threshold of around 2.25 °C before they influence the duration of the frozen period at Lake Ngoring.

3.2.3. The Maximum Possible Impact of Rising Air Temperature on the TP On Lake Nam Co and Lake Ngoring

Figures 10 and 11 show the impact of air temperature rises of 3.5 °C at the two lakes. Here, we use the differences between responses to air temperature increases of 0 °C and 3.5 °C to analyze the influence of the most significant warming on seasonal variations in five variables (H, LE, LST, MLT, and ice thickness) for the two lakes. In general, when the air temperature rises by 3.5 °C, annual LE and LST increase and H decreases at both lakes (Figure 10b,c, Figure 11b,c). The increase in LE and LST at Lake Ngoring is higher than at Lake Nam Co, and the decrease in H at Lake Nam Co is greater than at Lake Ngoring. The ice thickness at Lake Ngoring decreases more than at Lake Nam Co. The changes in the variables in response to the temperature increases are similar at the two lakes from the end of July to the beginning of December. From mid-December to the end of July, the two lakes respond differently to the temperature increase, due to their different frozen periods and the continuous impact on lake surface energy exchange after the lake ice melts. The difference between the responses of H to the temperature change at the two different lakes is greater than the difference between the LE responses at the two lakes. The maximum value for the increase in LE at Lake Nam Co is 39.52 W m^{-2}, which occurs at the beginning of February, and the maximum for Lake Ngoring is 47.37 W m^{-2}, which occurs at the end of December. The maximum decreases occur at the end of May and at the beginning of April, reaching -38.22 W m^{-2} and -23.95 W m^{-2}, respectively. The maximum values for the increases in H at Lake Nam Co and Lake Ngoring are also at the beginning of February and at the end of December, when they reach 18.08 W m^{-2} and 24.73 W m^{-2}, respectively. In contrast to the LE responses, the maximum decreases in H occur at the beginning of March and in mid-November, reaching -22.98 W m^{-2} and -21.39 W m^{-2} for Lake Nam Co and Lake Ngoring, respectively. Similar to H and LE, the maximum increases in surface temperature at Lake Nam Co and Lake Ngoring are at the beginning of February and the end of December, when they reach 5.06 °C and 6.58 °C, respectively. The maximum decrease in ice thickness at Lake Nam Co and Lake Ngoring occurs at the beginning of February and the end of March, reaching -0.24 m and -0.36 m, respectively. The possible reasons can be summarized as the following points. Firstly, the model lake depth is set to 25 m for Lake Ngoring and 40 m for Lake Nam Co. Secondly, the meteorological conditions of the two lakes are different, such as air temperature, precipitation and wind speed.

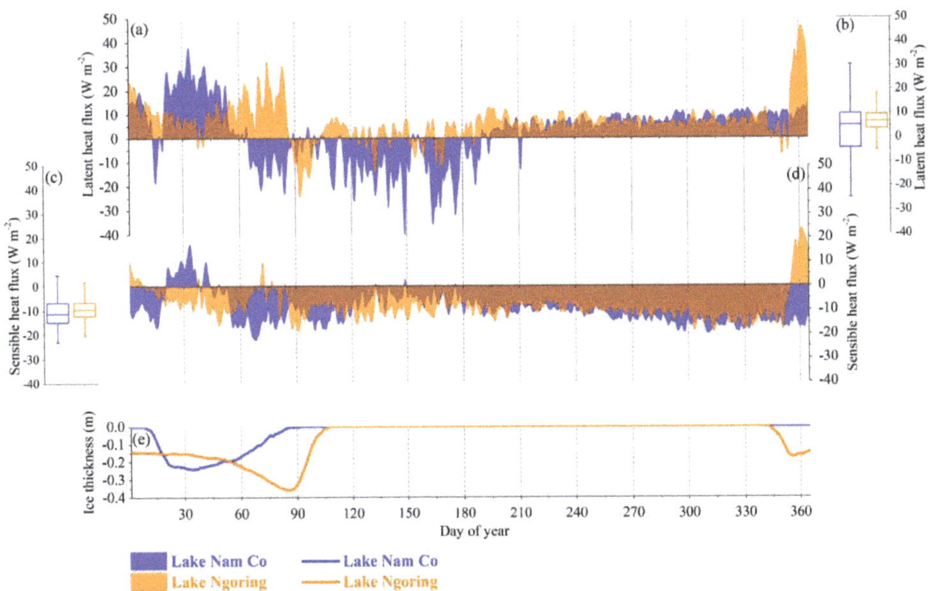

Figure 10. Comparison of changes in the differences, when the air temperature rises by 0 °C and by 3.5 °C (the latter minus the former), for H, LE and ice thickness for Lake Nam Co and Lake Ngoring (averaged from January 2012 to December 2016): (**a,b**) LE, (**c,d**) H, and (**e**) lake ice thickness.

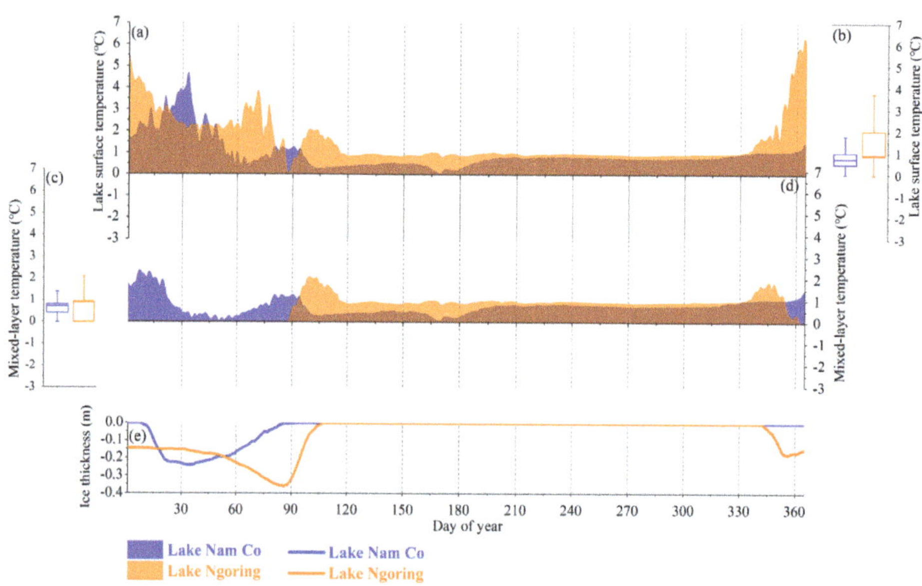

Figure 11. Comparison of changes in the differences, when the air temperature rises by 0 °C and by 3.5 °C (the latter minus the former), for LST, MLT and ice thickness, for Lake Nam Co and Lake Ngoring (averaged from January 2012 to December 2016): (**a**,**b**) LE, (**c**,**d**) H, and (**e**) for lake ice thickness.

4. Conclusions

The model simulations show that LST rises with increasing air temperature, the rate is much faster in winter than in summer, and the increase is greater at Lake Ngoring than at Lake Nam Co. When the air temperature rises, the number of frozen days is reduced, the maximum ice thickness decreases, the freeze-up date for the lake ice is delayed, and the break-up date advances. From the end of July to the beginning of December, the variables changed similarly for the two lakes, and from mid-December to the end of July, due to their different frozen periods and the continuous impact on the lake surface energy exchange after ice melt, the two lakes respond significantly differently to changes in temperature. The difference in the response of H is greater than the difference in the response of LE between Lake Ngoring and Lake Nam Co.

It is interesting that LST increases in summer and winter with rising air temperature, but it increases much faster in winter than in summer, especially as the lake begins to freeze, and it increases more at Lake Ngoring than at Lake Nam Co. This is completely consistent with the fact that the increase in surface temperature on the TP is higher in winter than in summer. The mechanism for surface warming on the TP has not yet been resolved, so in future work, we will explore a plateau warming mechanism that includes multiple freezing and thawing processes, such as glaciers, frozen soil, and snow. In combination, these phenomena may explain the freezing period response of lake ice.

Another interesting phenomenon is that when the air temperature rises from 2.25 °C to 2.50 °C, the freezing period at Lake Ngoring changes more obviously than following similarly sized increases between cooler air temperatures. There is no obvious rule to relate the duration of the frozen period at Lake Nam Co with rising air temperature. The physical processes that determine the duration of the frozen period are relatively complicated, especially during the times for ice freeze-up and break-up. Lake Ngoring has a relatively long and stable freezing period, and we speculate that temperature increases must reach a threshold of 2.25 °C before they influence the duration of the frozen period at Lake Ngoring. Our future work will try to investigate this by carrying out model simulations for other lakes on the TP. We will also focus on a wider range of climate change indicators, including wind speed, precipitation, and humidity, and try to analyze the impact of future climate change on lakes on the TP.

One aspect of this work that could be improved is that the diurnal pattern in the simulated albedo does not match the typical U-shaped curve seen in the observations. This is because the albedo parameterization scheme in the FLake model was developed from the sea ice albedo parameterization scheme, and its application to lakes requires further improvement. Our future work will establish long-term observation sites at multiple lakes to obtain sufficient observation data to improve the parameterization scheme for ice albedo, and thereby make it more suitable for lakes on the TP. Another aspect of this work that could be improved is that we use offline simulation, and only change the air temperature, mainly to compare the effects of different warming degrees on the lake. This method is commonly used, even though it has some obvious flaws, but the uncertainty is relatively small. In our future study, we will further use reginal climate models to comprehensively explore the effects of the offline and coupled climate models on the simulation results of the method.

Author Contributions: Y.M. conceived and designed the experiments; J.L., Z.L. and D.S. performed the experiments; Y.M. and J.L. analyzed the data; J.L. wrote the main manuscript. All authors have read and agreed to the published version of the manuscript.

Funding: This research was funded by the Second Tibetan Plateau Scientific Expedition and Research (STEP) program (2019QZKK0103), the Strategic Priority Research Program of Chinese Academy of Sciences (XDA20060101), the National Natural Science Foundation of China (91837208, 42075089), the Youth innovation Promotion Association CAS (QCH2019004).

Data Availability Statement: The input data used for FLake model simulations and the in situ observed data used for evaluate the simulated results presented in this study are available upon reasonable request from the corresponding author.

Acknowledgments: We acknowledge the hard work of Binbin Wang, Shaobo Zhang and Xiaohui Peng in the field observations.

Conflicts of Interest: The authors declare no conflict of interest.

References

1. *Climate Change 2013*; IPCC: Geneva, Switzerland, 2013; ISBN 9781107661820.
2. Lofgren, B.M. Simulated effects of idealized Laurentian Great Lakes on regional and large-scale climate. *J. Clim.* **1997**, *10*, 2847–2858. [CrossRef]
3. Long, Z.; Perrie, W.; Gyakum, J.; Caya, D.; Laprise, R. Northern lake impacts on local seasonal climate. *J. Hydrometeorol.* **2007**, *8*, 881–896. [CrossRef]
4. Dutra, E.; Stepanenko, V.M.; Balsamo, G.; Viterbo, P.; Miranda, P.M.A.; Mironov, D.; Schär, C. An offline study of the impact of lakes on the performance of the ECMWF surface scheme. *Boreal Environ. Res.* **2010**, *15*, 100–112.
5. Schindler, D.W. Lakes as sentinels and integrators for the effects of climate change on watersheds, airsheds, and landscapes. *Limnol. Oceanogr.* **2009**, *54*, 2349–2358. [CrossRef]
6. O'Reilly, C.M.; Sharma, S.; Gray, D.K.; Hampton, S.E.; Read, J.S.; Rowley, R.J.; Schneider, P.; Lenters, J.D.; McIntyre, P.B.; Kraemer, B.M.; et al. Rapid and highly variable warming of lake surface waters around the globe. *Geophys. Res. Lett.* **2015**, *42*, 10773–10781. [CrossRef]
7. Sharma, S.; Gray, D.K.; Read, J.S.; O'Reilly, C.M.; Schneider, P.; Qudrat, A.; Gries, C.; Stefanoff, S.; Hampton, S.E.; Hook, S.; et al. A global database of lake surface temperatures collected by in situ and satellite methods from 1985–2009. *Sci. Data* **2015**, *2*. [CrossRef]
8. Schneider, P.; Hook, S.J. Space observations of inland water bodies show rapid surface warming since 1985. *Geophys. Res. Lett.* **2010**, *37*, 1–5. [CrossRef]
9. Austin, J.A.; Colman, S.M. Lake Superior summer water temperatures are increasing more rapidly than regional temperatures: A positive ice-albedo feedback. *Geophys. Res. Lett.* **2007**, *34*, 1–5. [CrossRef]
10. Burnett, A.W.; Kirby, M.E.; Mullins, H.T.; Patterson, W.P. Increasing Great Lake-effect snowfall during the twentieth century: A regional response to global warming? *J. Clim.* **2003**, *16*, 3535–3542. [CrossRef]
11. Hostetler, S.W.; Bartlein, P.J. Simulation of lake evaporation with application to modeling lake level variations of Harney-Malheur Lake, Oregon. *Water Resour. Res.* **1990**, *26*, 2603–2612. [CrossRef]

12. Zhao, G.; Gao, H.; Cai, X. Estimating lake temperature profile and evaporation losses by leveraging MODIS LST data. *Remote Sens. Environ.* **2020**, *251*, 112104. [CrossRef]

13. Biermann, T.; Babel, W.; Ma, W.; Chen, X.; Thiem, E.; Ma, Y.; Foken, T. Turbulent flux observations and modelling over a shallow lake and a wet grassland in the Nam Co basin, Tibetan Plateau. *Theor. Appl. Climatol.* **2014**, *116*, 301–316. [CrossRef]

14. Duguay, C.R.; Prowse, T.D.; Bonsal, B.R.; Brown, R.D.; Lacroix, M.P.; Ménard, P. Recent trends in Canadian lake ice cover. *Hydrol. Process.* **2006**, *20*, 781–801. [CrossRef]

15. Lang, J.; Lyu, S.; Li, Z.; Ma, Y.; Su, D. An investigation of ice surface albedo and its influence on the high-altitude lakes of the Tibetan Plateau. *Remote Sens.* **2018**. [CrossRef]

16. Efremova, T.V.; Palshin, N.I. Ice phenomena terms on the water bodies of Northwestern Russia. *Russ. Meteorol. Hydrol.* **2011**, *36*, 559–565. [CrossRef]

17. Brown, L.C.; Duguay, C.R. The response and role of ice cover in lake-climate interactions. *Prog. Phys. Geogr.* **2010**, *34*, 671–704. [CrossRef]

18. Shimoda, Y.; Azim, M.E.; Perhar, G.; Ramin, M.; Kenney, M.A.; Sadraddini, S.; Gudimov, A.; Arhonditsis, G.B. Our current understanding of lake ecosystem response to climate change: What have we really learned from the north temperate deep lakes? *J. Great Lakes Res.* **2011**, *37*, 173–193. [CrossRef]

19. Arhonditsis, G.B.; Brett, M.T.; DeGasperi, C.L.; Schindler, D.E. Effects of climatic variability on the thermal properties of Lake Washington. *Limnol. Oceanogr.* **2004**, *49*, 256–270. [CrossRef]

20. Bernhardt, J.; Engelhardt, C.; Kirillin, G.; Matschullat, J. Lake ice phenology in Berlin-Brandenburg from 1947–2007: Observations and model hindcasts. *Clim. Chang.* **2012**, *112*, 791–817. [CrossRef]

21. Qiu, J. China: The third pole. *Nature* **2008**, *454*, 393–396. [CrossRef] [PubMed]

22. Gu, S.; Tang, Y.; Cui, X.; Kato, T.; Du, M.; Li, Y.; Zhao, X. Energy exchange between the atmosphere and a meadow ecosystem on the Qinghai-Tibetan Plateau. *Agric. For. Meteorol.* **2005**, *129*, 175–185. [CrossRef]

23. Yang, K.; Wu, H.; Qin, J.; Lin, C.; Tang, W.; Chen, Y. Recent climate changes over the Tibetan Plateau and their impacts on energy and water cycle: A review. *Glob. Planet. Change* **2014**, *112*, 79–91. [CrossRef]

24. Kuang, X.; Jiao, J.J. Review on climate change on the Tibetan Plateau during the last half century. *J. Geophys. Res. Atmos.* **2016**, 3979–4007. [CrossRef]

25. Zhang, G.; Yao, T.; Xie, H.; Yang, K.; Zhu, L.; Shum, C.K.; Bolch, T.; Yi, S.; Allen, S.; Jiang, L.; et al. Response of Tibetan Plateau lakes to climate change: Trends, patterns, and mechanisms. *Earth Sci. Rev.* **2020**, *208*, 103269. [CrossRef]

26. Wu, F.; You, Q.; Xie, W.; Zhang, L. Temperature change on the Tibetan Plateau under the global warming of 1.5 °C and 2 °C. *Clim. Chang. Res.* **2019**, *15*, 92–109. [CrossRef]

27. Messager, M.L.; Lehner, B.; Grill, G.; Nedeva, I.; Schmitt, O. Estimating the volume and age of water stored in global lakes using a geo-statistical approach. *Nat. Commun.* **2016**, *7*, 1–11. [CrossRef]

28. Ma, R.; Duan, H.; Hu, C.; Feng, X.; Li, A.; Ju, W.; Jiang, J.; Yang, G. A half-century of changes in China's lakes: Global warming or human influence? *Geophys. Res. Lett.* **2010**, *37*, 2–7. [CrossRef]

29. Zhang, G.; Yao, T.; Chen, W.; Zheng, G.; Shum, C.K.; Yang, K.; Piao, S.; Sheng, Y.; Yi, S.; Li, J.; et al. Regional differences of lake evolution across China during 1960s–2015 and its natural and anthropogenic causes. *Remote Sens. Environ.* **2019**, *221*, 386–404. [CrossRef]

30. Kropáček, J.; Maussion, F.; Chen, F.; Hoerz, S.; Hochschild, V. Analysis of ice phenology of lakes on the Tibetan Plateau from MODIS data. *Cryosphere* **2013**, *7*, 287–301. [CrossRef]

31. Bengtsson, L. Mixing in ice-covered lakes. *Hydrobiologia* **1996**, *322*, 91–97. [CrossRef]

32. Liu, X.; Chen, B. Climatic warming in the tibetan plateau during recent decades. *Int. J. Climatol.* **2000**, *1742*, 1729–1742. [CrossRef]

33. Gou, P.; Ye, Q.; Che, T.; Feng, Q.; Ding, B.; Lin, C.; Zong, J. Lake ice phenology of Nam Co, Central Tibetan Plateau, China, derived from multiple MODIS data products. *J. Great Lakes Res.* **2017**, *43*, 989–998. [CrossRef]

34. Ke, C.-Q.; Tao, A.-Q.; Jin, X. Variability in the ice phenology of Nam Co Lake in central Tibet from scanning multichannel microwave radiometer and special sensor microwave/imager: 1978 to 2013. *J. Appl. Remote Sens.* **2013**, *7*, 073477. [CrossRef]

35. Yang, R.; Zhu, L.; Wang, J.; Ju, J.; Ma, Q.; Turner, F.; Guo, Y. Spatiotemporal variations in volume of closed lakes on the Tibetan Plateau and their climatic responses from 1976 to 2013. *Clim. Change* **2017**, *140*, 621–633. [CrossRef]

36. Yao, J.; Zhao, L.; Gu, L.; Qiao, Y.; Jiao, K. The surface energy budget in the permafrost region of the Tibetan Plateau. *Atmos. Res.* **2011**, *102*, 394–407. [CrossRef]

37. Wen, L.; Lyu, S.; Kirillin, G.; Li, Z.; Zhao, L. Air-lake boundary layer and performance of a simple lake parameterization scheme over the Tibetan highlands. *Tellus A Dyn. Meteorol. Oceanogr.* **2016**, *68*, 1–15. [CrossRef]

38. Lenters, J.D.; Kratz, T.K.; Bowser, C.J. Effects of climate variability on lake evaporation: Results from a long-term energy budget study of Sparkling Lake, northern Wisconsin (USA). *J. Hydrol.* **2005**, *308*, 168–195. [CrossRef]

39. Oswald, C.J.; Rouse, W.R. Thermal characteristics and energy balance of various-size Canadian Shield lakes in the Mackenzie River basin. *J. Hydrometeorol.* **2004**, *5*, 129–144. [CrossRef]

40. Wang, B.; Ma, Y.; Wang, Y.; Su, Z.; Ma, W. Significant differences exist in lake-atmosphere interactions and the evaporation rates of high-elevation small and large lakes. *J. Hydrol.* **2019**, *573*, 220–234. [CrossRef]

41. Li, Z.; Lyu, S.; Ao, Y.; Wen, L.; Zhao, L.; Wang, S. Long-term energy flux and radiation balance observations over Lake Ngoring, Tibetan Plateau. *Atmos. Res.* **2015**, *155*, 13–25. [CrossRef]

42. Lazhu; Yang, K.; Wang, J.; Yanbin, L.; Chen, Y.; Zhu, L.; Ding, B.; Qin, J. Quantifying evaporation and its decadal change for Lake Nam Co, central Tibetan Plateau. *J. Geophys. Res. Atmos.* **2016**, *121*, 7578–7591. [CrossRef]

43. Wang, B.; Ma, Y.; Chen, X.; Ma, W.; Su, Z.; Menenti, M. Observation and simulation of lake-air heat and water transfer processes in a high-altitude shallow lake on the Tibetan Plateau. *J. Geophys. Res. Atmos.* **2015**, *120*, 12327–12344. [CrossRef]

44. Wu, Y.; Zheng, H.; Zhang, B.; Chen, D.; Lei, L. Long-term changes of lake level and water budget in the nam co lake basin, central tibetan plateau. *J. Hydrometeorol.* **2014**, *15*, 1312–1322. [CrossRef]

45. Mironov, D. *Parameterization of Lakes in Numerical Weather Prediction. Part 1: Description of a Lake Model*; German Weather Service: Offenbach am Main, Germany, 2005.

46. Li, Z.; Ao, Y.; Lyu, S.; Lang, J.; Wen, L.; Stepanenko, V.; Meng, X.; Zhao, L. Investigation of the ice surface albedo in the Tibetan Plateau lakes based on the field observation and MODIS products. *J. Glaciol.* **2018**, *64*, 506–516. [CrossRef]

47. Lei, Y.; Yao, T.; Bird, B.W.; Yang, K.; Zhai, J.; Sheng, Y. Coherent lake growth on the central Tibetan Plateau since the 1970s: Characterization and attribution. *J. Hydrol.* **2013**, *483*, 61–67. [CrossRef]

48. Zhou, S.; Kang, S.; Chen, F.; Joswiak, D.R. Water balance observations reveal significant subsurface water seepage from Lake Nam Co, south-central Tibetan Plateau. *J. Hydrol.* **2013**, *491*, 89–99. [CrossRef]

49. Li, Z.; Lyu, S.; Zhao, L.; Wen, L.; Ao, Y.; Wang, S. Turbulent transfer coefficient and roughness length in a high-altitude lake, Tibetan Plateau. *Theor. Appl. Climatol.* **2016**, *124*, 723–735. [CrossRef]

50. Fratini, G.; Mauder, M. Towards a consistent eddy-covariance processing: An intercomparison of EddyPro and TK3. *Atmos. Meas. Tech. Discuss.* **2014**, *7*, 2107–2126. [CrossRef]

51. Göckede, M.; Rebmann, C.; Foken, T. A combination of quality assessment tools for eddy covariance measurements with footprint modelling for the characterisation of complex sites. *Agric. For. Meteorol.* **2004**, *127*, 175–188. [CrossRef]

52. Schaaf, C.B.; Gao, F.; Strahler, A.H.; Lucht, W.; Li, X.; Tsang, T.; Strugnell, N.C.; Zhang, X.; Jin, Y.; Muller, J.; et al. First operational BRDF, albedo nadir reflectance products from MODIS. *Remote Sens. Environ.* **2002**, *83*, 135–148. [CrossRef]

53. Svacina, N.A.; Duguay, C.R.; King, J.M.L. Modelled and satellite-derived surface albedo of lake ice—Part II: Evaluation of MODIS albedo products. *Hydrol. Process.* **2014**, *28*, 4562–4572. [CrossRef]

54. Tavares, M.H.; Cunha, A.H.F.; Motta-Marques, D.; Ruhoff, A.L.; Cavalcanti, J.R.; Fragoso, C.R.; Bravo, J.M.; Munar, A.M.; Fan, F.M.; Rodrigues, L.H.R. Comparison of methods to estimate lake-surface-water temperature using landsat 7 ETM+ and MODIS imagery: Case study of a large shallow subtropical lake in Southern Brazil. *Water* **2019**, *11*, 168. [CrossRef]

55. Zhang, G.; Yao, T.; Xie, H.; Qin, J.; Ye, Q.; Dai, Y.; Guo, R. Estimating surface temperature changes of lakes in the Tibetan Plateau using MODIS LST data. *J. Geophys. Res. Atmos.* **2014**, *119*, 8552–8567. [CrossRef]

56. Westermann, S.; Langer, M.; Boike, J. Systematic bias of average winter-time land surface temperatures inferred from MODIS at a site on Svalbard, Norway. *Remote Sens. Environ.* **2012**, *118*, 162–167. [CrossRef]

57. Crosman, E.T.; Horel, J.D. MODIS-derived surface temperature of the Great Salt Lake. *Remote Sens. Environ.* **2009**, *113*, 73–81. [CrossRef]

58. Reinart, A.; Reinhold, M. Mapping surface temperature in large lakes with MODIS data. *Remote Sens. Environ.* **2008**, *112*, 603–611. [CrossRef]

59. Robinson, I.S.; Wells, N.C.; Charnock, H. The sea surface thermal boundary layer and its relevance to the measurement of sea surface temperature by airborne and spaceborne radiometers. *Int. J. Remote Sens.* **1984**, *5*, 19–45. [CrossRef]

60. Bowling, L.C.; Lettenmaier, D.P. Modeling the effects of lakes and wetlands on the water balance of arctic environments. *J. Hydrometeorol.* **2010**, *11*, 276–295. [CrossRef]

61. Subin, Z.M.; Riley, W.J.; Mironov, D. An improved lake model for climate simulations: Model structure, evaluation, and sensitivity analyses in CESM1. *J. Adv. Model. Earth Syst.* **2012**, *4*, 1–27. [CrossRef]

62. Mironov, D.V. *Parameterization of Lakes in Numerical Weather Prediction. Description of a Lake Model*; COSMO Tech. Rep. No.11; Deutscher Wetterdienst: Offrnbach am Main, Germany, 2008; Volume 47. [CrossRef]

63. Kirillin, G.; Hochschild, J.; Mironov, D.; Terzhevik, A.; Golosov, S.; Nützmann, G. FLake-Global: Online lake model with worldwide coverage. *Environ. Model. Softw.* **2011**, *26*, 683–684. [CrossRef]

64. Hamilton, D.P.; Schladow, S.G. Prediction of water quality in lakes and reservoirs. Part I—Model description. *Ecol. Model.* **1997**, *96*, 91–110. [CrossRef]

65. Goudsmit, G.H.; Burchard, H.; Peeters, F.; Wüest, A. Application of k-ε turbulence models to enclosed basins: The role of internal seiches. *J. Geophys. Res. C Oceans* **2002**, *107*, 23-1–23-13. [CrossRef]

66. Yao, H.; Samal, N.R.; Joehnk, K.D.; Fang, X.; Bruce, L.C.; Pierson, D.C.; Rusak, J.A.; James, A. Comparing ice and temperature simulations by four dynamic lake models in Harp Lake: Past performance and future predictions. *Hydrol. Process.* **2014**, *28*, 4587–4601. [CrossRef]

67. Stepanenko, V.; Jöhnk, K.D.; Machulskaya, E.; Perroud, M.; Subin, Z.; Nordbo, A.; Mammarella, I.; Mironov, D. Simulation of surface energy fluxes and stratification of a small boreal lake by a set of one-dimensional models. *Tellus A Dyn. Meteorol. Oceanogr.* **2014**, *66*. [CrossRef]

68. Martynov, A.; Sushama, L.; Laprise, R. Simulation of temperate freezing lakes by one-dimensional lake models: Performance assessment for interactive coupling with regional climate models. *Boreal Environ. Res.* **2010**, *15*, 143–164.

69. Pour, H.K.; Duguay, C.R.; Martynov, A.; Brown, L.C. Simulation of surface temperature and ice cover of large northern lakes with 1-D models: A comparison with MODIS satellite data and in situ measurements. *Tellus A Dyn. Meteorol. Oceanogr.* **2012**, *64*. [CrossRef]
70. Huang, A.; Lazhu; Wang, J.; Dai, Y.; Yang, K.; Wei, N.; Wen, L.; Wu, Y.; Zhu, X.; Zhang, X.; et al. Evaluating and Improving the Performance of Three 1-D Lake Models in a Large Deep Lake of the Central Tibetan Plateau. *J. Geophys. Res. Atmos.* **2019**, *124*, 3143–3167. [CrossRef]
71. Bennington, V.; Notaro, M.; Holman, K.D. Improving climate sensitivity of deep lakes within a regional climate model and its impact on simulated climate. *J. Clim.* **2014**, *27*, 2886–2911. [CrossRef]
72. Gu, H.; Jin, J.; Wu, Y.; Ek, M.B.; Subin, Z.M. Calibration and validation of lake surface temperature simulations with the coupled WRF-lake model. *Clim. Change* **2015**, *129*, 471–483. [CrossRef]
73. Xiao, C.; Lofgren, B.M.; Wang, J.; Chu, P.Y. Improving the lake scheme within a coupled WRF-lake model in the Laurentian Great Lakes. *J. Adv. Model. Earth Syst.* **2016**, *8*, 1969–1985. [CrossRef]
74. Kirillin, G.; Wen, L.; Shatwell, T. Seasonal thermal regime and climatic trends in lakes of the Tibetan highlands. *Hydrol. Earth Syst. Sci.* **2017**, *21*, 1895–1909. [CrossRef]
75. Li, Z.; Lyu, S.; Wen, L.; Zhao, L.; Ao, Y.; Meng, X. Study of freeze-Thaw cycle and key radiation transfer parameters in a Tibetan Plateau lake using LAKE2.0 model and field observations. *J. Glaciol.* **2021**, *67*, 91–106. [CrossRef]
76. Perroud, M.; Goyette, S.; Martynov, A.; Beniston, M.; Anneville, O. Simulation of multiannual thermal profiles in deep Lake Geneva: A comparison of one-dimensional lake models. *Limnol. Oceanogr.* **2009**, *54*, 1574–1594. [CrossRef]
77. Stepanenko, V.M.; Goyette, S.; Martynov, A.; Perroud, M.; Fang, X.; Mironov, D. First steps of a Lake Model intercomparison project: LakeMIP. *Boreal Environ. Res.* **2010**, *15*, 191–202.
78. Semmler, T.; Cheng, B.; Yang, Y.; Rontu, L. Snow and ice on Bear Lake (Alaska)—Sensitivity experiments with two lake ice models. *Tellus A Dyn. Meteorol. Oceanogr.* **2012**, *64*. [CrossRef]
79. Kirillin, G. Modeling the impact of global warming on water temperature and seasonal mixing regimes in small temperate lakes. *Boreal Environ. Res.* **2010**, *15*, 279–293.
80. Shatwell, T.; Thiery, W.; Kirillin, G. Future projections of temperature and mixing regime of European temperate lakes. *Hydrol. Earth Syst. Sci.* **2019**, *23*, 1533–1551. [CrossRef]
81. Mironov, D.; Heise, E.; Kourzeneva, E.; Ritter, B.; Schneider, N.; Terzhevik, A. Implementation of the lake parameterisation scheme FLake into the numerical weather prediction model COSMO. *Boreal Environ. Res.* **2010**, *15*, 218–230.
82. Mironov, D.; Kirillin, G.; Heise, E.; Golosov, S.; Terzhevik, A.; Zverev, I. Parameterization of lakes in numerical models for environmental applications. *Proc. 7th Work. Phys. Process. Nat. Waters* **2003**, 135–143.
83. Qu, B.; Kang, S.; Chen, F.; Zhang, Y.; Zhang, G. Lake Ice and Its Effect Factors in the Nam Co Basin, Tibetan Plateau. *Adv. Clim. Change Res.* **2012**, *8*, 327. [CrossRef]
84. Guo, P.; Ye, Q.; Wei, Q. Lake ice change at the Nam Co Lake on the Tibetan Plateau during 2000–2013 and influencing factors. *Prog. Geogr.* **2015**, *34*, 1241–1249. [CrossRef]

 *water*

Article

Simulating the Evolution of Da Anglong Glacier, Western Tibetan Plateau over the 21st Century

Wenqing Zhao [1,2,3], Liyun Zhao [1,*], Lide Tian [4,5], Michael Wolovick [1,6] and John C. Moore [1,5,7,*]

1   College of Global Change and Earth System Science, Beijing Normal University, Beijing 100875, China; 201721490035@mail.bnu.edu.cn (W.Z.); michael.wolovick@gmail.com (M.W.)
2   Key Laboratory of Tibetan Environment Changes and Land Surface Processes, Institute of Tibetan Plateau Research, Chinese Academy of Sciences, Beijing 100101, China
3   College of Earth and Planetary Sciences, University of Chinese Academy of Sciences, Beijing 100049, China
4   Institute of International Rivers and Eco-Security, Yunnan University, Kunming 600500, China; ldtian@ynu.edu.cn
5   CAS Center for Excellence in Tibetan Plateau Earth Sciences, Chinese Academy of Sciences, Beijing 100101, China
6   Alfred Wegener Institute, Helmholtz Centre for Polar and Marine Science, Glaciology Section, 27570 Bremerhaven, Germany
7   Arctic Centre, University of Lapland, 96101 Rovaniemi, Finland
*   Correspondence: zhaoliyun@bnu.edu.cn (L.Z.); john.moore.bnu@gmail.com (J.C.M.)

Abstract: We apply a three-dimensional (3D) full-Stokes model to simulate the evolution of Da Anglong Glacier, a large glacier in the western Tibetan Plateau from the year 2016 to 2098, using projected temperatures and precipitations from the 25-km-resolution RegCM4 nested within three Earth System Models (ESM) simulating the RCP2.6 and RCP8.5 scenarios. The surface mass balance (SMB) is estimated by the degree-day method using a quadratic elevation-dependent precipitation gradient. A geothermal flux of 60 mW m^{-2} produces a better fit to measured surface velocity than lower heat fluxes and represents a new datum in this region of sparse heat flux observations. The ensemble mean simulated glacier volume loss during 2016–2098 amounts to 38% of the glacier volume in the year 2016 under RCP2.6 and 83% under RCP8.5. Simulation from 2016 to 2098 without ice dynamics leads to an underestimation of ice loss of 22–27% under RCP2.6 and 16–24% under RCP8.5, showing that ice dynamics play an important amplifying factor in ice loss for this glacier, unlike for small Tibetan glaciers where SMB dominates glacier change.

Keywords: Tibetan Plateau; glacier modeling; mass balance; full-Stokes model

Citation: Zhao, W.; Zhao, L.; Tian, L.; Wolovick, M.; Moore, J.C. Simulating the Evolution of Da Anglong Glacier, Western Tibetan Plateau over the 21st Century. *Water* **2022**, *14*, 271. https://doi.org/10.3390/w14020271

Academic Editors: Yaoming Ma and Guoqing Zhang

Received: 12 November 2021
Accepted: 12 January 2022
Published: 17 January 2022

Publisher's Note: MDPI stays neutral with regard to jurisdictional claims in published maps and institutional affiliations.

1. Introduction

The Tibetan Plateau has tens of thousands of mountain glaciers, which act as important water resources supporting both local dry season irrigation and downstream rivers, such as the Indus, Brahmaputra, Ganges, Yellow, Yangtze, and Mekong, and, hence, the agricultural needs and economy of hundreds of millions of people. The Tibetan Plateau has undergone significant warming, at a faster rate than the global mean during recent decades, which is projected to continue in the future [1–3]. However, the response of the glaciers has not been spatially homogeneous, and, although many glaciers have experienced rapid retreats in recent decades, a minority of glaciers have advanced [4–6].

The response of glaciers in the Tibetan Plateau and the wider area of High Mountain Asia (HMA) to ongoing and future climate change is a topic of concern and cross-boundary disputes for several countries in Asia. Future glacier changes have been evaluated by the area-volume scaling method under various climate scenarios, such as the A1B emission scenario [7], the various Representative Concentration Pathway (RCP) scenarios [8–11], and for a global temperature rise of 1.5 °C [12]. In contrast with the area-volume scaling method, mechanistic models can investigate 3D velocity and thermal distribution, as well as simulating the dynamical evolution of individual glaciers [13–15]. Mechanistic models vary in complexity, for instance, a "flowline model" or the "shallow ice approximation" neglect some stress components to ease computation. A full-Stokes model can produce 3D velocity, stress and temperature distributions, and facilitates comparison between simulated glacier evolution and observations.

All ice dynamical models require estimates of surface and bed geometry, along with atmospheric boundary conditions such as surface mass balance (SMB), surface temperature, and precipitation records. Simplified dynamics models sometimes appear to require less data because they implicitly parameterize some aspects of the problem. Due to lack of such data, relatively little work has been done with mechanistic models of the Tibetan Plateau glacier responses to climate warming [13,16]. These have, to date, been on small glaciers, with no simulations of a large Tibetan glacier to our knowledge. Glaciers less than 1 km^2 represent ~83% of the total Tibetan glaciers but only occupy 20% of the total glacierized area. However, glaciers larger than 5 km^2 represent only ~3% of the Tibetan glaciers but occupy 51% of the total glacierized area. Thus, the response of large glaciers to climate warming is proportionally more important. However, their large area and remoteness usually make fieldwork challenging.

Da Anglong Glacier in the Ali Region of the western Tibetan Plateau is in a geologically interesting transition region between old cratonic basins and orogenic belts [17], but only sparse observations of geothermal heat observations exist. Since glacier dynamics allow inferences to be made of the basal geothermal heat flux that controls how fast the glacier flows, glaciological simulations can augment the heat flux map.

Because of both the practical difficulties involved in ground surveying glaciers in Tibet and the large numbers of relatively small glaciers, there is a reliance on remotely sensed information to infer the state and rate of change of the ice mass. However, we will show that, in the particular case of Da Anglong Glacier, there is large bias in both satellite radar surface elevations and remotely sensed ice velocity products comparing with field observations.

In this paper, we simulate the dynamic evolution and mass balance of Da Anglong Glacier from the year 2016 to 2098. The Ali Region is at the intersection of the Indian monsoon and westerlies; the inter-annual balance between these two climate patterns is the dominant factor affecting patterns of snow accumulation and ablation on glaciers in and around the Tibetan Plateau [6]. Advancing glaciers can be found to the north in West Kunlun and Karakoram, while all the observed glaciers in the Himalayas to its south and to the east in inner Tibet have been retreating [5,6]. Da Anglong Glacier is, to our knowledge, the only glacier in the Ali Region to have been studied in the field.

We combine an SMB parameterization using the degree-day model with a full-Stokes ice flow model, Elmer/Ice (available online: http://elmerice.elmerfem.org/ (accessed on 11 November 2021)), to simulate the evolution of Da Anglong Glacier from 2016 to 2098 with two climate scenarios representing the range from aggressive greenhouse gas mitigation (RCP2.6) to little effort at mitigation (RCP8.5). The study region and observation data are described in the next section. The surface mass balance parameterization and ice flow model setup are described in Sections 3 and 4. The simulations, projections, and discussions follow in the final sections.

2. Study Area and Observational Data

Da Anglong Glacier (32.84° N, 80.92° E) is large (6.66 km^2) by Tibetan standards where the average size is less than 1 km^2 [18]. It is in the Anglong Glacier Range, in the upper reaches of the Indus River, on the western Tibetan Plateau, which is one of the least studied areas in China (Figure 1). The glacier ranges between about 5620 and 6430 m a.s.l., and it is about 4.7 km long. The surface slope varies from 5.1° to 35.8° with a mean value of 12.6°. The terminus of Da Anglong Glacier retreated 78 m between 2000 and 2015 [19], at a mean rate of about 5.2 m a^{-1}.

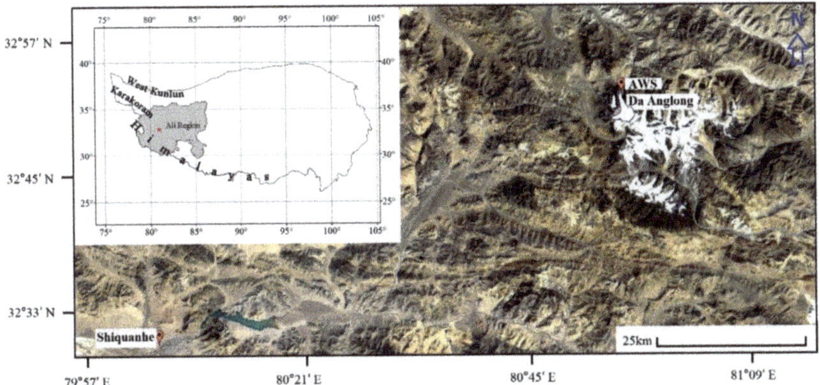

Figure 1. The location and image from Google Earth of Da Anglong Glacier, the Shiquanhe station and the AWS. The inset map displays the outline of the Tibetan Plateau, our study region (red box), locations of Ali region, West Kunlun, Karakoram, and Himalaya.

Data from the nearest meteorological station to the glacier, Shiquanhe (32.5° N, 80.08° E, 4279 m a.s.l., 85 km away from the glacier), show that, over the past three decades, the mean annual air temperature has increased by 0.063 °C a^{-1}, while the annual precipitation shows no trend (Figure 2). There is an automatic weather station (AWS; 32.87° N, 80.92° E, 5640 m a.s.l; Figure 1) near the terminus of Da Anglong Glacier. Daily air temperatures and precipitation at the AWS were measured from 1 August 2015 to 25 August 2016 (Figure 3). The mean annual temperature lapse rate is 0.65 °C (100 m)$^{-1}$ between the AWS and Shiquanhe station from August 2015 to July 2016. Daily temperatures at the AWS and Shiquanhe station are well correlated (r = 0.83) but daily precipitation less well (r = 0.45), as may be expected in mountainous regions. Precipitation is concentrated in summer on Da Anglong Glacier (Figure 3), meaning that both the main ablation and accumulation seasons are in summer, and temperatures at the AWS site are above the freezing point for around 5 months of the year.

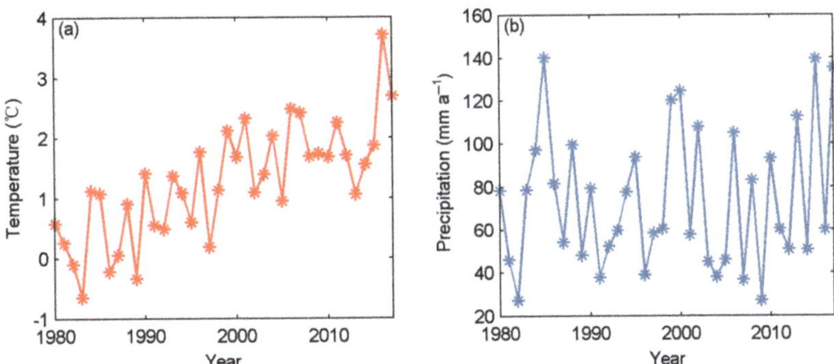

Figure 2. Annual mean temperature (**a**) and precipitation (**b**) records at the Shiquanhe station from 1980 to 2017.

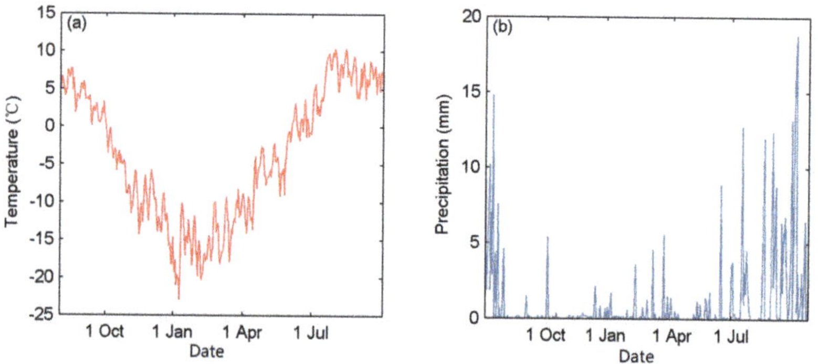

Figure 3. Daily mean temperature (**a**) and precipitation (**b**) records at the AWS from 1 August 2015 to 25 August 2016.

Surface topography and bed elevation were measured by differential GPS and ground penetrating radar (GPR). SMB and ice velocity were estimated by repeated stake observations in August in 2014 and 2015 and October in 2016 and 2017. The differential GPS used was Starfire E3050 in 2014 and 2015 and Starfire E3040 in 2016 and 2017. The GPS vertical precision is 0.1 m and the horizontal precision is 0.05 m. There is a supraglacial stream on the surface of the glacier flowing roughly along the center flow line, limiting surveys to only one side of the river (Figure 4). Elevations above 6150 m were not surveyed due to terrain hazards. Ice thickness was measured in 2016 by Pulse Ekko 100 GPR with antenna center frequency of 100 MHz, with a 4 m antenna spacing and measurement interval. The ice thickness data has an accuracy of 1–2 m. The maximum thickness measured was 216 m.

Figure 5 shows the measured SMB-elevation profile from the stakes on the glacier between 2014 and 2017. SMB was measured over the elevation range 5600–6100 m, and the Equilibrium Line Altitude (ELA) was at 5900–5950 m. The mean annual surface velocity measured at the stakes was 4.4 m a^{-1}, with the maximum of 6.0 m a^{-1} at 5834 m elevation. We also analyzed satellite velocity data at 240 m spatial resolution, generated using auto-RIFT [20] and provided by the NASA MEaSUREs ITS_LIVE project [21], to compare with field measurements. Annual velocities are provided at yearly intervals. The imagery used to create the velocity mosaics comes from USGS/NASA's Landsat 8 Operational Land Imager-Band 8 (15 m resolution).

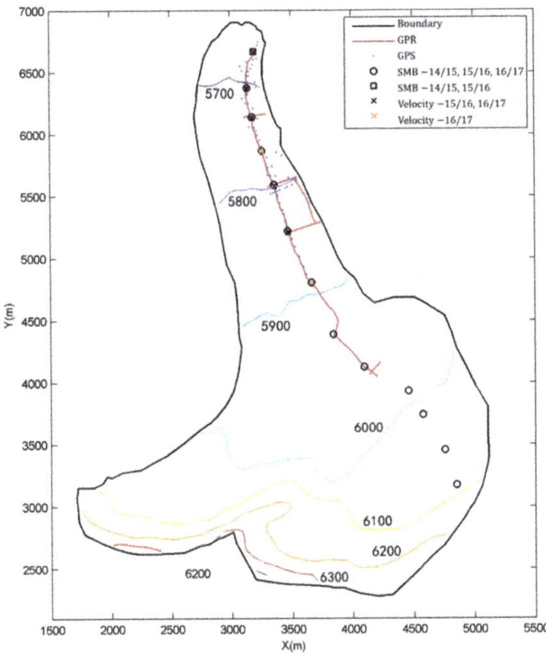

Figure 4. The glacier, with the outline marked in black. Ground Penetrating Radar (GPR) measurements are shown as the red track in the lower glacier and differential GPS measurement as blue points. Stakes are marked as black squares for SMB measured over 2 years (2014/2015, 2015/2016) and black circles for SMB measured for 3 years (2014/2015, 2015/2016, 2016/2017). Stakes are marked by black crosses for velocity measured in 2015/2016 and 2016/2017, and by orange crosses for velocity measured only between 2016 and 2017. Surface elevation contours in the year 2016 (Section 5). Recent Equilibrium Line Altitudes (ELA) from the year 2014 to 2017 are around 5900 m.

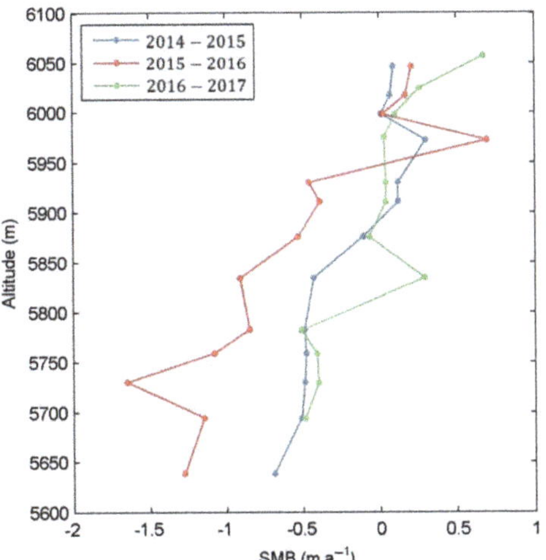

Figure 5. The observed SMB at stakes from August 2014 to August 2015 (blue), from August 2015 to October 2016 (red), and from October 2016 to October 2017 (green).

3. Surface Mass Balance Parameterization

We have SMB measurements (Figure 5) in three different mass balance years: from August 2014 to August 2015, from August 2015 to October 2016, and from October 2016 to October 2017. We parameterize glacier accumulation and ablation as a function of local temperature and precipitation based on the measurements at the AWS and Shiquanhe station.

3.1. Accumulation

Da Anglong Glacier is located at the upper reaches of the Indus River, which is influenced by westerly air flow and precipitation sources. Sun et al. [22] found that precipitation firstly increases to a maximum and then decreases with elevation in the basins influenced by westerlies over the Tibetan Plateau. They also estimate the average annual precipitation gradient in the upper Indus basin, 0.29 mm m^{-1}, which equals a daily mean precipitation gradient of 0.08 mm (100 m)$^{-1}$. We calculated the daily precipitation gradients using the observed daily precipitation at the AWS near the glacier and Shiquanhe station, from August 2015 to August 2016. We found that the daily precipitation gradients are concentrated between 0.1 mm (100 m)$^{-1}$ and 0.2 mm (100 m)$^{-1}$, see Figure 6, which is close to the above finding in [22], considering the natural variability within mountainous regions.

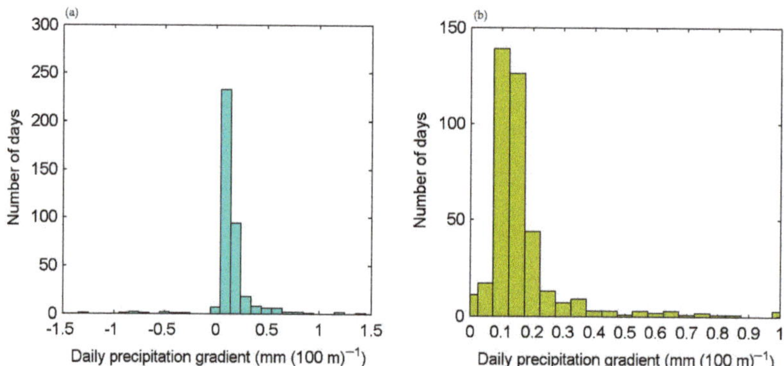

Figure 6. The histogram of measured daily precipitation gradients using observed daily precipitation at the AWS and Shiquanhe station. Intervals in plot (**a**) is 0.1 mm (100 m)$^{-1}$. Plot (**b**) shows the range of 0−1 mm (100 m)$^{-1}$ with an interval of 0.05 mm (100 m)$^{-1}$.

The precipitation gradient in reality varies with elevation. The profile of precipitation firstly increasing then decreasing with elevation in the basins influenced by westerly winds [22] implies that the precipitation gradient changes from positive to negative at the maximum precipitation height. We use a quadratic polynomial to capture this feature of precipitation gradient. Three pairs (elevation, precipitation gradient) fully determine the quadratic polynomial. We choose three elevations, the AWS, Shiquanhe station, and the maximum precipitation height. The observed SMB in both the years 2014/2015 and 2015/2016 reaches a maximum at 5975 m (Figure 5), which take to mean that the precipitation reaches a local maximum near that elevation. Therefore, we assume the daily precipitation gradient is zero at 5975 m.

Because the daily precipitation gradients are mostly in the range of 0.1–0.2 mm (100 m)$^{-1}$, we explore daily precipitation gradients at the Shiquanhe station and AWS by selecting from the set {0.1, 0.15, 0.2} mm (100 m)$^{-1}$, giving nine combinations of daily precipitation gradients at the Shiquanhe station and the AWS. Having fixed the zero of precipitation gradient at 5975 m, we fit a quadratic polynomial to the daily precipitation gradients (Figure 7a), then integrate it from 4278 m to all the elevations on the glacier surface and estimate precipitation there (Figure 7b).

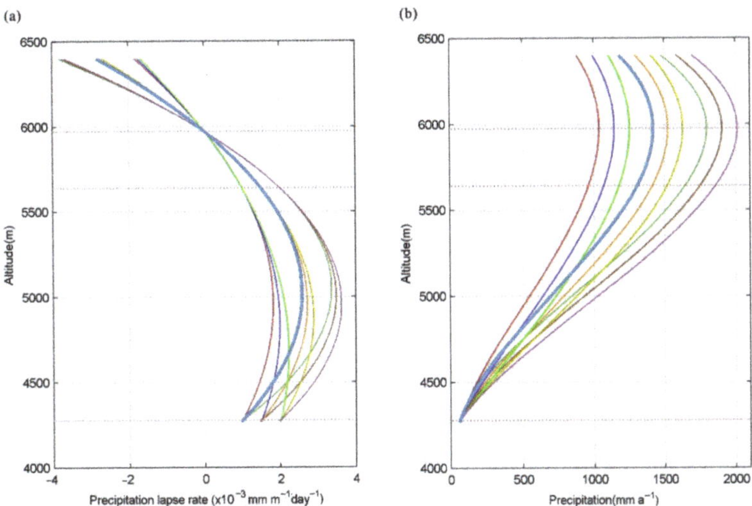

Figure 7. (**a**) The estimated quadratic polynomial profile of elevation−dependent daily precipitation gradients using 9 combinations (by 9 colors) of precipitation gradient values at the Shiquanhe station (4278 m) and AWS (5640 m) from the sets {0.1, 0.15, 0.2} mm (100 m day)$^{-1}$ and zero value at 5975 m; for instance, the thick light blue curve represents the best fit combination that using the precipitation gradient of 0.1 mm (100 m day)$^{-1}$ at 4278 m and 0.15 mm (100 m day)$^{-1}$ at 5640 m and 0 at 5975 m; (**b**) the estimated total precipitation using the estimated precipitation gradients in plot (**a**) and daily precipitation records at the Shiquanhe station from August 2015 to August 2016. The three elevations of 4278 m, 5640 m, and 5975 m are marked by red dotted lines.

The quadratic polynomials using the above nine combinations give total precipitation at the elevation of the AWS from August 2015 to August 2016 in the range of 961–1860 mm. The quadratic polynomial with daily precipitation gradient value of 0.1 mm (100 m)$^{-1}$ day^{-1} at the Shiquanhe station and 0.15 mm (100 m)$^{-1}$ day^{-1} at the AWS gives a total precipitation value of 1303.7 mm, closest to the observed 1302.7 mm, at the AWS over the same period. Therefore, we use the best-fit quadratic polynomials of daily precipitation gradients (Figure 7a) hereafter.

Next, we calculated (elevation-dependent) daily solid precipitation, P_s, as follows [23]:

$$P_s = \begin{cases} P & T \leq T_S \\ \frac{1-T}{T_L} \cdot P & T_S < T < T_L \\ 0 & T \geq T_L \end{cases} \tag{1}$$

where P represents daily total precipitation and T, daily mean temperature. T_L and T_S are the critical values for liquid and solid precipitation, taken as 0 °C and 4 °C, respectively. Elevation-dependent glacier accumulation is obtained by summing daily solid precipitation.

3.2. Ablation

We employ the widely used (e.g., [24]) positive degree-day model [25] to estimate the ablation at different elevations. The annual ablation rate is calculated by multiplying the sum of positive daily mean air temperatures (*PDD*) by a suitable degree-day factor (*DDF*) as follows:

$$Ablation = DDF \times PDD \tag{2}$$

$$PDD = \sum_{t=1}^{n} H_t \cdot T_t \tag{3}$$

T_t is daily temperature. H_t is logical function, set to 1 when $T_t \geq 0$ and else 0.

The *DDF* determines the amount of ice and snow ablation produced by unit positive accumulated temperature. The choice of *DDF* greatly affects the accuracy of the model. *DDF* ranged from 2.6 to 13.8 mm·d^{-1}·°C^{-1} on 15 glaciers across the Tibetan Plateau [24], increasing from northwest to southeast. Maritime (that is, comparatively warm and wet) glaciers have higher *DDF* than subcontinental and extremely continental glaciers, which may be characterized as cold and dry. Deng and Zhang [26] studied degree–day factors from 24 glaciers on the Tibetan Plateau, and also found a tendency for high values in the east and south and low values in the west and north.

We estimate daily precipitation on the glacier for the three mass balance periods (2014/2015, 2015/2016, 2016/2017) using daily precipitation records at the Shiquanhe station and the estimated best-fit precipitation gradient (Section 3.1).

We estimate the SMB using solid precipitations in Equation (1) and ablation calculated in Equation (2), with the *DDF* selected from the range of [2.6, 13.8] mm·d^{-1}·°C^{-1} to minimize misfit with observed SMB (Figure 5) in three mass balance periods. We find a *DDF* of 3.4 mm·d^{-1}·°C^{-1} gives the best fit (Figure 8) and is reasonable for glaciers in the region [24].

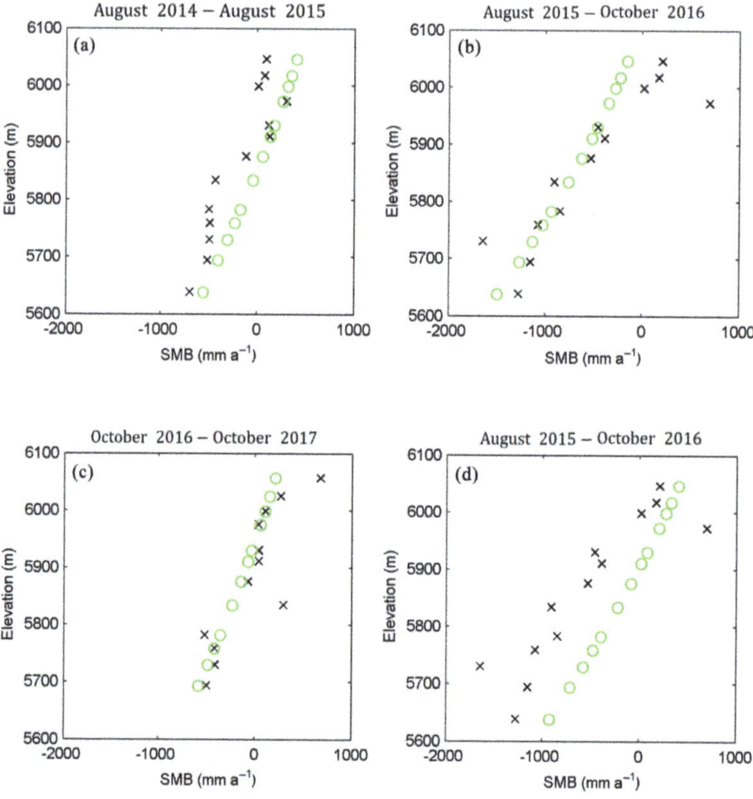

Figure 8. The observed SMBs (black crosses) compared with the estimate (green dots) using the best-fit precipitation gradients quadratic polynomial and DDF value of 3.4 mm·d^{-1} °C^{-1} for the periods from August 2014 to August 2015 (**a**), from August 2015 to October 2016 (**b,d**), and from October 2016 to October 2017 (**c**). Plots (**a–c**) are based on the observed precipitation at the Shiquanhe station, and plot (**d**) is based on the observed precipitation at the AWS.

These best SMB parameters are used to parametrize SMB for future simulations. Additionally, we impose a limit on the maximum SMB on the glacier of 0.5 m since observed SMB rarely exceeds 0.5 m even at the highest point of the glacier (Figure 5) due to frequent strong winds. Furthermore, the upper glacier surface and the side walls that constrain it are almost the same elevation, suggesting that the highest elevation parts of the glacier are close to, or at, maximum possible elevation.

4. Ice Flow Modeling

Da Anglong Glacier flow dynamics were modeled using the thermo-mechanically coupled full-Stokes model, Elmer/Ice, an open source finite element code that has been used to simulate several mountain glacier flow models successfully [16,27,28].

4.1. Field Equations

Elmer/Ice is a thermo-mechanically coupled full-Stokes model that includes the following equations:

$$\nabla \cdot v = 0 \tag{4}$$

$$\nabla \cdot \tau - \nabla p = \rho g, \tag{5}$$

$$\rho c \left(\frac{\partial T}{\partial t} + v \cdot \nabla T \right) = \nabla \cdot (\kappa T) + 4 \eta d_e^2 \tag{6}$$

Ice density, ρ, is set to 910 km·m^{-3}. $v = (u, v, w)$ is the vector of ice flow velocity. Equation (5) expresses the conservation of momentum, where τ is the deviatoric stress tensor, p is the isotropic pressure, and g is the acceleration due to gravity $(0,0,-9.81)$ m·s^{-2}. Via Glen's flow law, the deviatoric stress tensor is related to the deviatoric part of the strain rate tensor, D, which can be described by

$$\tau = 2 \eta D \tag{7}$$

where η denotes ice viscosity, given by

$$\eta = \frac{1}{2} A(T)^{-\frac{1}{n}} d_e^{\frac{(1-n)}{n}} \tag{8}$$

where the flow rate factor $A(T)$ is derived from an Arrhenius law [29]; d_e is the second invariant of the strain-rate, $(0.5 \mathrm{tr} D^2)^{1/2}$, and the Glen exponent n is taken as 3. Equation (6) is the heat transfer equation, and T is ice temperature. The heat capacity c and heat conductivity K are functions of ice temperature [29]. The ice temperature is bounded by an upper limit at the pressure melting point.

4.2. Boundary Conditions

Da Anglong Glacier is entirely above 5000 m elevation, so we initially assumed that the bottom of the glacier is frozen and non-sliding. The radargram along the flowline shows a consistent ice-bedrock interface that does not suggest any of the glacier is subject to basal melt [19], though that does not mean that basal melt is certainly excluded. We set a Neumann boundary condition at the ice bottom interface for temperature simulation with a geothermal heat flux. Jiang et al. [17] compiled geothermal heat flux data for China; there are no geothermal heat flux measurements in this region, and data are very sparse on the western Tibetan Plateau. Their map shows the glacier is located on a transition region between the Central China Orogen with low-medium heat fluxes and the Tibet-Saijiang Orogen with higher fluxes, and probably has a heat flux in the range 50–60 mW m^{-2}. Therefore, we tested values of 50 and 60 mW m^{-2} for geothermal heat flux and determine a likely value based on comparison between modeled and observed surface flow velocity (Section 5).

We prescribe a stress-free condition at the glacier surface, meaning that atmospheric pressure and its change over the glacier surface are neglected. Surface temperature is estimated by annual mean temperature and a lapse rate of 0.65 °C (100 m)$^{-1}$.−7.1 °C is the annual mean temperature from the Shiquanhe station:

$$T(z) = -7.1 - 0.0065 \cdot (z - 5640) \tag{9}$$

where T is the glacier surface temperature (°C) and z is surface elevation (m a.s.l.).

The evolution of the free surface in prognostic simulations was modeled by a kinematic boundary condition:

$$\frac{\partial z}{\partial t} + u\,\frac{\partial z}{\partial t} + v\,\frac{\partial z}{\partial t} - w = SMB \tag{10}$$

where z is surface elevation; u, v and w denote ice flow velocity components in the x, y, and z direction, respectively; SMB is calculated as described in the SMB parameterization section.

4.3. Climate Forcing Scenarios

For the period 2016–2098, we estimated the time series of SMB by the method in Section 3 based on output from the 25-km-resolution regional climate model RegCM4 ([30], available online: https://esgf-data.dkrz.de/search/cordex-dkrz/ (accessed on 11 November 2021)). RegCM4 was nested within three different Earth System Models (ESMs): HadGEM2−ES [31], MPI−ESM−MR [32], and NorESM1−M [33], all running the RCP2.6 and RCP8.5 scenarios.

The mean annual air temperature increases at the AWS during the 21st century range from 0.002–0.008 °C a^{-1} under RCP2.6 to 0.054–0.068 °C a^{-1} under RCP8.5 (Figure 9a, Table 1). HadGEM2ES projects the largest rise while MPI−ESM−MR, the lowest. The ensemble mean projected rise rates are 0.006 °C a^{-1} under RCP 2.6 and 0.059 °C a^{-1} under RCP 8.5. We bias correct RegCM4 modeled daily temperatures using observed temperatures over 1980–2005 at the Shiquanhe station, with an altitudinal temperature lapse rate of 0.65 °C (100 m)$^{-1}$ and correct monthly means with an offset from the 1980–2005 observations. Summer mean temperature dominates the glacier ablation in the whole year. The summer temperature rises slightly slower than mean annual temperature (Table 1). Again, HadGEM2ES projects the highest summer temperature rises.

Table 1. Modeled annual and summer mean temperature warming rate (°C·a^{-1}) at the elevation of AWS projected by RegCM4 driven by 3 ESMs under RCP2.6 and RCP8.5 from 2018 to 2098.

	HadGEM2ES	MPI−ESM−MR	NorESM1−M	Ensemble Mean
RCP2.6 annual	0.008	0.002	0.007	0.006
RCP8.5 annual	0.068	0.056	0.054	0.059
RCP2.6 summer	0.005	0.002	0.007	0.005
RCP8.5 summer	0.053	0.043	0.043	0.047

The RegCM4 modeled daily precipitation at the nearest grid point to the AWS (within about 1 km) has a similar distribution to observations during its measurement period, and, since there is only 1 year of data, we do not bias correct modeled daily precipitation. There is no trend in modeled precipitation by all models under RCP 2.6 but a slightly increasing trend after the year 2050 under RCP8.5. The ensemble mean projected precipitation at the elevation of AWS is in the range of 831–1116 mm a^{-1} under RCP 2.6 and 778–1202 mm a^{-1} under RCP 8.5.

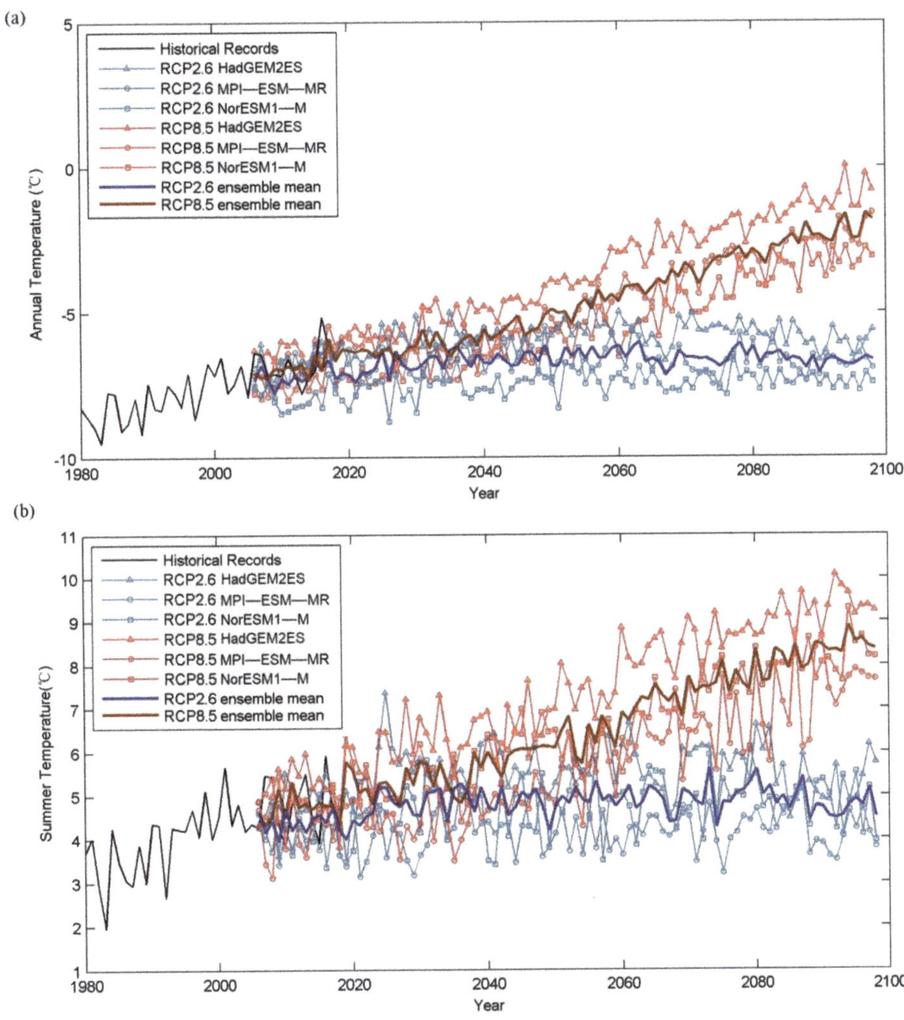

Figure 9. Historical annual mean temperature (**a**) and summer mean temperature (**b**) at the elevation of AWS from Shiquanhe station from 1980 to 2017 (black), and the future annual mean temperature projected by RegCM4 under RCP2.6 (blue) and RCP8.5 (red) from 2006 to 2098 driven by 3 ESMs (HadGEM2ES, MPI−ESM−MR, NorESM1−M). Thick lines represent ensemble means.

5. Glacier Geometry in the Year 2016

We produce the best-fit state of the glacier in the year 2016 from the detailed but spatially limited in situ observations, extended by inverse modeling of the glacier velocities to estimate errors in ice thickness based on the spatially complete but bias-prone Shuttle Radar Topography Mission (SRTM version 4.1 with resolution of 90 m) dataset measured in the year 2000.

The SRTM dataset specifies 90% of errors will be <16 m [34]. Since this error is not random either spatially or temporally, we must consider the possibility of a bias in the SRTM data. The glacier surface profile in 2000 would have been different from 2016 because of trends in SMB. These will not be uniform over the glacier: in a warming climate, the lower glacier will thin more than the upper parts. In contrast with this actual change in glacier geometry, a bias in SRTM data will produce an offset to elevations, for example, due to local errors in representing the geoid.

The surface elevation data measured by GPS in the year 2016 are about 30 m lower than those from the year 2000 SRTM dataset at the corresponding locations, with no trend with elevation, suggesting that a bias rather than an SMB signal dominates this offset. Therefore, we lower the surface elevation from SRTM dataset by 30 m everywhere to get a glacier surface elevation map in the year 2016 (Figure 10a). This 30 m includes both the error in the SRTM dataset and the elevation change from the year 2000 to 2016. If, instead of allowing this correction to the SRTM data, we assumed the SRTM elevations were correct, it would imply a surface lowering of 30 m between 2000 and 2016 and a mean lowering rate far in excess of any remote sensing observations in the region [35]. Shean et al. [35] generated a 30 km resolution elevation change trend map from 2000 to 2018 and found an estimated local mean mass balance of -0.2 ± 0.2 m a^{-1} in the hexagonal cell containing Da Anglong Glacier. Observed surface lowering rates below the ELA (0–0.6 m a^{-1}) corresponds to 0–10 m surface lowering between 2000 and 2016. Furthermore, simulations with the SRTM geometry do not lead to terminus retreat as observed [19]. Therefore, the SRTM data is likely to overestimate the surface by 20–30 m in the lower part of Da Anglong Glacier, which is more than estimated 90th percentile of SRTM errors.

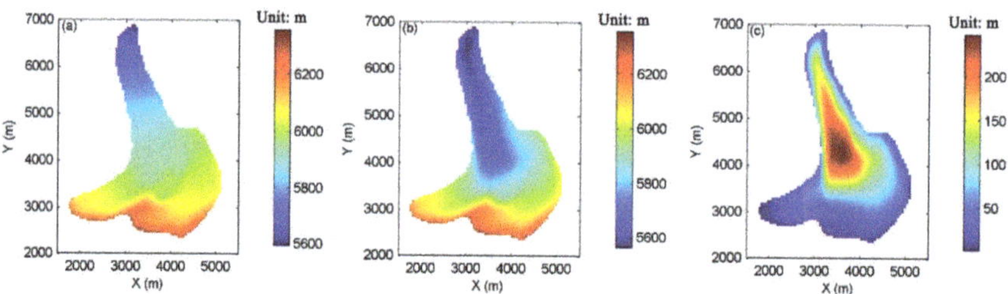

Figure 10. The estimated surface (**a**), bed (**b**), and ice thickness (**c**) distributions for the year 2016.

Next, we estimate the ice thickness in the year 2016. We measured ice thickness using GPR along several routes in 2015/2017 (Figure 4). However, the measurements are spatially limited and absent in the upper glacier. To get better spatial coverage of ice thickness, we supplement them with the modeled ice thickness for Da Anglong from the inverse model GlabTop2 [36]. GlabTop2 uses SRTM surface elevation data and empirical thickness–slope relations, a semi-elliptic cross-sectional geometry, and an estimate of basal shear stress from glacier vertical extent. GlabTop2 underestimates the mean ice thickness by 7% at six glaciers surveyed by radar measurements, but it also tends to overestimate small ice thicknesses [36]. Therefore, we cannot simply take GlabTop2 ice thicknesses as correct for the year 2016.

We do, however, have an ice dynamics model that can be used to give surface velocities for arbitrary glacier geometries. Hence, we can refine the ice thickness by manually doing an inversion process using the ice flow model to minimize misfit between modeled ice velocity and that measured by the GPS survey of the stakes on the glacier. For the upper part of the glacier, where radar data are absent, we take GlabTop2 ice thickness as an initial estimate and modify it to match the location of observed thickest ice. We manually generate thickness correction data based on the distances to the thickest part of the glacier, assuming the ice thickness in the lower glacier has high accuracy while the upper part has large uncertainty. These different ice thickness corrections are then used to produce simulations of surface velocity we can compare with the observations. We define a relative velocity error as

$$Error = \frac{1}{N} \sum \frac{|V_{obs} - V_{modeled}|}{V_{obs}} \tag{11}$$

where N is number of observations, V_{obs} and $V_{modeled}$ are observed and modeled velocities at each stake. The end results are ice thickness (Figure 10c) and surface velocity maps that best-fit observed stake velocities. We find that a geothermal heat flux of 60 mW m^{-2} produces a 10–20% better fit than with lower heat flux. We produce a bed geometry map by subtracting the best-fit ice thickness estimates from the glacier surface elevation in the year 2016, correcting SRTM elevations by 30 m (Figure 10b). This bed elevation is well correlated (r = 0.9) with the GPR measurements we made in the years 2015/2017. The RSME between the estimated bed elevation and GPR measured is 5.7 m.

We compare observed, modeled, and remotely sensed velocities at the stake sites in Figure 11. The uncertainty of imagery-derived velocity is quoted at 0.3 m a^{-1} for this glacier [21]. Figure 11 shows that the imagery-derived velocities are generally much lower (with differences far greater than the nominal satellite errors) than those from stake observations or our model, which is not entirely unexpected as Da Anglong Glacier is a relatively slow flowing (<10 m a^{-1}) glacier. Millan et al. [37] found that capturing fluctuations below 10 m a^{-1} remain challenging even for Sentinel-2, which they found is about twice as precise as that from Landsat 7/8 on some mountain glaciers. There are smaller errors between the stake and modeled velocities than between stake and satellite-derived velocities, showing that the image data is not especially helpful in extending the stake velocity observations, and, indeed, the remotely sensed velocities are so low as to be unphysical for an ice body of this observed thickness. Observed velocities are slightly larger than simulated, indicating that some sliding is occurring in addition to the pure ice deformation that we prescribed by specifying the no-slip condition at the bed.

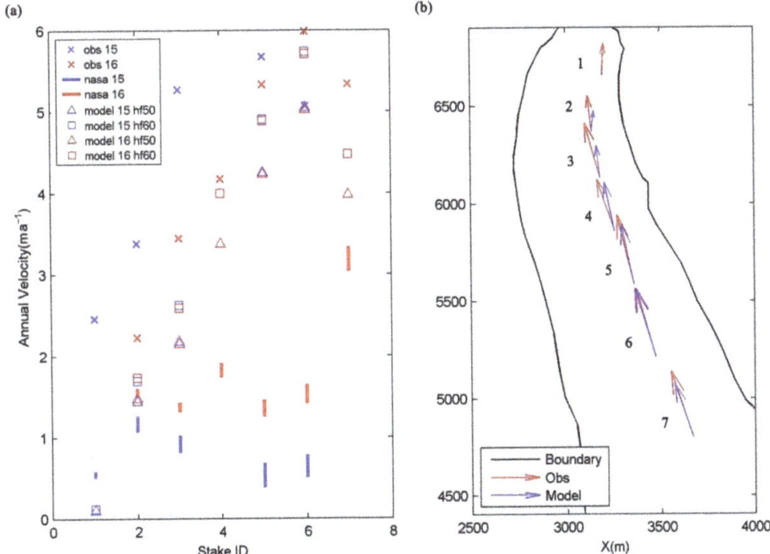

Figure 11. (a) The observed surface velocities at stakes (stars), best-fit modeled with geothermal heat flux of 50 (triangles) and 60 (squares) mW m^{-2} (see legend hf50, hf60), and annual mean surface velocities provided by the NASA MEaSUREs ITS_LIVE project [21] satellite images (lines, see legend nasa) in the year 2015/2016 (blue) and 2016/2017 (red); (b) the averaged observed (red arrow) and modeled (blue arrow) surface velocities in the years 2015 and 2016 as vectors on the glacier surface.

6. Model Projections from 2016 to 2098

Using the geometry for the year 2016, we build a 3D ice dynamic model with unstructured triangular elements at 50 m spatial resolution with 16 terrain-following layers between the bedrock and surface. We applied both diagnostic and prognostic simulations.

The former was done for a fixed geometry in the year 2016, with the purpose of building the steady-state temperature and velocity fields, and then to use it as an initial condition for prognostic simulations during the period 2017–2098. Simulations were done using geothermal heat fluxes of 50 and 60 mW m^{-2}.

We assumed that the glacier is not sliding on its bed, and the resulting basal temperature map (Figure 12) in the year 2016 is below pressure melting point by more than 2.2 °C everywhere for 50 mW m^{-2} and by 0.6 °C for 60 mW m^{-2} heat fluxes. Although this temperature field is consistent with no sliding, water can be present beneath a glacier at lower temperatures [38]. Furthermore, this is a steady state simulation, whereas we have argued that the glacier has been thinning at least over the 21st century. A thicker glacier in the 20th century may have had areas at the pressure melting point. This would lead to faster sliding velocities. We consider the result to be compatible with the limited sliding suggested by velocities in Figure 11. The modeled maximum ice velocities in the year 2016 are 5.6 m a^{-1} and 6.3 m a^{-1} with geothermal heat fluxes of 50 and 60 mW m^{-2}, respectively (Figure 13).

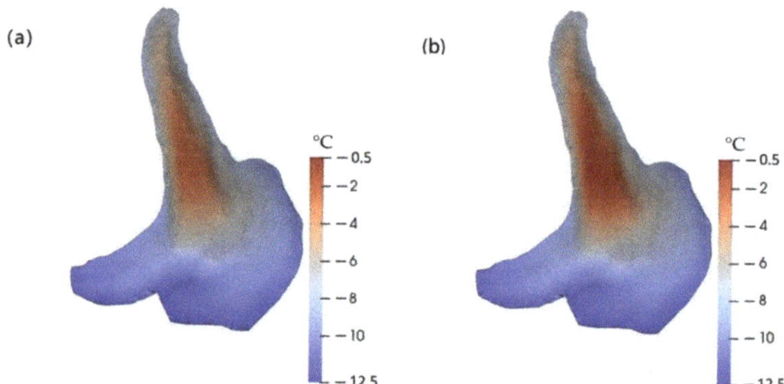

Figure 12. Distribution of temperature relative to pressure melting point at the bedrock with geothermal heat flux of 50 (**a**) and 60 (**b**) mW m^{-2}.

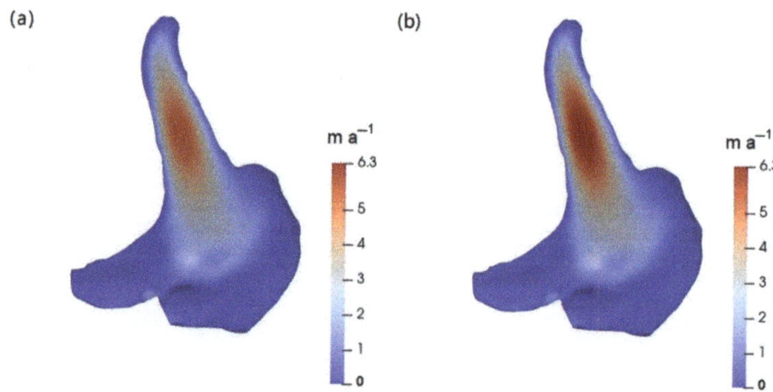

Figure 13. Modeled surface velocities in the year 2016 with geothermal heat flux of 50 (**a**) and 60 (**b**) mW m^{-2}.

The prognostic simulation from 2016 to 2098 was forced by projected temperatures and precipitations from RegCM4 driven by the three ESMs under the RCP2.6 and RCP8.5 scenarios (Section 4.3) with geothermal heat fluxes of 50 and 60 mW m^{-2}. While basal heat flux makes significant changes to surface velocities (Figure 11), it has little impact on glacier volume and area evolution.

Simulated glacier volume loss during 2016–2098 is equivalent to 16–62% of the glacier volume in the year 2016 under RCP2.6 and 75–91% under RCP8.5 (Figure 14). Therefore, the average annual volume loss rate is 0.19–0.75% a^{-1} under RCP2.6 and 1.0–1.1% a^{-1} under RCP8.5. Ensemble mean volume loss rate during 2016–2098 is 0.46% a^{-1} of the glacier volume in the year 2016 under RCP2.6 and 1.0% a^{-1} under RCP8.5.

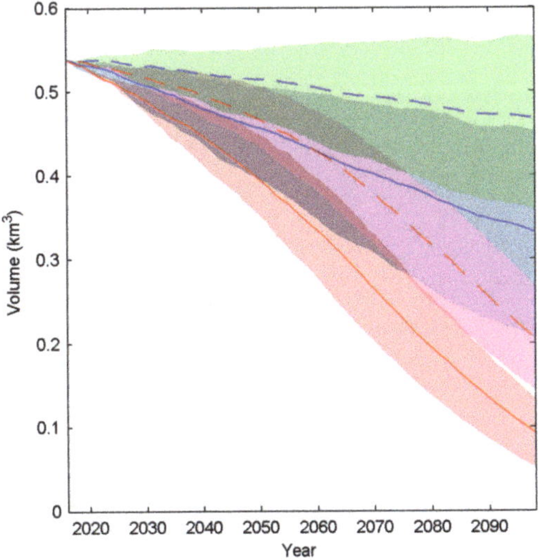

Figure 14. Modeled ensemble mean volume changes from 2016 to 2098 using RegCM4 outputs driven by the 3 ESMs (HadGEM2−ES, MPI−ESM−MR, and NorESM1−M) for the period 2017−2098 under RCP2.6 (blue) and RCP8.5 (red), with ice dynamics (solid line) and without dynamics (dashed line). The across-model spread with/without ice dynamics are denoted by shading under RCP2.6 (blue/green) and RCP8.5 (red/pink).

Summer mean temperature dominates glacier ablation. RegCM4 forced by HadGEM2ES projects the highest summer mean temperature among the ensemble and, thus, the largest volume loss, while RegCM4 forced by MPI−ESM−MR projects the lowest. Simulated glacier area loss during 2016–2098 is equivalent to 11–39% of the glacier area in the year 2016 under RCP2.6 and 60–83% under RCP8.5 (Figure 15). Therefore, the annual averaged area loss rate across ESMs is 0.13–0.47% a^{-1} under RCP2.6 and 0.72–1.00% a^{-1} under RCP8.5. Ensemble mean area loss rate during 2016–2098 is 0.22% a^{-1} of the glacier area in the year 2016 under RCP2.6 and 0.87% a^{-1} under RCP8.5.

We also compared simulated volume change in the final year 2098 from prognostic simulations with and without ice dynamics (Figure 14), and found relative difference of 24% (range 22–27%) under RCP2.6 and 20% (16–24%) under RCP8.5. Ice dynamics transports ice to the lower part of the glacier where more ablation occurs. Therefore, ice dynamics play a relatively larger role under RCP2.6 because surface temperature and ablation rate increases are lower than under RCP8.5. For small Tibetan glaciers like the Gurenhekou Glacier and Qiangtang No. 1 Glacier [13,16], SMB is the overwhelmingly dominant factor in mass loss. This is because the ice flow speed of glaciers scales strongly (theoretically, with the fourth power) of ice thickness and, thus, larger glaciers with greater mean thickness are much more dynamic than small glaciers. Changes to the dynamics of larger glaciers are relatively slow because the climate-driven changes in geometry that drive ice flow take longer to modify large glaciers and, thus, ice dynamics provide increasingly important positive feedback on larger glaciers in warming climates.

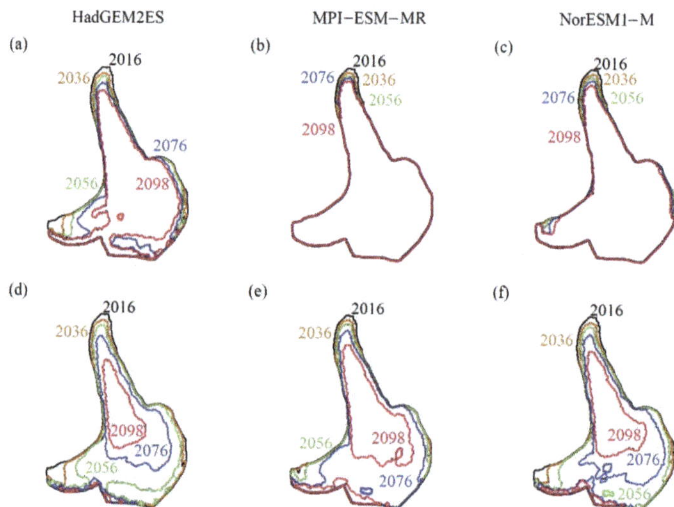

Figure 15. Modeled terminus retreat from 2016 to 2098 under RCP2.6 (upper row, plot (**a–c**)) and RCP 8.5 scenarios (lower row, plot (**d–f**)) using projected results from RegCM4 driven by 3 ESMs, HadGEM2ES (plot (**a,d**)), MPI−ESM−MR (plot (**b,e**)), NorESM1−M (plot (**c,f**)).

The measured annual mean terminus retreat rate from 2000 to 2015 is 5.2 m a^{-1} [19]. Modeled terminus retreats from 2016 to 2098 projected from RegCM4 displays differences between ESMs (Figure 15). Differences between models are much smaller than differences between scenarios, indicating the dominance of emissions scenario over ESM differences on the glacier evolution. Modeled glacier terminus retreat rate from 2016 to 2098 is 2.9–5.9 m a^{-1} under RCP2.6 and 7.7–12.5 m a^{-1} under RCP8.5. The simulated upper part of the glacier also shrinks under RCP8.5. The model suggests that only the central part of the glacier will exist by the end of this century under RCP8.5.

7. Discussion

The large relative difference in projected glacier volume changes when considering ice dynamics shows that both SMB and ice transport downslope are important factors in glacier evolution in the coming decades. This is different from previous studies on a few smaller glaciers on the Tibetan Plateau [13,16] where SMB is the only dominant factor. This finding shows that ice dynamics cannot be ignored in mass balance estimation for Tibetan glaciers, especially those large ones that dominate the sea level rise contribution.

Accurate modeling of ice dynamics requires knowledge of glacier geometry, which is derived from data on surface elevation and ice thickness. In the case of Da Anglong, it is difficult to measure ice thickness in the upper part of glacier due to the dangerous conditions. We only have observed surface elevation, ice thickness, and surface velocity at the lower part of the glacier. Ice thickness generation for large glaciers without full spatial coverage of measurements is challenging. We modify the SRTM dataset by 30 m to generate the glacier surface elevation in the year 2016 (Section 5). Applying this procedure gives a mean difference of 0.59 m from measured surface elevation with an RSME of 2.83 m. This seems reasonable given the seasonal fluctuations expected in glacier thickness. The 30 m correction, we argue, is plausible according to the stated uncertainties in SRTM data.

Various statistical models have been published that seek to estimate the ice thickness distribution of mountain glaciers from surface characteristics based on considerations of ice flow dynamics and mass conservation, including the 13 models that participated in the Ice Thickness Model Intercomparison eXperiment [39]. These are very useful where large numbers of glaciers allow for cancellation of random errors in volume but will not always produce adequate results for any single glacier. In this study, we developed a novel way to accommodate these data limitations by using an inverse procedure with the full-Stokes ice flow model generating surface velocities that can be compared with the observed surface velocity to deduce ice thickness in the un-surveyed upper glacier. Using the optimized ice thickness in the lower glacier, the estimated bed elevation is well correlated (r = 0.9) with the GPR measurements in the lower part of the glacier. The RSME between the estimated bed elevation and GPR measured is 5.7 m. The uncertainty from the radar data is expected to be 1–2 m, but there is an additional uncertainty due to positioning errors and the underlying roughness of the bed.

Velocity was only measured in the ablation zone of Da Anglong Glacier, where velocities are generally higher than those simulated, with an averaged relative error of 30–50% with geothermal heat flux of 50 mW m^2 and 10–40% with 60 mW m^2. The modeled underestimation might be because we simulate ice deformation alone, consistent with the assumption of no basal sliding and basal ice temperatures about 0.6 °C below the pressure melting point with the present-day ice thickness. The slightly faster velocities might be the result of melting at the base in the recent past when the glacier was thicker, and sliding over the bed would also generate frictional heat that could sustain a supply of lubricating water as the glacier cooled slowly by thinning. The absence of a complete surface velocity map of the glacier is one large limitation and uncertainty source on the modeled results. When and if an accurate map of surface velocity for the glacier becomes available, we can do an inversion for the basal friction coefficient in basal sliding law, resulting in a better fit between the modeled ice velocity and the observed.

An obvious source of surface velocities might be from remote sensing images. We also analyzed satellite velocity data at 240 m spatial resolution generated using auto-RIFT [20] and provided by the NASA MEaSUREs ITS_LIVE project [21] to compare with field measurements. Annual velocities are provided at yearly intervals. The imagery used to create the velocity mosaics comes from USGS/NASA's Landsat 8 Operational Land Imager-Band 8 (15 m resolution). We compared observed stake velocities on the glacier and those derived from satellite images and find the images underestimate the stake observation velocities, with twice the misfit as for the modeled velocities and with errors much larger than estimated for the derived satellite velocities [21]. This contrasts with experience from other glaciers in HMA: Dehecq et al. [40] found satellite derived annual velocity compared well with measurements in 2003 [41] from the ablation area for the Chhota Shigri Glacier, in Himachal Pradesh. However, very few measurements of glacier velocity exist in HMA and the utility of satellite-derived annual velocity would probably benefit from more widespread field measurements. Gardner et al. [21] acknowledge that the formal errors they quote seem too low and suggest that they should be used rather as qualitative metrics for assessing error. This might be so, but, with few verifications of the satellite product, it is hard to know how to interpret the given error. The ice velocity on Da Anglong Glacier is only slightly higher than pure internal ice deformation expected from its geometry and, hence, is flowing at nearly the slowest rate possible, since any basal lubrication would increase surface velocities. The internal deformation rate of glaciers should provide a minimum bound on glacier speeds, and, with statistical methods of estimating ice volume from surface observations as discussed earlier, would perhaps be usefully included in large-scale multi-glacier velocity estimates.

Precipitation gradient and *DDF* are crucial parameters in the SMB estimation, but they have large spatial and temporal variability over Tibetan glaciers [22,24,42]. Due to lack of measurements, many previous studies use the simple assumption of a linear relationship between precipitation with elevation, but this assumption is rather different from the sparse observations in the region [22,43–45] and, thus, likely introduces systematic errors in mountainous regions. This motivated us to use a quadratic elevation-dependent precipitation gradient parameterization based on the actual observed precipitation and SMB measurements. The very limited data available do not justify a detailed analysis with Monte Carlo searching of parameter space, and, thus, we can only estimate gradients to the nearest 0.05 mm (100 m)$^{-1}$ day^{-1}. Although we accounted for spatial variability, the paucity of data also means we cannot consider temporal variability due to the short observational period. In a region where the influence of both westerlies and monsoonal precipitation can be expected, the changes in precipitation gradient over time may well be important. Furthermore, the possible impact of climate warming on the climatic setting of the glacier might change the gradients in the future. However, the DDF factor, we deduce, is consistent with the observed variation of DDF across the region. The DDF is at the low end of the range on the Tibetan Plateau, indicating continental conditions for the glacier, as might be expected from its location.

A future step in simulating SMB might be a more sophisticated surface energy and mass balance model [46–48] or application of an operational physical model such as WRF [49]. However, these models have challenging data requirements and require many climate fields as drivers and sources of validation. These observations are very rare on the Tibetan Plateau, and reanalysis products are typically in error by an order of magnitude in snow accumulation [50].

Geothermal heat flux is an important boundary condition for ice flow models. It is essential to determine the ice temperature and, hence, ice viscosity and ice flow velocity. To match modeled ice velocity with measurements, we compared modeled ice velocity under different geothermal heat flux with observation, finding better fits with a geothermal heat flux of 60 mW m^{-2} than 50 mW m^{-2}. We experimented with heat fluxes as low as 30 mW m^{-2} and these produce ice velocities well below observed, and, conversely, higher heat fluxes that produce extensive regions of the glacier bed at the melting point lead to surface velocities much faster than observed. Thus, in this glacier, the ice velocity data provide a fairly strong constraint on geothermal heat flux—which, in a sparsely surveyed region, can provide a useful datum. However, the heat flux makes only a negligible difference to ice volume loss, despite the importance of ice dynamics for Da Anglong Glacier. Therefore, derivation of the geothermal heat flux from glaciological velocity observations requires both sufficient observations of surface velocity, and sufficient ice geometrical data to simulate a full-Stokes flow model. The procedure would not be feasible on typical small glaciers where ice dynamics are slower, and mass wastage rates due to SMB are relatively larger than on the exceptionally large Da Anglong Glacier.

One way of testing the glacier dynamics model is to compare the modeled surface lowering with observations. We do this with a simulation for the period 2001–2015 starting from the steady state results in the year 2000. Our averaged modeled surface elevation lowering rates from 2000 to 2015 is about 0.4 m a^{-1} over the whole glacier. This can be compared with estimates from satellite altimetry. Brun et al. [5] estimated glacier elevation change rate in HMA from the year 2000 to 2016 and found surface lowering of 0.7 m a^{-1} in Himalaya but increases in elevation of close to 0.2 m a^{-1} in western Kunlun and eastern Pamir, with a gap in data coverage over our region of interest, consistent with Shean et al. [35] estimates of -0.2 ± 0.2 m a^{-1}. The quoted uncertainty is only the estimated standard deviation with a necessarily incomplete knowledge of unknown systematic biases. Therefore, we judge that this satellite derived lowering rate is not significantly different from our modeled results.

Simulated ensemble mean volume loss rate during 2016–2098 is 0.46% a^{-1} of the glacier volume in the year 2016 under RCP2.6 and 1.0% a^{-1} under RCP8.5. Both area and volume loss rates have wider across-ESM model spread under RCP2.6 than RCP8.5, for example, volume loss rates range 0.19–0.75% a^{-1} under RCP 2.6 and 0.91–1.1% a^{-1} under RCP8.5. This is due to the much larger warming under RCP8.5 with SMB relatively more important than ice dynamics.

Our ensemble mean mass loss in the region from the year 2016 to 2098 is 38% under RCP2.6 and 83% under RCP8.5. This estimate under RCP2.6 is close to the ensemble mean mass loss estimates for the subregion South Asia West from the year 2015 to 2100 of 35% by Marzeion et al. [15] and 42% by Hock et al. [14] from glacier models forced with General Circulation Models. The estimate under RCP8.5 is larger than the ensemble mean mass loss for the subregion South Asia West from the year 2015 to 2100 of 62% by Marzeion et al. [15] and 65% by Hock et al. [14]. However, their models give a very wide range of estimates: 48–80% by Marzeion et al. [15] and 42–83% by Hock et al. [14].

8. Conclusions

Suites of simulations for Da Anglong Glacier have been made for the period 2016–2098 based on SMB parameterizations with projected temperature and precipitation from the 25-km-resolution RegCM4 nested within three ESMs running the RCP2.6 and RCP8.5 scenarios. We parameterized SMB using the positive degree-day method. We fit quadratic polynomials for an elevation-dependent precipitation gradient, using daily precipitation gradients at different elevations estimated by measured precipitation at both a local AWS and a regional meteorological station. The key factor, DDF, in the degree-day model for Tibetan glaciers in literature is expected to lie within the range 2.6 to 13.8 mm·d^{-1}·°C^{-1}. The best-fit DDF to the field observations is 3.4 mm·d^{-1}·°C^{-1} for Da Anglong Glacier, which is comfortably at the continental climate end of the range of glaciers and typical of the northwest Tibetan Plateau [24].

Da Anglong Glacier had an area of 6.92 km^2 in the year 2000 and 6.66 km^2 in 2016. Observations on the upper glacier are missing because of dangerous conditions. To deal with this, we developed a novel inverse procedure using GlabTop2 output as the initial ice thickness and the full-Stokes Elmer flow model to determine ice velocities for a given map of ice thickness. We modified ice thickness to minimize misfit between modeled steady-state surface velocity and the measured surface velocity.

By using field observations within the ice dynamics model, we showed that Da Anglong Glacier ice thickness is beyond the 90th percentile of uncertainty in SRTM elevation data in the year 2000. SRTM ice elevations imply a glacier too thick to undergo its documented terminus retreat. Furthermore, NASA MEaSUREs ITS_LIVE satellite velocity data are unphysically slow for a glacier as thick and steep as ground-based data show it to be. Hence, the remotely sensed data suggest simultaneously thicker and thinner ice than are observed from ground measurements on the glacier.

The modeled glacier loses 16–62% (ensemble mean of 38%) of its 2016 volume during 2016–2098 under the aggressive mitigation RCP2.6 scenario and 75–91% (ensemble mean of 83%) under the "business-as-usual" RCP8.5 scenario. The Paris 2015 agreement on greenhouse gas emissions produces a climate intermediate between these two extremes, thus we might expect Da Anglong to lose 1/2 to 3/4 of its mass by 2100 under the existing greenhouse gas emissions agreement.

Simulation from 2016 to 2098 neglecting ice dynamics show an underestimation of 23–29% under RCP2.6 and 19–26% under RCP8.5 compared with including ice dynamics. This is different from the result found for small Tibetan glaciers [13,16] for which SMB is the only significant factor in mass loss. Thus, mass loss estimates for large glaciers, which dominate the total sea level rise commitment, may need to be estimated individually when evaluating large scale mass balance estimation for High Mountain Asia.

Author Contributions: Conceptualization, L.Z. and J.C.M.; methodology, L.Z. and W.Z.; software, W.Z. and L.Z.; validation, L.Z., W.Z. and J.C.M.; formal analysis, W.Z. and L.Z.; investigation, L.Z.; resources, L.Z.; data curation, L.T., W.Z., L.Z.; writing—original draft preparation, W.Z., L.Z.; writing—review and editing, L.Z., J.C.M., M.W., L.T.; visualization, W.Z.; supervision, L.Z. and J.C.M.; project administration, L.Z.; funding acquisition, L.Z., L.T. All authors have read and agreed to the published version of the manuscript.

Funding: This research was funded by the National Key Research and Development Program of China (2021YFB3900105) and the National Natural Science Foundation of China (No. 41941006, 41530748).

Institutional Review Board Statement: Not applicable.

Informed Consent Statement: Not applicable.

Data Availability Statement: The SRTM version 4.1 dataset was obtained from https://cgiarcsi. community/data/srtm-90m-digital-elevation-database-v4-1/ (accessed on 11 November 2021). The output of regional climate model RegCM4 was obtained from https://esgf-data.dkrz.de/search/ cordex-dkrz/ (accessed on 11 November 2021). The historical temperature and precipitation data at Shiquanhe meteorological station was downloaded from http://data.cma.cn/ (accessed on 11 November 2021). The ice thickness determined using GlabTop2 model was downloaded from https://mountainhydrology.org/data-nature-2017/ (accessed on 11 November 2021). The in-situ observation data, Elmer/Ice model scripts and modelled results are available from the authors.

Acknowledgments: This study has been supported by the National Key Research and Development Program of China (2021YFB3900105) and the National Natural Science Foundation of China (No. 41941006, 41530748). We thank Gao Xuejie for providing the output of regional climate model RegCM4.

Conflicts of Interest: The authors declare no conflict of interest.

References

1. Kang, S.; Xu, Y.; You, Q.; Flügel, W.-A.; Pepin, N.; Yao, T. Review of climate and cryospheric change in the Tibetan Plateau. *Environ. Res. Lett.* **2010**, *5*, 015101. [CrossRef]
2. Pepin, N.; Bradley, R.S.; Diaz, H.F.; Baraer, M.; Caceres, E.B.; Forsythe, N.; Fowler, H.; Greenwood, G.; Hashmi, M.Z.; Liu, X.D.; et al. Elevation-dependent warming in mountain regions of the world. *Nat. Clim. Chang.* **2015**, *5*, 424–430.
3. Zhu, X.; Wei, Z.; Dong, W.; Wen, X.; Liu, Y. Projected temperature and precipitation changes on the Tibetan Plateau: Results from dynamical downscaling and CCSM4. *Theor. Appl. Climatol.* **2019**, *138*, 861–875. [CrossRef]
4. Azam, M.F.; Wagnon, P.; Berthier, E.; Vincent, C.; Fujita, K.; Kargel, J.S. Review of the status and mass changes of Himalayan-Karakoram glaciers. *J. Glaciol.* **2018**, *64*, 61–74. [CrossRef]
5. Brun, F.; Berthier, E.; Wagnon, P.; Kaab, A.; Treichler, D. A spatially resolved estimate of high mountain Asia glacier mass balances, 2000–2016. *Nat. Geosci.* **2017**, *10*, 668–673. [CrossRef] [PubMed]
6. Yao, T.; Thompson, L.; Yang, W. Different glacier status with atmospheric circulations in Tibetan Plateau and surroundings. *Nat. Clim. Chang.* **2012**, *2*, 663–667. [CrossRef]
7. Nakićenović, N.; Sward, R. *Special Report on Emission Scenarios*; Cambridge University Press: Cambridge, UK, 2000; pp. 570–599.
8. Marzeion, B.; Jarosch, A.H.; Hofer, M. Past and future sea level change from the surface mass balance of glaciers. *Cryosphere* **2012**, *6*, 1295–1322. [CrossRef]
9. Radic, V.; Bliss, A.; Beedlow, C.A.; Hock, R.; Miles, E.; Cogley, J.G.. Regional and global projections of the 21st century glacier mass changes in response to climate scenarios from GCMs. *Clim. Dyn.* **2014**, *42*, 37–58. [CrossRef]
10. Zhao, L.; Ding, R.; Moore, J.C. The High Mountain Asia glacier contribution to sea-level rise from 2000 to 2050. *Annu. Glaciol.* **2016**, *57*, 223–231. [CrossRef]
11. Zhao, L.; Yang, Y.; Ji, D.; Moore, J.C. Glacier evolution in high mountain Asia under stratospheric sulfate aerosol injection geoengineering. *Atmos. Chem. Phys.* **2017**, *17*, 6547–6564. [CrossRef]
12. Kraaijenbrink, P.D.A.; Bierkens, M.F.P.; Lutz, A.F.; Immerzeel, W.W. Impact of a global temperature rise of 1.5 degrees Celsius on Asia's glaciers. *Nature* **2017**, *549*, 257–260. [CrossRef]
13. Zhao, L.; Tian, L.; Zwinger, T.; Ding, R.; Zong, J.; Ye, Q.; Moore, J.C. Numerical simulations of Gurenhekou glacier on the Tibetan Plateau. *J. Glaciol.* **2014**, *60*, 71–82. [CrossRef]
14. Hock, R.; Bliss, A.; Marzeion, B.; Giesen, R.H.; Hirabayashi, Y.; Huss, M.; Radić, V.; Slangen, A.B.A. GlacierMIP—A model intercomparison of global-scale glacier mass-balance models and projections. *J. Glaciol.* **2019**, *65*, 453–467. [CrossRef]
15. Marzeion, B.; Hock, R.; Anderson, B.; Bliss, A.; Champollion, N.; Fujita, K.; Huss, M.; Immerzeel, W.W.; Zekollari, H. Partitioning the uncertainty of ensemble projections of global glacier mass change. *Earth's Future* **2020**, *8*, e2019EF001470. [CrossRef]
16. Li, Y.; Tian, L.; Yang, Y.; John, C.M.; Sun, S.; Zhao, L. Simulating the evolution of Qiangtang No. 1 Glacier in the central Tibetan Plateau to 2050. *Arct. Antarct. Alp. Res.* **2017**, *49*, 1–12. [CrossRef]
17. Jiang, G.; Hu, S.; Shi, Y.; Zhang, C.; Wang, Z.; Hu, D. Terrestrial heat flow of continental China: Updated dataset and tectonic implications. *Tectonophysics* **2019**, *753*, 36–48. [CrossRef]
18. Guo, W.; Liu, S.; Xu, J.; Wu, L.; Shangguan, D.; Yao, X.; Wei, J.; Bao, W.; Yu, P.; Liu, Q.; et al. The second Chinese glacier inventory: Data, methods and results. *J. Glaciol.* **2015**, *61*, 357–372. [CrossRef]
19. Chen, Y.; Tian, L.; Zong, J.; Zhu, D.; Wang, C.; Jin, S. Variation of the large and small Anglong Glaciers in the Ali district, Tibet autonomous region. *J. Glaciol. Geocryol.* **2019**, *41*, 14–23, (In Chinese with English Summary). [CrossRef]
20. Gardner, A.S.; Mohold, G.; Scambos, T.; Fahnstock, M.; Ligtenberg, S.; Van Den, B.M.; Nilsson, J. Increased West Antarctic and unchanged East Antarctic ice discharge over the last 7 years. *Cryosphere* **2018**, *12*, 521–547. [CrossRef]
21. Gardner, A.S.; Fahnestock, M.A.; Scambos, T.A. ITS_LIVE Regional Glacier and Ice Sheet Surface Velocities. Data Archived at National Snow and Ice Data Center; 2019. Available online: https://its-live.jpl.nasa.gov/ (accessed on 15 January 2020).
22. Sun, H.; Su, J.; Huang, J.; Yao, T.; Luo, Y.; Chen, D. Contrasting precipitation gradient characteristics between westerlies and monsoon dominated upstream river basins in the Third Pole. *Chin. Sci. Bull.* **2020**, *65*, 91–104. [CrossRef]

23. Sakai, A.; Nuimural, T.; Fujita, K.; Takenaka, S.; Nagail, H.; Lamsal, D. Climate regime of asian glaciers revealed by gamdam glacier inventory. *Cryosphere* **2015**, *9*, 865–880. [CrossRef]

24. Zhang, Y.; Liu, S.; Ding, Y. Spatial variation of degree-day factors on the observed glaciers in Western China. *Acta Geogr. Sinica* **2006**, *61*, 89–98. [CrossRef]

25. Braithwaite, R.J. Calculation of degree-days for glacier–climate research. *Z. Gletsch. Glazialgeol.* **1984**, *20*, 1–20.

26. Deng, C.; Zhang, W. Spatial distribution pattern of degree–day factors of glaciers on the Qinghai–Tibetan Plateau. *Environ. Monit. Assess.* **2018**, *190*, 475. [CrossRef]

27. Adhikari, S.; Marshall, S.J. Influence of high-order mechanics on simulation of glacier response to climate change: Insights from Haig Glacier, Canadian Rocky Mountains. *Cryosphere* **2013**, *7*, 1527–1541. [CrossRef]

28. Gagliardini, O.; Zwinger, T.; Gillet-Chaulet, F.; Durand, G. Capabilities and performance of elmer/ice, a new generation ice-sheet model. *Geosci. Model Dev.* **2013**, *6*, 1299–1318. [CrossRef]

29. Paterson, W.S.B. *The Physics of Glaciers*, 3rd ed.; Elsevier: Oxford, UK, 1994.

30. Gao, X.J.; Wu, J.; Shi, Y.; Wu, J.; Han, Z.Y.; Zhang, D.F.; Tong, Y.; Li, R.K.; Xu, Y.; Giorgi, F. Future changes in thermal comfort conditions over China based on multi-RegCM4 simulations. *Atmos. Ocean. Sci. Lett.* **2018**, *11*, 291–299. [CrossRef]

31. Martin, G.M.; Ringer, M.A.; Pope, V.D.; Jones, A.; Dearden, C.; Hinton, T.J. The physical properties of the atmosphere in the new Hadley Centre Global Environmental Model (HadGEM1). Part I: Model description and global climatology. *J. Clim.* **2006**, *19*, 1274–1301. [CrossRef]

32. Marsland, S.J.; Haak, H.; Jungclaus, J.H.; Latif, M.; Röske, F. The Max-Planck-Institute global ocean/sea ice model with orthogonal curvilinear coordinates. *Ocean Model.* **2003**, *5*, 91–127. [CrossRef]

33. Bentsen, M.; Bethke, I.; Debernard, J.B.; Iversen, T.; Kirkevåg, A.; Seland, Ø; Drange, H.; Roelandt, C.; Seierstad, I.A.; Hoose, C.; et al. The Norwegian Earth System Model, NorESM1-M—Part 1: Description and basic evaluation of the physical climate. *Geosci. Model Dev.* **2013**, *6*, 687–720. [CrossRef]

34. Falorni, G.; Teles, V.; Vivoni, E.R.; Bras, R.L.; Amaratunga, K.S. Analysis and characterization of the vertical accuracy of digital elevation models from the Shuttle Radar Topography Mission. *J. Geophys. Res.* **2005**, *110*, F02005. [CrossRef]

35. Shean, D.E.; Bhushan, S.; Montesano, P.; Rounce, D.R.; Arendt, A.; Osmanoglu, B. A systematic, regional assessment of high mountain Asia glacier mass balance. *Front. Earth Sci.* **2020**, *7*, 363. [CrossRef]

36. Frey, H.; Machguth, H.; Huss, M.; Huggel, C.; Bajracharya, S.; Bolch, T.; Kulkarni, A.; Linsbauer, A.; Salzmann, N.; Stoffel, M. Estimating the volume of glaciers in the Himalayan–Karakoram region using different methods. *Cryosphere* **2014**, *8*, 2313–2333. [CrossRef]

37. Millan, R.; Mouginot, J.; Rabatel, A.; Jeong, S.; Cusicanqui, D.; Derkacheva, A.; Chekki, M. Mapping surface flow velocity of glaciers at regional scale using a multiple sensors approach. *Remote Sens.* **2019**, *11*, 2498. [CrossRef]

38. Dash, J.G.; Rempel, A.W.; Wettlaufer, J.S. The physics of premelted ice and its geophysical consequences. *Rev. Mod. Phys.* **2006**, *78*, 695–741. [CrossRef]

39. Farinotti, D.; Brinkerhoff, D.J.; Fürst, J.J.; Gantayat, P.; Gillet-Chaulet, F.; Huss, M.; Leclercq, P.W.; Maurer, H.; Morlighem, M.; Pandit, A.; et al. Results from the Ice Thickness Models Intercomparison eXperiment phase 2 (ITMIX2). *Front. Earth Sci.* **2021**, *8*, 571923. [CrossRef]

40. Gourmelen, N.; Gardner, A.S.; Brun, F.; Goldberg, D.; Nienow, P.W..; Berthier, E.; Vincent, C.; Wagnon, P.; Trouvé, E.; Gourmelen, N. Twenty-first century glacier slowdown driven by mass loss in High Mountain Asia. *Nat. Geosci.* **2019**, *12*, 22–27. [CrossRef]

41. Azam, M.F.; Wagnon, P.; Berthier, E.; Vincent, C.; Fujita, K.; Kargel, J.S. From balance to imbalance: A shift in the dynamic behaviour of Chhota Shigri glacier, western Himalaya, India. *J. Glaciol.* **2012**, *58*, 315–324. [CrossRef]

42. Singh, V.; Goyal, M.K. Analysis and trends of precipitation lapse rate and extreme indices over north Sikkim eastern Himalayas under CMIP5ESM-2M RCPs experiments. *Atmos. Res.* **2016**, *167*, 34–60. [CrossRef]

43. Yang, K.; Guyennon, N.; Ouyang, L.; Tian, L.; Tartari, G.; Salerno, F. Impact of summer monsoon on the elevation-dependence of meteorological variables in the south of central Himalaya. *Int. J. Climatol.* **2018**, *38*, 1748–1759. [CrossRef]

44. Sevruk, B.; Mieglitz, K. The effect of topography, season and weather situation on daily precipitation gradients in 60 Swiss valleys. *Water Sci. Technol.* **2002**, *45*, 41–48. [CrossRef] [PubMed]

45. Sevruk, B. Regional dependency of precipitation-altitude relationship in the Swiss Alps. *Clim. Chang.* **1997**, *36*, 355–369. [CrossRef]

46. Klok, E.J.; Oerlemans, J. Model study of the spatial distribution of the energy and mass balance of Morteratschgletscher, Switzerland. *J. Glaciol.* **2002**, *48*, 505–518. [CrossRef]

47. Jiang, X.; Wang, N.; He, J.; Wu, X.; Song, J. A distributed surface energy and mass balance model and its application to a mountain glacier in China. *Chin. Sci. Bull.* **2010**, *55*, 2079–2087. [CrossRef]

48. Krapp, M.; Robinson, A.; Ganopolski, A. SEMIC: An efficient surface energy and mass balance model applied to the Greenland ice sheet. *Cryosphere* **2017**, *11*, 1519–1535. [CrossRef]

49. Turton, J.N.; Mölg, T.; Collie, E. High-resolution (1 km) Polar WRF output for 79o N Glacier and the northeast of Greenland from 2014 to 2018. *Earth Syst. Sci. Data* **2020**, *12*, 1191–1202. [CrossRef]

50. Wang, W.; Yang, K.; Zhao, L.; Zheng, Z.; Lu, H.; Mamtimin, A.; Ding, B.; Li, X.; Zhao, L.; Li, H.; et al. Characterizing surface albedo of shallow fresh snow and its importance for snow ablation on the interior of the Tibetan Plateau. *J. Hydrometeorol.* **2020**, *21*, 815–827. [CrossRef]

Review

Review on Per- and Poly-Fluoroalkyl Substances' (PFASs') Pollution Characteristics and Possible Sources in Surface Water and Precipitation of China

Fan Wang [1], Yiru Zhuang [1], Bingqi Dong [1] and Jing Wu [1,2,*]

1   The MOE Key Laboratory of Resource and Environmental System Optimization, College of Environmental Science and Engineering, North China Electric Power University, Beijing 102206, China; wf881228@163.com (F.W.); xiezizyr@163.com (Y.Z.); dbq971013@163.com (B.D.)
2   Institute of Transport Energy and Environment, Beijing Jiaotong University, Beijing 100044, China
*   Correspondence: wujing.108@163.com

Citation: Wang, F.; Zhuang, Y.; Dong, B.; Wu, J. Review on Per- and Poly-Fluoroalkyl Substances' (PFASs') Pollution Characteristics and Possible Sources in Surface Water and Precipitation of China. *Water* **2022**, *14*, 812. https://doi.org/10.3390/w14050812

Academic Editor: Maurizio Barbieri

Received: 10 February 2022
Accepted: 2 March 2022
Published: 4 March 2022

Publisher's Note: MDPI stays neutral with regard to jurisdictional claims in published maps and institutional affiliations.

Abstract: In recent years, due to the production and use of per- and poly-fluoroalkyl substances (PFASs), the research on the pollution characteristics and sources of PFASs in surface water and precipitation in China has attracted increasing attention. In this study, the related published articles with sampling years from 2010 to 2020 were reviewed, and the concentration levels, composition characteristics and possible sources of PFASs in surface water (rivers and lakes) and precipitation in China were summarized, including those in the Tibetan Plateau region. The results show that the concentrations of PFASs in surface water in different areas of China vary greatly, ranging from 0.775 to 1.06×10^6 ng/L. The production processes of fluorinated manufacturing facilities (FMFs) and sewage discharge from wastewater treatment plants (WWTPS) were the main sources of PFASs in surface water in China, and the concentrations of PFASs in water flowing through cities with high urbanization increased significantly compared with those before water flowed through cities with high urbanization. The compositions of PFASs in surface water gradually changed from long-chain PFASs, such as per-fluoro-octanoic acid (PFOA) and per-fluoro-octanesulfonic acid (PFOS) to short-chain PFASs, such as per-fluorobutanoic acid (PFBA), per-fluorobutanesulfonic acid (PFBS), perfluorohexanoic acid (PFHxA) and per-fluoropentanoic acid (PFPeA). The concentrations of PFASs in precipitation in China ranged from 4.2 to 191 ng/L, which were lower than those of surface water. The precipitation concentrations were relatively high around a fluorination factory and in areas with high urbanization levels. PFASs were detected in the surface water and precipitation in the Tibetan Plateau (TP), which is the global "roof of the world", but the concentrations were low (0.115–6.34 ng/L and 0.115–1.24 ng/L, respectively). Local human activities and surface runoff were the main sources of PFASs in the surface water of the Tibetan Plateau. In addition, under the influence of the Southeast Asian monsoon in summers, marine aerosols from the Indian Ocean and air pollutants from human activities in Southeast Asia and South Asia will also enter the water bodies through dry and wet depositions. With the melting of glaciers caused by global warming, the concentration of PFASs in the surface water of the TP was higher than that before the melting of glaciers flowed into the surface water of the TP. Generally, this study summarized the existing research progress of PFAS studies on surface water and precipitation in China and identified the research gaps, which deepened the researchers' understanding of this field and provided scientific support for related research in the future. The concentrations of PFASs in the water bodies after flowing through FMFs were significantly higher than those before water flowed through FMFs, so the discharge of the FMF production process was one of the main sources of PFASs in surface water.

Keywords: i-PFASs; China; river; lake; precipitation; the Tibetan Plateau

1. Introduction

Per- and poly-fluoroalkyl substances (PFASs) refer to a class of compounds in which all or part of the hydrogen atoms linked to carbon atoms in alkane molecules are replaced by fluorine atoms [1]. Because of the strong polarity of the C-F bond, PFASs have more stable and more excellent properties than other hydrocarbons (such as a remarkably high chemical stability and excellent hydrophobicity and oleophobicity), so they have been used in various fields of production, such as plastic wrap, paper, coatings, poly-tetrafluoroethylene products and foam fire-extinguishing agents [2]. According to different functional groups and physicochemical properties, PFASs can be divided into (1) ionic PFASs (i-PFASs), such as per-fluoroalkyl carboxylic acids (PFCAs) and per-fluoroalkane sulfonic acids (PFSAs), and (2) neutral PFASs (n-PFASs), such as fluorotelomer alcohols (FTOHs) and per-fluoro-octane sulfo-namidoethanols (FOSEs) [3]. The PFASs referred to in this paper are i-PFASs. Due to the mass production and use of some PFASs, PFASs have been widely detected in various environmental media, animals and plants in recent years [4–6], and have also been found in the human body [7,8]. With intensive studies, the persistence, bioaccumulation, long-distance transportation and biohazard of some long-chain PFASs (long-chain PFASs refer to PFCAs with seven or more perfluorinated carbons and PFSAs with six or more perfluorinated carbons) have been gradually confirmed [9]. Subsequently, some countries and organizations have successively issued series of rules and regulations to restrict the use of such substances [10], and PFOS, PFOA and their salts were listed under Annex B and Annex A of the Stockholm Convention on Persistent Organic Pollutants in 2009 and 2019, respectively [11,12]. Meanwhile, in 2019, the Persistent Organic Pollutants Review Committee (POPRC) recommended that per-fluorohexane sulfonate acid (PFHxS) and its salts be listed in Annex A of the Convention [13]. In addition, at the 17th meeting of the POPRC, held in January 2022, it was suggested that long-chain PFCA (involving carbon-chain lengths from 9 to 21), its salts and related compounds should be listed in Annexes A, B and/or C of the Stockholm Convention on Persistent Organic Pollutants [14]. With the restricted use of PFOS, PFOA (Table S2) and other substances, some short-chain per-fluoroalkyl acids and new polyfluoride substitutes [such as 6:2 chlorinated polyfluorinated ether sulphonic acid (F-53B)] are becoming research focuses [15,16]. The international community is paying increasing attention to PFASs.

Since 2000, some major manufacturers have phased out the production of PFOS, PFOA and their salts [17]. The production of these substances has gradually shifted from developed countries, such as North America and Europe, to developing countries, especially China [18]. China has become the largest manufacturer and supplier of PFOS and PFOA in the world since 2004 [3]. Some studies have estimated the emissions of PFOA and its salts in China from 2004 to 2012. The results show that the cumulative emissions reached 250 tons, and China became the largest PFOA emission site at that time [19]. Therefore, the environmental monitoring of and risk research on PFASs in China have attracted international attention. Studies have shown that PFASs can exist stably in water environments [20], and the ocean is the final sink for PFASs [21]. Therefore, it is necessary to explore the pollution characteristics and possible sources of PFASs in China's waters. Most of the related studies have been carried out for a single water body [22,23], but there were a few comprehensive literature reviews. In this study, the related published articles with sampling years from 2010 to 2020 were reviewed, and the concentration levels, composition characteristics and possible sources of PFASs in surface water (rivers and lakes) and precipitation in China were summarized (Table S1 shows the PFASs involved in the paper). The research area also covered the Tibetan Plateau. The purpose is to summarize the existing research progress, identify the existing research gaps and provide scientific support for future research in this field.

2. PFASs in Surface Water

This study summarized the concentration data from monitoring studies of PFASs in surface water in China over the past 10 years, as shown in Figure 1 and Table 1 for details. It was revealed that the concentrations of PFASs in surface water varied greatly in different areas in China (from 0.775 to 1.06×10^6 ng/L). The average concentrations in the Daling River, tested in 2018, and Xiaoqing River, tested in 2013, were highest (2.31×10^3 ng/L and 2.14×10^3 ng/L, respectively) [24,25], followed by that in the Daling River in 2011 (1.04×10^3 ng/L) [26]. The concentrations in other rivers were relatively low. It was found that the concentrations and composition characteristics of PFASs were mainly related to the emissions of fluorinated manufacturing facilities (FMFs), wastewater treatment plants (WWTPs) and urbanization development. The concentrations of PFASs in the water bodies after they flowed through FMFs were significantly higher than those before they flowed through FMFs, so the discharge from the FMF production process was one of the main sources of PFASs in surface water. For example, the concentration of PFASs in the Daling River flowing through the Fluorochemical Industrial Zone in Fuxin City was significantly higher than the concentration in other watersheds by an order of magnitude. With the development of the fluorochemical industry, the PFASs from manufacturing processes caused more pollution to enter the Daling River, and the concentrations detected in 2018 increased by 50% compared to 2011 [25,26]. Liaohe River was also affected by the Fluorochemical Industrial Zone in Fuxin City with the maximum concentration of 781 ng/L [27]. Similarly, Xiaoqing River was influenced by FMFs, such as metallurgy, electronics, and firefighting, with the maximum concentration of 1.06×10^6 ng/L detected in 2013 [24]. The north of Taihu Lake was close to FMFs, such as the manufacturing of paint and plastic products, with concentrations of 56.1–120 ng/L, while concentrations in other parts of the study area ranged from 10.0 to 79.4 ng/L [28]. The concentration in Guanlan River during the abundant water period was six times higher than that in the dry water period, probably due to the large amount of surface runoff and rainwater flushing during the abundant water period that carried pollutants from the periphery of the industrial area into the water body [29]. East China was also a concentrated area for FMFs, and the PFAS concentration in Huangpu River flowing through this region increased (from about 300 to 380 ng/L) [30]. Moreover, the variation tendencies of PFAS concentrations in different water bodies were also different; for example, the concentration of PFASs in the Daling River increased from 2011 (average of 1.04×10^3 ng/L [26]) to 2018 (average of 2.31×10^3 ng/L [25]), and that in Xiaoqing River decreased from 2013 (average of 2.14×10^3 ng/L) [24] to 2014 (average of 457 ng/L) [31].

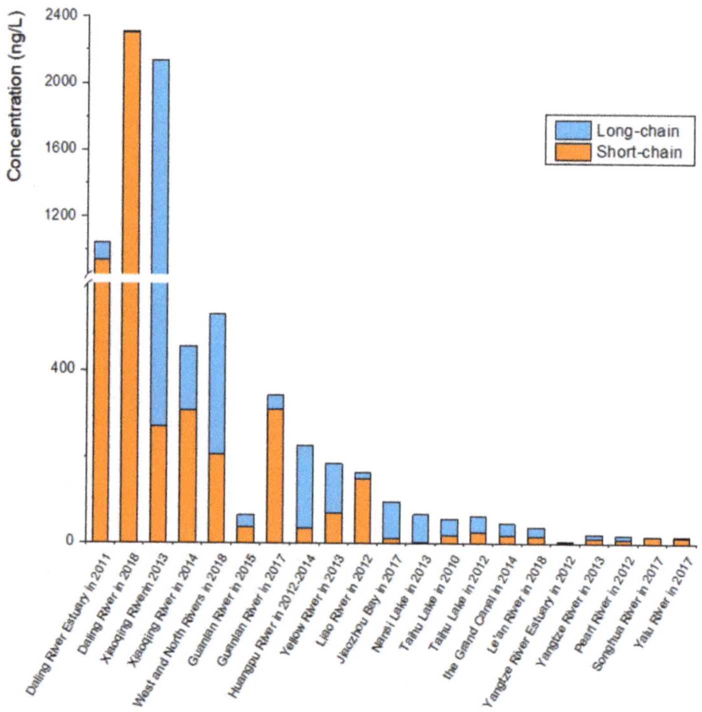

Figure 1. Levels and compositions in surface water in China in past ten years.

In addition, the concentrations of PFASs in the water bodies after flowing through WWTPs were obviously higher than those before the water flowed through WWTPs, so the sewage discharge of WWTPs was another main source of PFASs in surface water. For example, after flowing through WWTPs, the concentration of PFASs in Huangpu River, mentioned above, rose from ~200 to ~250 ng/L [30]. The peak concentration of the Yongding River occurred at 108 ng/L downstream of the WWTP in its watershed and 12.4 ng/L upstream [32]. Due to the sewage discharged by WWTPs entering into the north of Dianchi Lake directly, concentrations in the north (35.8–135.9 ng/L) were higher than those in the south (less than 25 ng/L) [33]. It was found that the concentrations of PFASs in water flowing through cities with high degrees of urbanization had significant rises because of traffic, commercial activities and so on. For example, the West and North Rivers flow through the cities of Shaoguan, Heze and Qingyuan, and rapid urban development had a remarkable impact on the PFASs in the water bodies with the concentrations up to 1.06×10^3 ng/L [34]. Yanghe River (one of the tributaries of the Yongding River) ran through the highly urbanized city of Zhangjiakou with a high concentration of 197 ng/L, and Sangan River, another tributary of the Yongding River, ran through less densely populated and less urbanized areas with relatively smaller concentrations ranging from 6.67 to 9.74 ng/L [32]. Likewise, the Songhua River and the Yalu River, which were relatively less densely populated, also had relatively low concentrations of PFASs, with a maximum of only 32 ng/L [35].

Table 1. PFAS concentrations in surface water in China in past ten years.

Sampling Year	Area	Analytes	Concentration (ng/L)	Reference
2010	Caohai (north of Dianchi Lake)	C4–C12 PFCAs, C8 PFSAs	35.8–135.9	[33]
2010	Taihu Lake	C4–C12 PFCAs, C6, C8 PFSAs	10.0–119.8	[28]
2011	Daling River estuary	C4–C10 PFCAs, C4, C6, C8 PFSAs	$5.26–4.74 \times 10^3$	[26]
2011	Tangxun Lake	C4–C13 PFCAs, C4, C6, C8 PFSAs	4570–11,890	[38]
2012	Liao River	C6–C12 PFCAs, C4, C6, C8 PFSAs and FOSA	44.4–781	[27]
2012	Pearl River	C4–C11 PFCAs, C4, C6–C8, C10 PFSAs	3.0–52	[36]
2012	Taihu Lake	C6–C12 PFCAs, C4, C6, C8 PFSAs and FOSA	17.2–94.3	[27]
2012	Yangtze River estuary	C4–C10 PFCAs, C4, C6, C8 PFSAs and FOSA	1.7–12	[39]
2012–2014	Huangpu River	C3–C12, C14 PFCAs, C4, C6, C8 PFSAs	39.8–596.2	[30]
2013	Nansi Lake	C5, C7–C11 PFCAs, C4, C6, C8 PFSAs	38.4–91.4	[40]
2013	Pearl River delta	C5–C14, C16, C18 PFCAs, C4, C6, C8, C10 PFSAs	1.53–33.5	[41]
2013	Xiaoqing River	C4–C11 PFCAs, C4, C6–C8, C10 PFSAs	$32.2–1.06 \times 10^6$	[24]
2013	Yangtze River	C4–C11 PFCAs, C4, C6, C8, C10 PFSAs and FOSA	2.2–74.56	[42]
2013	Yellow River	C4–C12 PFCAs, C4, C8 PFSAs	$44.7–1.52 \times 10^3$	[43]
2014	Xiaoqing River	C4–C12 PFCAs, C8 PFSAs	$36.5–4.96 \times 10^5$	[31]
2014	Grand Canal	C4–C11, C14 PFCAs, C4, C6, C8 PFSAs	7.8–218	[44]
2014–2015	Jiulong River estuary	C4–C14 PFCAs, C4, C6, C8, C10 PFSAs	3.30–110	[37]
2015	Guanlan River	C6–C12, C14 PFCAs, C4, C6, C8 PFSAs	37.04–103.7	[29]
2016	Sanggan River	C4–C12 PFCAs, C4, C6, C8 PFSAs	6.67–9.74	[32]
2016	Yanghe River	C4–C12 PFCAs, C4, C6, C8 PFSAs	2.10–197	[32]
2016	Yongding River	C4–C12 PFCAs, C4, C6, C8 PFSAs	12.4–108	[32]
2017	Guanlan River	C6–C12, C14 PFCAs, C4, C6, C8 PFSAs	179.15–613.68	[29]
2017	Jiaozhou Bay	C4–C12 PFCAs, C4, C6, C8, C10 PFSAs and FOSA	35.00–205.34	[45]
2017	Poyang Lake	C4–C11 PFCAs, C4, C6, C8 PFSAs	12.9–56.2	[46]
2017	Songhua River	C4–C14 PFCAs, C4, C6, C8, C10 PFSAs	6.4–32	[35]
2017	Yalu River	C4–C14 PFCAs, C4, C6, C8, C10 PFSAs	6.3–28	[35]
2017	Yangtze River	C4–C11 PFCAs, C4, C6, C8 PFSAs	7.8–586.2	[46]
2018	Daling River	C4–C13 PFCAs, C4, C6, C8, C10 PFSAs, HFPO-DA and 6:2 Cl-PFESA	$48.4–4.58 \times 10^3$	[25]
2018	Le'an River	C4–C14 PFCAs, C4, C6, C8, C10 PFSAs	14.71–114.72	[47]
2018	West and North Rivers	C4, C6–C10 PFCAs, C4, C6, C8, PFSAs, 6:2 Cl-PFESA and FHUEA	$0.775–1.06 \times 10^3$	[34]

In the past, long-chain PFASs made the greatest contributions, especially PFOA and PFOS. In recent years, short-chain PFASs have gradually become the main compounds in water pollution, such as PFBA, PFBS, PFHxA and PFPeA. For instance, in the Daling River, the proportions of PFBS (35.9% in 2011 and 48% in 2018) and PFBA (32.8% in 2011 and 41% in 2018) have gradually increased [25,26]. On the contrary, PFOA (17.5% in 2011 and 2.3% in 2018) showed a decreasing trend [25,26]. In the main stem of the Pearl River, PFBS contributed 36% while PFOS contributed 21% [36], and the same phenomenon was found in the tributaries of the Pearl River (52% PFBA and 25% PFOA) [34]. Similarly, PFHxA accounted for 46% in spring and 53% in summer for PFPeA, with short-chain PFASs predominating in the Jiulongjiang estuary [37].

3. PFASs in Precipitation

The monitoring results of studies of PFASs in precipitation in China over the past 10 years were summarized as shown in Table 2. The monitored concentrations of PFASs in precipitation from the available studies in China, except for TFA (trifluoroacetic acid), ranged from 4.2 to 191 ng/L, which were higher than those in precipitation from developed countries, such as Germany (1.6–48.6 ng/L) [48], Japan (8.16–37.2 ng/L) [49] and France (2.59–3.76 ng/L) [50] during the same period, but lower than the concentrations in the United States (50–850 ng/L) [51].

Table 2. PFAS concentrations in precipitation in past ten years.

Sampling Year	Region	Analytes	Concentration (ng/L)	Reference
2010	Tianjin	C4–C12 PFCAs, C4, C6, C8, C10 PFSAs, 6:2 FTUCA and 8:2 FTUCA	22.5–147	[53]
2016	28 cities in mainland	C2 PFCA (TFA)	8.80–1.8×10^3	[52]
		C3–C12 PFCAs, C4, C6, C8 PFSAs, 8:2 FTUCA, 6:2 FTSA, 6:2 Cl-PFESA, 6:2 diPAP and 8:2 diPAP	5.37–191	
2017	Jiaozhou Bay	C4–C12 PFCAs, C4, C6, C8, C10 PFSAs and FOSA	4.20–66.1	[45]

The PFAS concentrations in precipitation were generally lower than those in surface water in the same area. The average concentration in precipitation in Jiaozhou Bay was 22.0 ng/L [45], which was lower than that in surface water (60.5 ng/L) [45]. In the Nanchang section of Yangtze River, the average PFAS concentration in precipitation was 9.2 ng/L [52], which was lower than that in surface water (44.6 ng/L) [46]. Similarly, in the Xiamen section of Jiulong River, the average precipitation concentration of PFASs (22.0 ng/L) [52] was lower than that of surface water (38.2 ng/L) [37].

PFAS precipitation concentrations tended to be higher in areas with concentrated distribution of FMFs. High PFAS concentrations (80.6 ng/L) were found in urban areas in Northeast China (such as Fuxin, Dalian and Harbin) with a high distribution of FMFs [52]. Similarly, the PFAS precipitation concentration in Tianjin was also high (63.1 ng/L) [53]. Moreover, higher precipitation concentrations were found in inland and coastal cities with higher urbanization levels. The average concentration was 84.8 ng/L in more urbanized inland cities, such as Zhengzhou and Chengdu, with high traffic volumes and commercial activities [52]. In coastal cities, such as Shantou, Xiamen and Weifang, the average concentration of PFASs in precipitation reached 182 ng/L [53]. Ship maintenance in the port contributed greatly to PFAS pollution.

A precipitation monitoring study conducted by Chen et al. [52] in 28 cities in China found that TFA contributed the largest proportion, accounting for 78.1%, followed by PFOA, PFBA and PFOS, accounting for 5.08%, 3.16% and 3.12%, respectively. The proportions of short-chain PFASs (C4–C7) and long-chain PFASs ($\geq$C8) were almost the same [52]. Han et al. [45] found that PFOA occupied 61.2% of the precipitation monitored in Jiaozhou Bay, followed by PFOS and PFHxA with 9.20% and 8.21%, respectively, and long-chain PFASs represented the largest proportion (87.3%).

4. PFASs in the Tibetan Plateau (TP)

Compared to the above studies, concentrations of PFASs in surface waters (including rivers and meltwater runoff) in the TP were relatively low (0.115–6.34 ng/L) as shown in Table 3. The hydrographic network in the Tibetan Plateau area were shown in Figure 2. A monitoring study of river water on the eastern edge of the TP found that pollutants in the upstream area entered the river water directly with snow melt, with the PFAS concentration at about 2 ng/L [54]. The midstream area was located in a high-altitude area with no surrounding inhabitants, and the highest PFAS concentration was 1.12 ng/L. The downstream area was densely populated, and the river water was greatly affected by the direct discharge of domestic garbage and sewage into the river, with the concentration of PFASs at about 1.8 ng/L [54]. Chen et al. [55] studied runoff from Nam Co Lake and its surroundings in the TP and found that PFAS concentrations in glacial runoff (0.892–1.94 ng/L) were generally higher than those in non-glacial runoff (0.443–1.12 ng/L), and the average PFAS concentration was generally higher in the southern part of Nam Co Lake (1.28 ng/L) than in the northern part (0.901 ng/L). There is more glacial runoff in the south of Nam Co Lake. Under the trend of global warming, glacial meltwater in the south flows into the lake along with the runoff, which makes PFAS-enriched glaciers release PFASs into the lake, raising the PFAS concentration of Nam Co Lake in the south. In terms of components, PFBA and PFOS were the major pollutants in lake water, with concentrations accounting for 19% and 17%, respectively. The proportion of PFBA was significantly higher in glacial runoff (61%) than in non-glacial runoff (10%). PFPeA was widely present in both glacial runoff (14%) and non-glacial runoff (23%) [55].

Table 3. PFAS concentrations in surface water and precipitation in TP in past ten years.

Sampling Year	Region	Analytes	Concentration (ng/L)	Reference
Surface water				
2010	Central Tibetan Plateau	C4–C12 PFCAs, C2–C4, C6, C8, C10 PFSAs	0.146–4.39	[56]
2010	Gongga Mountain	C4–C12 PFCAs, C2–C4, C6, C8, C10 PFSAs	0.115–6.34	[56]
2015–2016	Eastern Tibetan Plateau	C4–C14, C16, C18 PFCAs, C4, C6, C8, C10 PFSAs	0.272–5.15	[54]
2017	Nam Co Lake	C4–C13 PFCAs, C4, C6, C8 PFSAs	0.353–2.17	[55]
Precipitation				
2017	Nam Co basin	C4–C13 PFCAs, C4, C6, C8 PFSAs	0.115–1.24	[55]
2017	Tibetan Plateau	C4–C12 PFCAs, C4, C6, C8 PFSAs	0.212–0.548	[57]

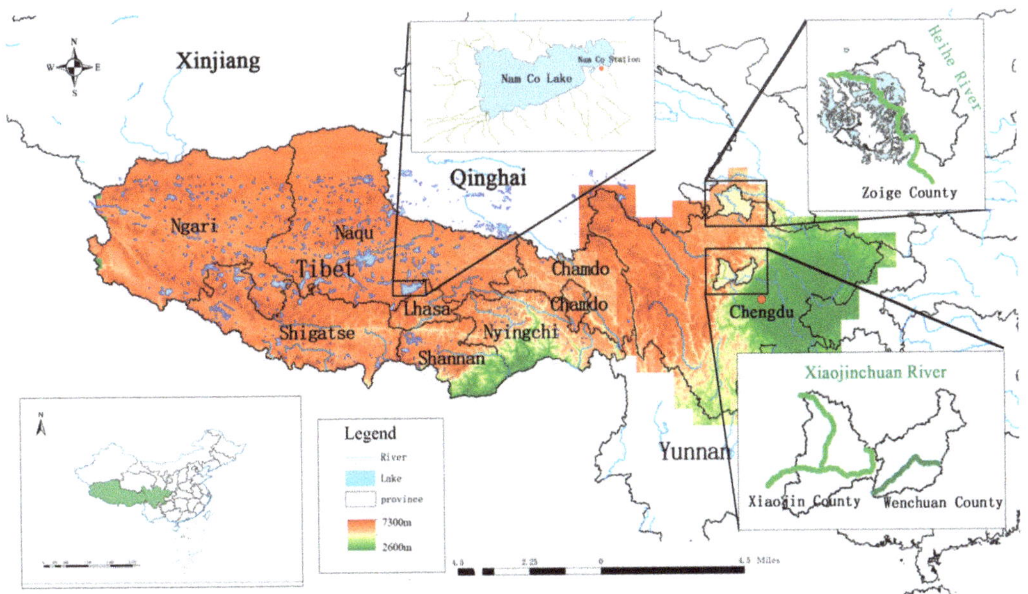

Figure 2. The hydrographic network in the Tibetan Plateau area.

In addition to pollution from local human activities and surface runoff, dry and wet depositions of atmospheric pollutants were also two of the main sources of PFASs in the water bodies of the TP. Influenced by the East Asian monsoon (from the east and southeast) and the Indian monsoon (from the south), PFAS pollutants in the atmosphere of the TP were increased [56]. It had been suggested that PFASs found in the waters of pristine areas in the eastern and central TP were greatly under the influence of the Southeast Asian monsoon in summers. Marine aerosols transported over long distances from the Indian Ocean and atmospheric pollutants from anthropogenic activities in Southeast and South Asia entered the water bodies through long-distance transportation and dry and wet depositions [55].

PFASs were also detected in precipitation in the TP, but at low concentrations (0.212–1.24 ng/L), with a 100-fold difference (0.212–190 ng/L) compared to precipitation concentrations from other regions. The average precipitation concentration of the Nam Co basin (0.616 ng/L) was higher than that of other areas of the TP (Motuo, Lulang, Lhasa and Muztagh Ata; 0.367 ng/L) [55,57]. The composition of PFASs in TP precipitation was similar to that of surface water with the main compounds PFOA and PFBA. Long- and short-chain PFASs in the Nam Co basin accounted for half of each, while short-chain PFASs were predominant in other regions of the TP (66%) [55]. One of the main sources of PFASs in precipitation in the TP may be atmospheric long-distance transportations from Northeast India, Southern Nepal and Northern Pakistan. Moreover, local emissions of PFAS-related manufacturing and the degradation of PFAS precursors in Eastern China also contributed to PFAS contaminations in precipitation [57].

5. Conclusions

The research on PFASs in surface water and precipitation in China has attracted increasing attention. Generally, the concentrations of PFASs in the surface water in different areas of China varied greatly, ranging from 0.775 ng/L in the West and North Rivers in 2018 to 1.06×10^6 ng/L in the Daling River in 2013. With the production and consumption of long-chain PFASs gradually changing to short-chain PFASs, the pollutant composition of PFASs in China's surface water also gradually changed from long-chain PFASs (such as PFOS and PFOA) to short-chain PFASs (such as PFBA, PFBS and PFPeA). The emissions from FMFs and WWTPS were considered the main sources of PFASs in China's surface water. It was also found that the concentrations of PFASs in water flowing through cities with high degrees of urbanization had significant rises. The concentration of PFASs in precipitation in China was lower than that in surface water, ranging from 4.2 ng/L in Jiaozhou Bay in 2017 to 191 ng/L in Weifang–Linqu County in 2016. PFAS concentrations were relatively high in precipitation around a fluorination plant and in areas with high urbanization levels. The release of PFASs from factories and human activities were assumed to be the main sources of PFASs in precipitation. Compared with other areas, the concentrations of PFASs in surface water and precipitation in the Tibetan Plateau were lower (0.115–6.34 ng/L and 0.115–1.24 ng/L, respectively). In addition to the emissions from local human activities, PFASs in the surface water of the Tibetan Plateau came from the inflow of surface runoff and dry and wet depositions of PFASs in the atmosphere. Under the general trend of global warming, the accelerated melting of glaciers will lead to further increases in PFAS concentrations. Moreover, influenced by the southeast monsoon in summers, PFASs can reach the Tibetan Plateau through long-distance transportation and finally enter the surface water of this area through depositions.

We find that there is a lack of continuous PFAS-monitoring research based on the PFASs in the same water body, and the research on novel PFAS substances (such as PFECAs and PFESAs) and their substitutes (such as HFPO-DA, ADONA and F-53B) in surface water is relatively limited. In addition, the research on PFASs in Tibetan Plateau surface water environments is extremely insufficient and needs further development. Overall, this study can supply researchers with a deeper understanding of the PFAS research progress on China's surface water and precipitation and provide scientific support for researchers to further grasp the research direction in this field.

Supplementary Materials: The following supporting information can be downloaded at: https://www.mdpi.com/article/10.3390/w14050812/s1, Table S1. Compound list of PFASs in the text. Table S2. Published administrative guidelines for PFOA and PFOS in water. References [34,58,59] are cited in the Supplementary Materials.

Author Contributions: F.W.: conceptualization, methodology, investigation, writing—original draft, validation; Y.Z.: conceptualization, methodology, investigation, writing—original draft; B.D.: conceptualization, methodology, investigation; J.W.: conceptualization, methodology, supervision, writing—reviewing and editing, project administration, validation. All authors have read and agreed to the published version of the manuscript.

Funding: This work was supported by the Second Tibetan Plateau Scientific Expedition and Research Program (STEP, grant no. 2019QZKK0103) and the National Natural Science Foundation of China (NSFC, grant no. 21976053).

Institutional Review Board Statement: Not applicable.

Informed Consent Statement: Not applicable.

Data Availability Statement: Not applicable.

Acknowledgments: The authors thank all individuals involved in this work.

Conflicts of Interest: The authors declare that they have no known competing financial interests or personal relationships that could have appeared to influence the work reported in this paper.

References

1. Pan, Y.; Zhang, H.; Cui, Q.; Shen, N.; Yeung, L.W.Y.; Sun, Y.; Guo, Y.; Dai, J. Worldwide Distribution of Novel Perfluoroether Carboxylic and Sulfonic Acids in Surface Water. *Environ. Sci. Technol.* **2018**, *52*, 7621–7629. [CrossRef] [PubMed]
2. Jian, J.M.; Guo, Y.; Zeng, L.; Liu, L.Y.; Lu, X.; Wang, F.; Zeng, E.Y. Global distribution of perfluorochemicals (PFCs) in potential human exposure source—A review. *Environ. Int.* **2017**, *108*, 51–62. [CrossRef] [PubMed]
3. Wang, Q.; Ruan, Y.; Lin, H.; Lam, P. Review on perfluoroalkyl and polyfluoroalkyl substances (PFASs) in the Chinese atmospheric environment. *Sci. Total Environ.* **2020**, *737*, 139804. [CrossRef] [PubMed]
4. Tian, Y.; Yao, Y.; Chang, S.; Zhao, Z.; Zhao, Y.; Yuan, X.; Wu, F.; Sun, H. Occurrence and Phase Distribution of Neutral and Ionizable Per- and Polyfluoroalkyl Substances (PFASs) in the Atmosphere and Plant Leaves around Landfills: A Case Study in Tianjin, China. *Environ. Sci. Technol.* **2018**, *52*, 1301–1310. [CrossRef]
5. Cui, Q.; Pan, Y.; Zhang, H.; Sheng, N.; Wang, J.; Guo, Y.; Dai, J. Occurrence and Tissue Distribution of Novel Perfluoroether Carboxylic and Sulfonic Acids and Legacy Per/Polyfluoroalkyl Substances in Black-Spotted Frog (Pelophylax nigromaculatus). *Environ. Sci. Technol.* **2018**, *52*, 982–990. [CrossRef]
6. Cao, X.; Wang, C.; Lu, Y.; Zhang, M.; Khan, K.; Song, S.; Wang, P.; Wang, C. Occurrence, sources and health risk of polyfluoroalkyl substances (PFASs) in soil, water and sediment from a drinking water source area. *Ecotoxicol. Environ. Saf.* **2019**, *174*, 208–217. [CrossRef]
7. Worley, R.R.; Moore, S.M.; Tierney, B.C.; Ye, X.; Calafat, A.M.; Campbell, S.; Woudneh, M.B.; Fisher, J. Per- and polyfluoroalkyl substances in human serum and urine samples from a residentially exposed community. *Environ. Int.* **2017**, *106*, 135–143. [CrossRef]
8. Jin, H.; Zhu, J.; Chen, Z.; Hong, Y.; Cai, Z. Occurrence and Partitioning of Bisphenol Analogues in Adults' Blood from China. *Environ. Sci. Technol.* **2018**, *52*, 812–820. [CrossRef]
9. Wang, Z.; Cousins, I.T.; Scheringer, M.; Hungerbuehler, K. Hazard assessment of fluorinated alternatives to long-chain perfluoroalkyl acids (PFAAs) and their precursors: Status quo, ongoing challenges and possible solutions. *Environ. Int.* **2015**, *75*, 172–179. [CrossRef]
10. Stockholm Convention. Available online: http://www.pops.int/TheConvention/POPsReviewCommittee/Meetings/POPRC3/POPRC3documents/tabid/77/Default.aspx (accessed on 1 March 2022).
11. USEPA. Drinking Water Contaminant Candidate List 3 (CCL3)—Final. *Fed. Regist.* **2009**, *74*, 51850–51862.
12. Stockholm Convention. Available online: http://www.pops.int/TheConvention/POPsReviewCommittee/Meetings/POPRC13/Overview/tabid/5965/Default.aspx (accessed on 1 March 2022).
13. Stockholm Convention. Available online: http://www.pops.int/TheConvention/POPsReviewCommittee/Meetings/POPRC15/Overview/tabid/8052/Default.aspx (accessed on 1 March 2022).
14. Stockholm Convention. Available online: http://www.pops.int/TheConvention/POPsReviewCommittee/Meetings/POPRC17/Overview/tabid/8900/Default.aspx (accessed on 1 March 2022).
15. Chen, F.; Yin, S.; Kelly, B.; Liu, W. Isomer-specific transplacental transfer of perfluoroalkyl acids: Results from a survey of paired maternal, cord sera, and placentas. *Environ. Sci. Technol.* **2017**, *51*, 5756–5776. [CrossRef]
16. Lee, J.W.; Lee, H.K.; Lim, J.E.; Moon, H.B. Legacy and emerging per- and polyfluoroalkyl substances (PFASs) in the coastal environment of Korea: Occurrence, spatial distribution, and bioaccumulation potential. *Chemosphere* **2020**, *251*, 126633. [CrossRef] [PubMed]
17. Buck, R.; Franklin, J.; Berger, U.; Conder, J.; Cousins, I.; De Voogt, P.; Jensen, A.; Kannan, K.; Mabury, S.; Van Leeuwen, S. Perfluoroalkyl and polyfluoroalkyl substances in the environment: Terminology, classification, and origins. *Integr. Environ. Assess. Manag.* **2011**, *7*, 513–541. [CrossRef]
18. Chen, C.; Lu, Y.; Zhang, X.; Geng, J.; Wang, T.; Shi, Y.; Hu, W.; Li, J. A review of spatial and temporal assessment of PFOS and PFOA contamination in China. *Chem. Ecol.* **2009**, *25*, 163–177. [CrossRef]
19. Li, L.; Zhai, Z.; Liu, J.; Hu, J. Estimating industrial and domestic environmental releases of perfluorooctanoic acid and its salts in China from 2004 to 2012. *Chemosphere* **2015**, *129*, 100–109. [CrossRef] [PubMed]

20. Remucal, C.K. Spatial and temporal variability of perfluoroalkyl substances in the Laurentian Great Lakes. *Environ. Sci. Process. Impacts* **2019**, *21*, 1816–1834. [CrossRef] [PubMed]

21. Prevedouros, K.; Cousins, I.; Buck, R.; Korzeniowski, S. Sources, fate and transport of perfluorocarboxylates. *Environ. Sci. Technol.* **2006**, *40*, 32–44. [CrossRef]

22. Shao, M.; Ding, G.; Zhang, J.; Wei, L.; Xue, H.; Zhang, N.; Li, Y.; Chen, G.; Sun, Y. Occurrence and distribution of perfluoroalkyl substances (PFASs) in surface water and bottom water of the Shuangtaizi Estuary, China. *Environ. Pollut.* **2016**, *216*, 675–681. [CrossRef]

23. Zhao, Z.; Cheng, X.; Hua, X.; Jiang, B.; Tian, C.; Tang, J.; Li, Q.; Sun, H.; Lin, T.; Liao, Y.; et al. Emerging and legacy per- and polyfluoroalkyl substances in water, sediment, and air of the Bohai Sea and its surrounding rivers. *Environ. Pollut.* **2020**, *263*, 114391. [CrossRef]

24. Wang, P.; Lu, Y.; Wang, T.; Meng, J.; Li, Q.; Zhu, Z.; Sun, Y.; Wang, R.; Giesy, J.P. Shifts in production of perfluoroalkyl acids affect emissions and concentrations in the environment of the Xiaoqing River Basin, China. *J. Hazard. Mater.* **2016**, *307*, 55–63. [CrossRef]

25. Gao, L.; Liu, J.; Bao, K.; Chen, N.; Meng, B. Multicompartment occurrence and partitioning of alternative and legacy per- and polyfluoroalkyl substances in an impacted river in China. *Sci. Total Environ.* **2020**, *729*, 138753. [CrossRef] [PubMed]

26. Wang, P.; Lu, Y.; Wang, T.; Zhu, Z.; Li, Q.; Zhang, Y.; Fu, Y.; Xiao, Y.; Giesy, J.P. Transport of short-chain perfluoroalkyl acids from concentrated fluoropolymer facilities to the Daling River estuary, China. *Environ. Sci. Pollut. Res.* **2015**, *22*, 9626–9636. [CrossRef] [PubMed]

27. Chen, X.; Zhu, L.; Pan, X.; Fang, S.; Zhang, Y.; Yang, L. Isomeric specific partitioning behaviors of perfluoroalkyl substances in water dissolved phase, suspended particulate matters and sediments in Liao River Basin and Taihu Lake, China. *Water Res.* **2015**, *80*, 235–244. [CrossRef] [PubMed]

28. Guo, C.; Zhang, Y.; Zhao, X.; Du, P.; Liu, S.; Lv, J.; Xu, F.; Meng, W.; Xu, J. Distribution, source characterization and inventory of perfluoroalkyl substances in Taihu Lake, China. *Chemosphere* **2015**, *127*, 201–207. [CrossRef] [PubMed]

29. Wang, Z.; Liang, X.; Zhan, B.; Wu, J.; Gao, Y.; Xu, N. Pollution Characteristics and Ecological Risk of Perfluorinated Compounds in a Rapidly Urbanizing Catchment. *Acta Sci. Nat. Univ. Pekin.* **2019**, *55*, 543–552. (In Chinese)

30. Sun, Z.; Zhang, C.; Yan, H.; Han, C.; Chen, L.; Meng, X.; Zhou, Q. Spatiotemporal distribution and potential sources of perfluoroalkyl acids in Huangpu River, Shanghai, China. *Chemosphere* **2017**, *174*, 127–135. [CrossRef]

31. Shi, Y.; Vestergren, R.; Xu, L.; Song, X.; Niu, X.; Zhang, C.; Cai, Y. Characterizing direct emissions of perfluoroalkyl substances from ongoing fluoropolymer production sources: A spatial trend study of Xiaoqing River, China. *Environ. Pollut.* **2015**, *206*, 104–112. [CrossRef]

32. Meng, J.; Zhou, Y.; Liu, S.; Chen, S.; Wang, T. Increasing perfluoroalkyl substances and ecological process from the Yongding Watershed to the Guanting Reservoir in the Olympic host cities, China. *Environ. Int.* **2019**, *133*, 105224. [CrossRef]

33. Zhang, Y.; Meng, W.; Guo, C.; Xu, J.; Yu, T.; Fan, W.; Li, L. Determination and partitioning behavior of perfluoroalkyl carboxylic acids and perfluorooctanesulfonate in water and sediment from Dianchi Lake, China. *Chemosphere* **2012**, *88*, 1292–1299. [CrossRef]

34. Chen, C.; Yang, Y.; Zhao, J.; Liu, Y.; Hu, L.; Li, B.; Li, C.; Ying, G. Legacy and alternative per- and polyfluoroalkyl substances (PFASs) in the West River and North River, south China: Occurrence, fate, spatio-temporal variations and potential sources. *Chemosphere* **2021**, *283*, 131301. [CrossRef]

35. Zhang, X.; Hu, T.; Yang, L.; Guo, Z. The Investigation of Perfluoroalkyl Substances in Seasonal Freeze—Thaw Rivers During Spring Flood Period: A Case Study in Songhua River and Yalu River, China. *Bull. Environ. Contam. Toxicol.* **2018**, *101*, 166–172. [CrossRef] [PubMed]

36. Zhang, Y.; Lai, S.; Zhao, Z.; Liu, F.; Chen, H.; Zou, S.; Xie, Z.; Ebinghaus, R. Spatial distribution of perfluoroalkyl acids in the Pearl River of Southern China. *Chemosphere* **2013**, *93*, 1519–1525. [CrossRef] [PubMed]

37. Cai, Y.; Wang, X.; Wu, Y.; Zhao, S.; Li, Y.; Ma, L.; Chen, C.; Huang, J.; Yu, G. Temporal trends and transport of perfluoroalkyl substances (PFASs) in a subtropical estuary: Jiulong River Estuary, Fujian, China. *Sci. Total Environ.* **2018**, *639*, 263–270. [CrossRef]

38. Zhou, Z.; Liang, Y.; Shi, Y.; Xu, L.; Cai, Y. Occurrence and Transport of Perfluoroalkyl Acids (PFAAs), Including Short-Chain PFAAs in Tangxun Lake, China. *Environ. Sci. Technol.* **2013**, *47*, 9249–9257. [CrossRef] [PubMed]

39. Zhao, Z.; Tang, J.; Mi, L.; Tian, C.; Zhong, G.; Zhang, G.; Wang, S.; Li, Q.; Ebinghaus, R.; Xie, Z.; et al. Perfluoroalkyl and polyfluoroalkyl substances in the lower atmosphere and surface waters of the Chinese Bohai Sea, Yellow Sea, and Yangtze River estuary. *Sci. Total Environ.* **2017**, *599–600*, 114–123. [CrossRef] [PubMed]

40. Cao, Y.; Cao, X.; Wang, H.; Wan, L.; Wang, S. Assessment on the distribution and partitioning of perfluorinated compounds in the water and sediment of Nansi Lake, China. *Environ. Monit. Assess.* **2015**, *187*, 611. [CrossRef]

41. Liu, B.; Zhang, H.; Xie, L.; Li, J.; Wang, X.; Zhao, L.; Wang, Y.; Yang, B. Spatial distribution and partition of perfluoroalkyl acids (PFAAs) in rivers of the Pearl River Delta, southern China. *Sci. Total Environ.* **2015**, *524–525*, 1–7. [CrossRef]

42. Pan, C.; Ying, G.; Zhao, J.; Liu, Y.; Jiag, Y.; Zhang, Q. Spatiotemporal distribution and mass loadings of perfluoroalkyl substances in the Yangtze River of China. *Sci. Total Environ.* **2014**, *493*, 580–587. [CrossRef]

43. Zhao, P.; Xia, X.; Dong, J.; Xia, N.; Jiang, X.; Li, Y.; Zhu, Y. Short- and long-chain perfluoroalkyl substances in the water, suspended particulate matter, and surface sediment of a turbid river. *Sci. Total Environ.* **2016**, *568*, 57–65. [CrossRef]

44. Piao, H.; Jiao, X.; Gai, N.; Chen, S.; Lu, G.; Yin, X.; Yamazaki, E.; Yamashita, N.; Tan, K.; Yang, Y.; et al. Perfluoroalkyl substances in waters along the Grand Canal, China. *Chemosphere* **2017**, *179*, 387–394. [CrossRef]

45. Han, T.; Gao, L.; Chen, J.; He, X.; Wang, B. Spatiotemporal variations, sources and health risk assessment of perfluoroalkyl substances in a temperate bay adjacent to metropolis, North China. *Environ. Pollut.* **2020**, *265*, 115011. [CrossRef] [PubMed]
46. Tan, K.; Lu, G.; Yuan, X.; Zheng, Y.; Shao, P.; Cai, J.; Zhao, Y.; Zhu, X.; Yang, Y. Perfluoroalkyl Substances in Water from the Yangtze River and Its Tributaries at the Dividing Point Between the Middle and Lower Reaches. *Bull. Environ. Contam. Toxicol.* **2018**, *101*, 598–603. [CrossRef] [PubMed]
47. Zhang, H.; Wang, S.; Yu, Y. Concentrations of Typical Perfluoroalkyl Acids and Contributions of Their Precursors in the Water of the Le'an River in China. *Environ. Sci.* **2020**, *41*, 3204–3211. (In Chinese)
48. Dreyer, A.; Matthias, V.; Weinberg, I.; Ebinghaus, R. Wet deposition of poly- and perfluorinated compounds in Northern Germany. *Environ. Pollut.* **2010**, *158*, 1221–1227. [CrossRef] [PubMed]
49. Taniyasu, S.; Yamashita, N.; Moon, H.; Kwok, K.; Lam, P.; Horii, Y.; Petrick, G.; Kannan, K. Does wet precipitation represent local and regional atmospheric transportation by perfluorinated alkyl substances? *Environ. Int.* **2013**, *55*, 25–32. [CrossRef] [PubMed]
50. Kwok, K.; Taniyasu, S.; Yeung, L.; Murphy, M.; Lam, P.; Horii, Y.; Kannan, K.; Petrick, G.; Sinha, R.; Yamashita, N. Flux of Perfluorinated Chemicals through Wet Deposition in Japan, the United States, And Several Other Countries. *Environ. Sci. Technol.* **2010**, *44*, 7043–7049. [CrossRef]
51. Pike, K.; Edmiston, P.; Morrison, J.; Faust, J. Correlation Analysis of Perfluoroalkyl Substances in Regional U.S. Precipitation Events. *Water Res.* **2021**, *190*, 116685. [CrossRef]
52. Chen, H.; Zhang, L.; Li, M.; Yao, Y.; Zhao, Z.; Munoz, G.; Sun, H. Per- and polyfluoroalkyl substances (PFASs) in precipitation from mainland China: Contributions of unknown precursors and short-chain (C2-C3) perfluoroalkyl carboxylic acids. *Water Res.* **2019**, *153*, 169–177. [CrossRef]
53. Zhao, L.; Zhou, M.; Zhang, T.; Sun, H. Polyfluorinated and Perfluorinated Chemicals in Precipitation and Runoff from Cities Across Eastern and Central China. *Arch. Environ. Contam. Toxicol.* **2013**, *64*, 198–207. [CrossRef]
54. Zheng, Y.; Lu, G.; Shao, P.; Gai, N.; Yang, Y.; Feng, J.; Zhao, Q. Level and distribution of perfluorinated compounds in snow and water samples from the transition zone in eastern Qinghai-Tibet. *Environ. Chem.* **2020**, *39*, 1192–1201. (In Chinese)
55. Chen, M.; Wang, C.; Wang, X.; Fu, J.; Gong, P.; Yan, J.; Yu, Z.; Yan, F.; Nawab, J. Release of Perfluoroalkyl Substances from Melting Glacier of the Tibetan Plateau: Insights into the Impact of Global Warming on the Cycling of Emerging Pollutants. *J. Geophys. Res. Atmos.* **2019**, *8*, 7442–7456. [CrossRef]
56. Yamazaki, E.; Falandysz, J.; Taniyasu, S.; Hui, G.; Jurkiewicz, G.; Yamashita, N.; Yang, Y.; Lam, P. Perfluorinated carboxylic and sulphonic acids in surface water media from the regions of Tibetan Plateau: Indirect evidence on photochemical degradation? *J. Environ. Sci. Health A* **2016**, *51*, 63–69. [CrossRef] [PubMed]
57. Chen, M.; Wang, C.; Gao, K.; Wang, X.; Fu, J.; Gong, P.; Wang, Y. Perfluoroalkyl substances in precipitation from the Tibetan Plateau during monsoon season: Concentrations, source regions and mass fluxes. *Chemosphere* **2021**, *282*, 131105. [CrossRef] [PubMed]
58. Barisci, S.; Suri, R. Occurrence and removal of poly/perfluoroalkyl substances (PFAS) in municipal and industrial wastewater treatment plants. *Water Sci. Technol.* **2021**, *84*, 3442–3468. [CrossRef] [PubMed]
59. Xiao, F. Emerging poly- and perfluoroalkyl substances in the aquatic environment: A review of current literature. *Water Res.* **2017**, *124*, 482–495. [CrossRef] [PubMed]

MDPI

St. Alban-Anlage 66

4052 Basel

Switzerland

Tel. +41 61 683 77 34

Fax +41 61 302 89 18

www.mdpi.com

Water Editorial Office

E-mail: water@mdpi.com

www.mdpi.com/journal/water